CURRENT TOPICS IN

DEVELOPMENTAL BIOLOGY

VOLUME 13

IMMUNOLOGICAL APPROACHES TO EMBRYONIC DEVELOPMENT AND DIFFERENTIATION PART I

CONTRIBUTORS

RICHARD AKESON

EDWARD A. BOYSE

EDWARD P. COHEN

SARAH C. R. ELGIN

KAY L. FIELDS

FRANÇOIS JACOB

CAROL JONES

BARBARA B. KNOWLES

WEITZE LIANG

DAVID R. McCLAY

CLARKE F. MILLETTE

ALBERTO MONROY

GARY W. MOY

GARTH L. NICOLSON

THEODORE T. PUCK

FLORIANA ROSATI

MELITTA SCHACHNER

LEE M. SILVER

DAVOR SOLTER

VICTOR D. VACQUIER

LYNN M. WILEY

CURRENT TOPICS IN DEVELOPMENTAL BIOLOGY

EDITED BY

A. A. MOSCONA
DEPARTMENTS OF BIOLOGY AND PATHOLOGY
THE UNIVERSITY OF CHICAGO
CHICAGO, ILLINOIS

ALBERTO MONROY
STAZIONE ZOOLOGICA
NAPLES, ITALY

VOLUME 13

Immunological Approaches to Embryonic Development and Differentiation
Part I

VOLUME EDITOR

MARTIN FRIEDLANDER

THE ROCKEFELLER UNIVERSITY
NEW YORK, NEW YORK

1979

ACADEMIC PRESS

A Subsidiary of Harcourt Brace Jovanovich, Publishers

New York   London   Toronto   Sydney   San Francisco

COPYRIGHT © 1979, BY ACADEMIC PRESS, INC.
ALL RIGHTS RESERVED.
NO PART OF THIS PUBLICATION MAY BE REPRODUCED OR
TRANSMITTED IN ANY FORM OR BY ANY MEANS, ELECTRONIC
OR MECHANICAL, INCLUDING PHOTOCOPY, RECORDING, OR ANY
INFORMATION STORAGE AND RETRIEVAL SYSTEM, WITHOUT
PERMISSION IN WRITING FROM THE PUBLISHER.

ACADEMIC PRESS, INC.
111 Fifth Avenue, New York, New York 10003

United Kingdom Edition published by
ACADEMIC PRESS, INC. (LONDON) LTD.
24/28 Oval Road, London NW1 7DX

LIBRARY OF CONGRESS CATALOG CARD NUMBER: 66–28604

ISBN 0–12–153113–9

PRINTED IN THE UNITED STATES OF AMERICA

79 80 81 82   9 8 7 6 5 4 3 2 1

CONTENTS

List of Contributors	ix
Preface	xi
Conspectus by Edward A. Boyse	xv

CHAPTER 1. **Cell Surface Antigens during Mammalian Spermatogenesis**
CLARKE F. MILLETTE

I.	Introduction	1
II.	Isolation of Mammalian Seminiferous Cells	6
III.	Temporal Appearance of Surface Antigens during Spermatogenesis	10
IV.	Membrane Mobility of Spermatogenic Cells	15
V.	Selective Partitioning of Membrane Antigens during Spermiogenesis	17
VI.	Isolation of Germ Cell Plasma Membranes	24
VII.	Concluding Remarks	27
	References	28

CHAPTER 2. **Immunoperoxidase Localization of Bindin during the Adhesion of Sperm to Sea Urchin Eggs**
GARY W. MOY AND VICTOR D. VACQUIER

I.	Introduction	31
II.	Preparation of Antigen and Antisera	33
III.	Characterization of Antibindin by Immunodiffusion Assay	34
IV.	Ultrastructural Immunohistochemical Localization of Bindin	36
V.	Concluding Remarks	42
	References	44

CHAPTER 3. **Cell Surface Differentiations during Early Embryonic Development**
ALBERTO MONROY AND FLORIANA ROSATI

I.	Introduction	45
II.	The Problem of Polarity	46
III.	The Segregation of Cell Lines	49
IV.	The Sperm–Egg Interaction in Fertilization	59
V.	Concluding Remarks	65
	References	67

CHAPTER 4. Immunofluorescent Analysis of Chromatin Structure in Relation to Gene Activity: A Speculative Essay
LEE M. SILVER AND SARAH C. R. ELGIN

I.	Introduction	71
II.	The Technique of Immunofluorescence Antibody Staining	72
III.	Patterns of Chromosomal Protein Distribution in Relation to Gene Activity	76
IV.	Concluding Remarks	86
	References	87

CHAPTER 5. Immunogenetics of Mammalian Cell Surface Macromolecules
CAROL JONES AND THEODORE T. PUCK

I.	Introduction	89
II.	Human–Chinese Hamster Ovary Cell Hybrids	90
III.	Immunologic Specificity of Human and Chinese Hamster Ovary Cell Surface Structures	92
IV.	Human Antigens (A_L, B_L, C_L, etc.) Expressed on Human–Chinese Hamster Somatic Cell Hybrids	93
V.	Analysis in Depth of the A_L Antigenic System	95
VI.	Use of A_L System to Detect Environmental Mutagens	107
VII.	Cell Surface Antigens of Wide and of Limited Distribution among Different Tissues	112
VIII.	Discussion	113
IX.	Concluding Remarks	114
	References	115

CHAPTER 6. Cell Surface and Early Stages of Mouse Embryogenesis
FRANÇOIS JACOB

I.	Introduction	117
II.	The Teratocarcinoma System	118
III.	Glycopeptides	119
IV.	Lectin Receptors	123
V.	Surface Antigens	125
VI.	Surface Structures and Cellular Interactions in Early Development	132
VII.	Concluding Remarks	134
	References	135

CHAPTER 7. Developmental Stage-Specific Antigens during Mouse Embryogenesis
DAVOR SOLTER AND BARBARA B. KNOWLES

I.	Introduction	139
II.	Immunological Analysis of Stage-Specific Antigens	141
III.	Conclusions and Perspectives	160
IV.	Concluding Remarks	162
	References	163

CONTENTS vii

CHAPTER 8. Early Embryonic Cell Surface Antigens as Developmental Probes
LYNN M. WILEY

I. Introduction 167
II. Cell Surface Antigens of Gametes and Preimplantation Embryos 169
III. Future Studies on Embryonic Cell Surface Antigens as Developmental Probes 187
IV. Concluding Remarks 192
References 193

CHAPTER 9. Surface Antigens Involved in Interactions of Embryonic Sea Urchin Cells
DAVID R. MCCLAY

I. Introduction 199
II. Adhesive Specificity: Demonstrations of Recognition Capabilities 200
III. Immunochemical Specificity: Changes Associated with Gastrulation 203
IV. Relationship between Cell Surface Antigens and Cell–Cell Interactions 209
V. Synthesis and Conclusions 211
VI. Concluding Remarks 213
References 213

CHAPTER 10. Cell Surface Antigens of Cultured Neuronal Cells
RICHARD AKESON

I. Introduction 215
II. Neural-Specific Antigens of Cultured Neuronal Cells 216
III. Do Mouse Neuroblastoma Cells Differentiate *in Vitro*? 224
IV. Approaches to Neural Cell–Cell Interaction 226
V. Perspectives on Future Research Directions 230
VI. Concluding Remarks 232
References 233

CHAPTER 11. Cell Type-Specific Antigens of Cells of the Central and Peripheral Nervous System
KAY L. FIELDS

I. Introduction 237
II. Specific Antigens for Different Neural Cell Types 238
III. Thy-1: A Developmental Antigen on Neuronal Cells 245
IV. Present and Potential Applications of Antibodies 252
V. Concluding Remarks 254
References 255

CONTENTS

CHAPTER 12. Cell Surface Antigens of the Nervous System
MELITTA SCHACHNER

I.	Introduction	259
II.	Cell Type-Specific Cell Surface Antigens	261
III.	Developmental Expression of Antigens	268
IV.	Subcellular Distribution of Antigens	273
V.	Concluding Remarks	276
	References	277

CHAPTER 13. Antibody Effects on Membrane Antigen Expression
EDWARD P. COHEN AND WEITZE LIANG

I.	Introduction	281
II.	Antigenic Modulation as a Model System for Studying Ligand-Induced Effects on Receptor Expression	283
III.	The Metabolism of TL Antigens in the Presence (and Absence) of TL Antiserum	285
IV.	Membrane Antigens of Somatic Hybrids of TL(+) and TL(−) Cells	290
V.	Failure of TL Antigens of Hybrid Cells to Undergo Modulation	293
VI.	Metabolism of Membrane Antigens of Hybrid Cells	294
VII.	The Quantities of TL Antigens of Parental and Hybrid Cells	295
VIII.	Discussion	297
IX.	Similar Features of Down Regulation and Antigenic Modulation	300
X.	Concluding Remarks	301
	References	303

CHAPTER 14. Topographic Display of Cell Surface Components and Their Role in Transmembrane Signaling
GARTH L. NICOLSON

I.	Introduction	305
II.	Plasma Membrane Organization	306
III.	Dynamics of Cell Membrane Components	310
IV.	Topographic Control of Cell Surface Receptors	318
V.	Transmembrane-Mediated Communication	326
VI.	Concluding Remarks	333
	References	333

Subject Index	339
Contents of Previous Volumes	343

LIST OF CONTRIBUTORS

Numbers in parentheses indicate the pages on which the authors' contributions begin.

RICHARD AKESON, *Departments of Pediatrics and Biological Chemistry, Children's Hospital Research Foundation, Cincinnati, Ohio* 45229 (215)

EDWARD A. BOYSE, *Memorial Sloan-Kettering Cancer Center, New York, New York* 10021 (xv)

EDWARD P. COHEN,* *La Rabida-University of Chicago Institute and Department of Microbiology, University of Chicago, Chicago, Illinois* 60649 (281)

SARAH C. R. ELGIN, *Harvard University, The Biological Laboratories, Cambridge, Massachusetts* 02138 (71)

KAY L. FIELDS, *Department of Neurology, Albert Einstein College of Medicine, Bronx, New York* 10461 (237)

FRANÇOIS JACOB, *Service de Génétique cellulaire du Collège de France et de l'Institut Pasteur, 25 rue du Dr. Roux, 75015 Paris, France* (117)

CAROL JONES, *Eleanor Roosevelt Institute for Cancer Research, Florence R. Sabin Laboratories for Genetic and Developmental Medicine, Department of Biochemistry, Biophysics and Genetics, University of Colorado Medical Center, Denver, Colorado* 80262 (89)

BARBARA B. KNOWLES, *The Wistar Institute of Anatomy and Biology, Philadelphia, Pennsylvania* 19104 (139)

WEITZE LIANG, *La Rabida-University of Chicago Institute, University of Chicago, Chicago, Illinois* 60649 (281)

DAVID R. MCCLAY, *Department of Zoology, Duke University, Durham, North Carolina* 27706 (199)

CLARKE F. MILLETTE, *Department of Anatomy, Laboratory of Human Reproduction and Reproductive Biology, Harvard Medical School, Boston, Massachusetts* 02115 (1)

ALBERTO MONROY, *Stazione Zoologica, Napoli, Italy* (45)

* Present address: Department of Microbiology and Immunology, University of Illinois at the Medical Center, Chicago, Illinois 60612.

GARY W. MOY, *Marine Biology Research Division, Scripps Institution of Oceanography, University of California-San Diego, La Jolla, California* 92093 (31)

GARTH L. NICOLSON, *Departments of Developmental and Cell Biology and Physiology, College of Medicine, University of California, Irvine, California* 92717 (305)

THEODORE T. PUCK, *Eleanor Roosevelt Institute for Cancer Research, Florence R. Sabin Laboratories for Genetic and Developmental Medicine, Department of Biochemistry, Biophysics and Genetics, University of Colorado Medical Center, Denver, Colorado* 80262 (89)

FLORIANA ROSATI, *Stazione Zoologica, Napoli, Italy* (45)

MELITTA SCHACHNER, *Department of Neurobiology, Heidelberg University, Im Neuenheimer Feld 320, D-6900 Heidelberg, Federal Republic of Germany* (259)

LEE M. SILVER, *Section on Developmental Genetics, Sloan-Kettering Institute, New York, New York* 10021 (71)

DAVOR SOLTER, *The Wistar Institute of Anatomy and Biology, Philadelphia, Pennsylvania* 19104 (139)

VICTOR D. VACQUIER, *Marine Biology Research Division, Scripps Institution of Oceanography, University of California-San Diego, La Jolla, California* 92093 (31)

LYNN M. WILEY, *Department of Anatomy, University of Virginia, Charlottesville, Virginia* 22908 (167)

PREFACE

The most fundamental processes underlying the orderly progression from a single, fertilized egg to a complex, multicellular organism remain largely unknown today, in spite of significant progress in many areas of developmental biology. Several recurring themes have emerged from a plethora of information generated by recent studies of developing systems. Cellular differentiation is frequently preceded by division and proliferation within cell sets, although the precise relationship between mitotic events and molecular specialization remains unknown. The transduction of metabolic (and developmental) cues from the environment to the level of the genome is probably mediated by the cell surface and may involve subplasmalemmal complexes that somehow join membrane receptors with nuclear components. The basis for the extensive repertoires of selective cell associations and morphogenetic patterns observed during development may arise from the assembly of basic molecular or cellular components into informational mosaics. While impressive strides have been made toward understanding the fundamental questions and theories set forth by classical embryology, it is only recently that we have been able to understand the molecular mechanisms underlying basic developmental events. The approaches to these problems have been greatly facilitated by the interaction of cell, developmental, and molecular biology at both the conceptual and methodological levels. A common denominator has been the exquisitely sensitive technical achievements and elegant conceptual frameworks provided by work in the field of immunology.

The objective of Volumes 13 and 14 of *Current Topics in Developmental Biology* is to provide a survey of concepts, achievements, and prospects concerning the application of immunological methodologies to the analysis of various aspects of cell differentiation and morphogenesis—molecular, cellular, and histogenic—in embryonic, fetal, and postnatal development. Contributions to Volume 13 focus on early embryonic development, the nervous system, and the structural and conceptual basis of transmembrane signaling. The contents of Volume 14 include studies of the immune system, connective tissue, non-muscle contractile proteins, and slime mold. Thus, the important feature of these volumes is the diversity of research interests unified by the theme of immunological approaches to the analysis of development and differentiation. It is certain that such approaches will be useful in continued investigations; the intent of these volumes is to focus attention on, and stimulate interest in, the many problems of development that may be studied by the application of immunological techniques and concepts. As Dr. Boyse observes in his conspectus, the field of immunology provides far more than methodological resources for students of development: As an "exemplar of ontogeny," the immune system should prove a rich conceptual resource as well.

We could not hope to provide contributions from every laboratory active in the research areas discussed in this volume. Rather, representative work was solicited and, in several areas, manuscripts from two or three laboratories in the same field were brought together to provide a broader perspective. A general review of spermatogenesis, including data on the use of immunological probes specific for developmentally regulated sperm cell surface antigens, is presented by Millette. Moy and Vacquier's discussion is more focused; ultrastructural immunocytochemical methods are used to corroborate the distribution of bindin with its proposed role in mediating sperm–egg adhesion. The chapter by Monroy and Rosati addresses the fundamental issues of acquisition of polarity in the fertilized egg and the segregation of somatic from germ cell lines.

Silver and Elgin discuss the use of immunofluorescent techniques to investigate the distribution of nonhistone chromosomal proteins in relation to established banding patterns in *Drosophila* polytene chromosomes. Genetic control of cell membrane differentiation antigen expression may be analyzed through immunogenetic approaches as detailed in the chapter by Jones and Puck.

A number of laboratories have been examining early embryos with the intent of identifying developmental stage specific antigens which may be used to monitor embryonic development. The chapters by Jacob, by Solter and Knowles, and by Wiley provide a broad discussion of advances in this area, with particular attention to the teratocarcinoma system. The generation of antisera specific to a number of developmental stage-specific embryonic antigens should prove useful in determining what biological role, if any, these molecules play during development. Substantial data have been accumulated in support of a role for surface antigens in the regulation of morphogenesis during embryonic development of the sea urchin, as discussed in McClay's chapter.

As the field of developmental neuroscience continues to expand, it is hoped that the development of specific markers for the various cell components in the nervous system will be useful in understanding the mechanisms that underlie the establishment of neural pathways. The chapters by Akeson, by Fields, and by Schachner review recent progress in the analysis of neural cell antigenic markers and offer the hope that immunological approaches may eventually permit the fractionation of heterologous cell types and reconstruction of neuronal networks. The final chapters by Cohen and Liang and by Nicolson discuss antigenic modulation and more general phenomena of transmembrane signaling.

Martin Friedlander

CONSPECTUS

PROBLEMS OF DEVELOPMENT: IMMUNOLOGY AS MASTER AND SERVANT

*Edward A. Boyse**

MEMORIAL SLOAN-KETTERING CANCER CENTER
NEW YORK, NEW YORK

It is a happy time for immunology, especially in its relation to developmental biology, for in this relationship immunology can be both master and servant—both a prime exemplar of ontogeny and a methodological resource to students of ontogeny.

The purpose of this conspectus is to suggest a context—evolution—in which the chapters in this volume, and other matters that escaped the editor's net, can be viewed as parts of a coherent whole. For the most part, brevity has not been the soul of latter-day writings on evolution, but readers of this note will have nothing to complain of on that score.

Any thorough treatment of developmental biology must include phylogenesis. In this context immunogenetics and cellular immunology have suggested a series of propositions that perhaps allow a clearer view of how higher evolution came about (Boyse and Cantor, 1978). These propositions can be stated only briefly here, and they are presented in the form of axioms or assertions.

First, a word about cellular differentiation. This is defined here as the capacity of a cell to commit itself irrevocably to a subset of its genes. Cellular differentiation is taken to be the biological novelty that inaugurated metazoan evolution. By a series of divergent and sequential differentiative steps, each perhaps restricted to a binary choice (Abbott *et al.*, 1974), and each producing a distinct cell set, a complete organism is generated from the zygote. These processes, though most dramatic during embryogenesis, continue throughout life, in some organs and tissues more than in others, for replenishment and repair.

(i) Each cell set has a unique surface phenotype composed of a particular set of genes. The number of genes assigned to these programs may account, in large measure, for the great expansion of the genome during phylogeny.

(ii) Commitment to a program and manifestation of the pro-

* American Cancer Society Research Professor of Cell Surface Immunogenetics.

gram on induction are distinct and separable features of a single step that conveys a cell from one compartment of differentiation to the next.

(iii) Surface phenotype and prescribed cellular functions are coordinately programmed in each cell set.

(iv) Fulfillment of a program consists in premitotic expression of a gene set for surface phenotype and postmitotic expression of a gene set for function.

(v) The surface topography of a cell, meaning the ordered disposition of the components of the surface phenotype, is inherent in its program.

(vi) The supramolecular configurations of the surface phenotype are susceptible to precise reorganization, called "phenotypic adaptation," in response to extraneous information from outside the cell (Flaherty and Zimmerman, 1979). Forthcoming reports by Lorraine Flaherty will be of great interest in this connection.

(vii) The derivation of cooperating cell sets, with different but related programs, from a phylogenetically antecedent cell set with one antecedent program, is a major feature of metazoan evolution, and parallels ontogeny. New communication networks are derived from old communication networks.

(viii) The great complexity of programs (i–vii) speaks for a common ancestral machinery of programming essential to metazoan evolution. New programs are derived from old programs. The remarkable phenomenon of transdetermination in transplanted imaginal disks of *Drosophila*, whereby a disk of cells destined on induction to generate a structure such as a leg spontaneously "mutates" to become a different potential structure such as a wing (reviewed by Kauffman, 1973), can be viewed as a pointer to relatedness in the machinery and origin of programs.

On the basis of these postulates one may ponder two questions: What is the nature of program commitments? How may whole programs undergo duplication and variation en masse, to account, among other things, for those perplexing spurts in evolution that appear to have achieved so much in so short a time?

1. The nature of commitment is the crux of developmental biology. A fine introduction for students of commitment is provided by *Sl* and *W*, two unlinked pleiotropic mouse genes with mutants which cause similar abnormal developmental syndromes which may rep-

resent, respectively, defects in the commitment-effector and commitment-receptor apparatus of the affected cell sets (Trentin, 1970; Russell, 1970). But we still lack a tangible model of commitment *in vitro* in which cells can be instructed to take up one of two or more alternative program options, rather than simply to manifest a foreclosed option (Scheid *et al.*, 1978). This is one of several reasons for widespread interest in the work of Wachtel and Ohno (1979; Ohno, 1978), indicating that the cell-surface molecule bearing H-Y antigen is the determinant of primary sex, because this system is yielding to investigation *in vitro* with cells from the sexually indifferent embryonic gonad.

2. The facts of nuclear transplantation, indicating that an enucleated ovum combined with the nucleus of a differentiated cell can undergo substantial development, point to the participation of a cytoplasmic entity in the process of commitment; hypotheses of program duplication should not be limited to conventional genetic explanations. Evidence that immune cells equip themselves by rearranging their immunoglobulin genes (Gilbert, 1978; Tonegawa *et al.*, 1978) is a challenge to the dogma that the genetic endowment of all cell sets is as a rule the same. What if the reported limitation in development of lymphocyte nuclei in enucleated eggs (Wabl *et al.*, 1975) were due to this editing of genes? And what if ontogenetic limitations of nuclei transplanted from cells of different sets were caused in each case by maldevelopment in the cell lineage from which the nucleus originated?

Tetraploidization, in which the entire genome is doubled, and which can be induced experimentally by cold shock with relative ease in some vertebrate species, has been proposed as the prime source of program duplication (Boyse and Cantor, 1978) because it should provide a veritable abundance of program variations available for the derivation of new cooperating cell sets in all cell lineages. Evidence for periodic tetraploidization during phylogeny is given by Ohno (1970), and its possible relevance to program duplication and variation is discussed elsewhere (Boyse and Cantor, 1978).

One of the attractions of an evolutionary device as powerful as the partitioning of variant duplicate programs into new cell sets is that it enables one to envisage how a systematically generated series of markers for individual cells of a given set, a "dictionary" which itself may have been a major achievement of evolution, might be handed on to another cell set as a ready-made system for detailed associative recognition. Associative recognition, in this sense, is one of several starting points for theorists seeking useful comparisons between the nervous and immune systems (Boyse and Cantor, 1978; Boyse et al., 1979; Cohn, 1970; Edelman, 1978; Hood et al., 1977; Moscona, 1976). Sperry's hypothesis that the surfaces of optic neurons are individually coded for precise pathfinding (Sperry, 1963) should hardly come as a surprise to immunologists who anticipate that the production of a particular antibody will prove to require the cooperation of a clone of T cells with a clone of B cells. A coding system of such range and precision is more than sufficient to label optic neurons and their partners. And there are ways in which immunobiochemical techniques might be used to investigate the reality and nature of Sperry's optic markers (Boyse and Cantor, 1978).

Another set of primary sensory neurons has recently come to the attention of immunologists. These are the olfactory neurons, which perceive a range of chemical signals that surely rivals that perceived by lymphocytes, and which are regularly replaced by freshly differentiated neurons every few weeks. The major histocompatibility complex (MHC), which is involved in many facets of chemical recognition and response by lymphocytes, includes genes that are involved in chemical recognition and response by neurons, manifest in the ability of mice to scent each other's MHC types, and to respond according to their own MHC types (Yamazaki et al., 1976; Boyse et al., 1979). It is not certain to what extent the great precision and range of odor recognition is due to diversity of receptors (see Cagan and Zeiger, 1978) rather than combination perception (Amoore, 1977), but small wonder that immunologists and people working on chemical sensory communication are seeking a rendezvous.

It has been remarked that the rapid evolution of higher cerebral structure and function, particularly among mammals, and spectacularly evident in man, at least by the slippery criteria of intelligence, is explicable in terms of multiple integrations of basic self-contained functional units, without a corresponding advance in fundamental complexity (Boyse and Cantor, 1978; Szentagothai, 1978). This view harmonizes with the general hypothesis that the unit of progression in higher evolution is the cell set. Neuroanatomists seem to be saying much the same thing when they speak, for example, of "exceptional qualities of the human brain [which] would rest primarily in the number of columnar modules" (Szentagothai, 1978). Or may we say "cell sets"?

ACKNOWLEDGMENTS

This work was supported in part by grants from the National Institutes of Health, the American Cancer Society, and the Rockefeller Foundation.

REFERENCES

Abbott, J., Schiltz, J., Dienstman, S., and Holtzer, H. (1974). *Proc. Natl. Acad. Sci. U.S.A.* **71,** 1506–1510.
Amoore, J. E. (1977). *Chem. Senses Flavor* **2,** 267–281.
Boyse, E. A., and Cantor, H. (1978). *In* "The Molecular Basis of Cell–Cell Interaction" (R. A. Lerner and D. Bergsma, eds.), Vol. XIV, pp. 249–283. Liss, New York.
Boyse, E. A., Yamazaki, K., Yamaguchi, M., and Thomas, L. (1979). *In* "The Immune System: Functions and Therapy of Dysfunction." Academic Press, New York (in press).
Cagan, R. H., and Zeiger, W. N. (1978). *Proc. Natl. Acad. Sci. U.S.A.* **75,** 4679–4683.
Cohn, M. (1970). *Ciba Found. Symp. Control Processes Multicell. Org., 1969* pp. 255–297.
Edelman, G. M. (1978). *In* "The Mindful Brain" (G. M. Edelman and V. B. Mountcastle, eds.), pp. 51–100. MIT Press, Cambridge, Mass.
Flaherty, L., and Zimmerman, D. (1979). *Proc. Natl. Acad. Sci. U.S.A.* **76,** 1990–1993.
Gilbert, W. (1978). *Nature (London)* **271,** 501.
Hood, L., Huang, H. V., and Dreyer, W. (1977). *J. Supramolec. Structure* **7,** 531–559.
Kauffman, S. A. (1973). *Science* **181,** 310–318.
Moscona, A. A. (1976). *In* "Neuronal Recognition" (S. H. Barondes, ed.), pp. 205–226. Plenum, New York.
Ohno, S. (1970). "Evolution by Gene Duplication." Springer, New York.
Ohno, S. (1978). "Major Sex-Determining Genes." Springer-Verlag, Berlin and New York.
Russell, E. S. (1970). *In* "Regulation of Hematopoiesis. Vol. I. Red Cell Production" (A. S. Gordon, ed.), pp. 649–675. Appleton, New York.
Scheid, M. P., Goldstein, G., and Boyse, E. A. (1978). *J. Exp. Med.* **147,** 1727–1743.
Sperry, R. W. (1963). *Proc. Natl. Acad. Sci. U.S.A.* **50,** 703–710.
Szentagothai, J. (1978). *Proc. R. Soc. London, Ser. B* **201,** 219–248.
Tonegawa, S., Maxam, A. M., Tizard, R., Bernard, O., and Gilbert, W. (1978). *Proc. Natl. Acad. Sci. U.S.A.* **75,** 1485–1489.
Trentin, J. J. (1970). *In* "Regulation of Hematopoiesis. Vol. I. Red Cell Production" (A. S. Gordon, ed.), pp. 159–186. Appleton, New York.
Wabl, M. R., Brun, R. B., and DuPasquier, L. (1975). *Science* **190,** 1310–1312.
Wachtel, S. S., and Ohno, S. (1979). *Prog. Med. Genet.* (in press).
Yamazaki, K., Boyse, E. A., Mike, V., Thaler, H. T., Mathieson, B. J., Abbott, J., Boyse, J., Zayas, Z. A., and Thomas, L. (1976). *J. Exp. Med.* **144,** 1324–1335.

CHAPTER 1

CELL SURFACE ANTIGENS DURING MAMMALIAN SPERMATOGENESIS

Clarke F. Millette

DEPARTMENT OF ANATOMY
LABORATORY OF HUMAN REPRODUCTION AND REPRODUCTIVE BIOLOGY
HARVARD MEDICAL SCHOOL
BOSTON, MASSACHUSETTS

I.	Introduction	1
II.	Isolation of Mammalian Seminiferous Cells	6
III.	Temporal Appearance of Surface Antigens during Spermatogenesis	10
IV.	Membrane Mobility of Spermatogenic Cells	15
V.	Selective Partitioning of Membrane Antigens during Spermiogenesis	17
VI.	Isolation of Germ Cell Plasma Membranes	24
VII.	Concluding Remarks	27
	References	28

I. Introduction

The role of the plasma membrane in the regulation of mammalian sperm differentiation remains obscure despite the fact that spermatogenesis offers a unique opportunity for the analysis of both mitotic and meiotic cell proliferation, for the study of cell–cell communication via direct intercellular contact as well as via hormonal interactions, for the investigation of cell differentiation, and for studies of cell motility. Furthermore, spermatogenesis provides an advantage in that all of these problems may be examined in a single developmental lineage, all of whose constituent cells have been well defined morphologically. Progenitor stem cells of the seminiferous epithelium, for example, may be recognized readily in both the light and electron microscope, thus facilitating studies on the temporal regulation of the mitotic phases of spermatogenesis (cf. Courot *et al.,* 1970). This may be contrasted with other systems of differentiation, such as erythropoiesis, where stem cell populations are currently difficult to analyze with precision. Moreover, all of the later stages of spermatocyte and spermatid differentiation in the testis are (1) also easily defined using routine microscopic techniques, (2) readily located in a single anatomical compartment of the body, and (3) usually isolable in sufficiently high purity to allow both immunological and biochemical investigation. Most other pathways of cell development do not possess these advantages. Even studies of lymphoid cells which have been highly successful in characterizing subtle molecular differences defining functional subclasses of lymphocytes must rely on externally applied cell surface labels or upon elaborate biological assays in order to identify with certainty a particular cell type. Mammalian spermatogenesis presents little need for indirect means of cell identification and is, therefore, ideally suited for detailed experimentation designed to analyze cell growth, cell differentiation, cell movement, and even cell death.

There are a variety of cellular interactions during spermatogenesis which may be regulated at the level of the cell surface. Some of these possible involvements of plasma membranes are listed in Table I. A brief discussion of these points should serve to illustrate the necessity for detailed immunological and biochemical analysis of spermatogenic cell surface antigens.

The interactions between Sertoli cells and developing male germ cells are dynamic and highly ordered with respect to the timing and spatial arrangements of successive waves of cellular differentiation. Sertoli cells are connected by an extensive series of tight junctions which form the blood–testis barrier (Dym and Fawcett, 1970). As a result the mammalian testis is an immunologically privileged site. Gap junctions (Gilula et al., 1976) and septate junctions (Connell, 1977) also interconnect adjacent Sertoli cells and presumably provide the coordination required for the translocation of germ cells from the basal to the luminal aspect of the seminiferous tubule. In addition to Sertoli–Sertoli cell interactions, the sustentacular cells also have specialized junctional complexes in the plasma membrane responsible for continued adhesion to developing spermatogenic cells. Hypertonic perfusion experiments, for example, demonstrate that spermatogonia are attached to Sertoli cells at focal regions of membrane contact (Gilula et

TABLE I

CELL MEMBRANES AND SPERMATOGENESIS

1. Germ cell–Sertoli cell interactions
2. Regulation of spermatogonial proliferation
3. Initiation of meiosis
4. Restriction of membrane mobility during spermiogenesis
5. Membrane partitioning during spermiogenesis
6. Localized antigenic topography on spermatozoa
7. Selective sperm survival (haploid gene expression)
8. Recognition of the egg membrane at fertilization

al., 1976; Russell, 1977b). Spermatocytes and spermatids are also firmly associated with the lateral and apical Sertoli cell membranes after ligation of the efferent ductule and the resultant distortion of the seminiferous tubule (Ross, 1974; Ross and Dobler, 1975). At the ultrastructural level, submembranous junctional specializations are found both in Sertoli cells and in germ cells (Kaya and Harrison, 1976; Russell, 1977a). Some surface specializations, similar in appearance to desmosomes, exist only transiently in early spermatids, but are retained by the Sertoli cell up to spermiation, when spermatozoa are released into the tubule lumen (Flickinger and Fawcett, 1967; Kaya and Harrison, 1976; Ross, 1976; Cooper and Bedford, 1976). The molecular nature of the cell surface constituents comprising these different membrane junctional complexes has not yet been examined, but such elements could play vital roles in the regulation of germ cell translocations during spermatogenesis.

During the initial phase of spermatogenesis, termed spermatocytogenesis, spermatogonia proliferate mitotically. It is generally accepted that mouse spermatogonia undergo six successive divisions prior to the initiation of the prolonged meiotic prophase (Clermont and Bustos-Obregon, 1968; Oakberg and Huckins, 1976). The timing and spatial orientation of each of these mitotic proliferation steps is strictly regulated and is distinguished from proliferative events in somatic cells by incomplete cytokinesis and the retention of wide intercellular bridges between germ cells (Fawcett *et al.*, 1959; Schleiermacher and Schmidt, 1973). The intercellular bridges may connect hundreds of spermatogonia and persist throughout spermatogenesis until the last stages of spermiation. Some spermatogonia appear to be degraded selectively during the later spermatogonial divisions so that only 25% of the theoretically possible number of preleptotene spermatocytes are finally produced (Huckins, 1978). None of the regulatory events controlling these processes have been established, but it is feasible that alterations in the plasma membranes of spermatogonia relate directly to the regulation of mitosis in the testis, in a manner analogous to that postulated for control of the normal cell cycle (Nicolson, 1976) and the induction of cell division by antibodies or lectins (Edelman, 1976).

The second phase of spermatogenesis is meiosis. Meiosis in mammals is a very complex process characterized by an ordered series of chromosomal rearrangements outside the scope of this chapter. Although intracellular factors most probably affect these events to the highest degree, recent data indicate that external controls may initiate meiosis in the mammal. The rete ovarii, for example, seems to be required for the initiation of meiosis in the female cat, mink, and ferret

(Byskov, 1975); and studies in the mouse suggest that diffusible substances from the fetal ovary cause premature meiotic induction in either sex (Byskov and Saxén, 1976). Other workers have obtained similar results in the hamster (O and Baker, 1976), although a possible requirement for direct cell–cell contact has not been definitively excluded. Whether mediated by specific receptors for diffusible materials or by direct intercellular communication, these inductive events strongly implicate the plasma membrane of seminiferous cells in the regulation of meiosis. Further analysis requires the isolation and characterization of cell surface components responsible for triggering the switch from mitotic cell division to meiotic proliferation in the testis.

Spermiogenesis is the third and final phase of spermatogenesis. During spermiogenesis, haploid round spermatids undergo a remarkable series of morphological changes resulting in the formation of a mature spermatozoon having a distinct architectural polarity more extensive than any seen in somatic cells (Phillips, 1974). These alterations occur in the absence of DNA synthesis or cytokinesis and are known to be accompanied by major changes in plasma membrane constituents. Mammalian spermatozoa exhibit many unusual features at the cell surface which suggest that male gametes, and by inference earlier spermatogenic cells, have evolved membranes in some ways dissimilar from most somatic cell types.

First, mammalian spermatozoa have on their surfaces a series of serologically defined antigens not common to most tissues of the body. Examples of these restricted membrane markers include antigens shared only with the nervous system (NS-3; NS-4; Cb1) (Schachner et al., 1975; Weeds, 1975), the F9 antigen (Gachelin et al., 1976), and PCC4 antigen (Gachelin et al., 1977). In addition, spermatozoa contain LDH-X, a germ cell-specific isozyme of lactate dehydrogenase which is unusual in at least two regards. The enzyme may be located in germ cell mitochondria and not in the cytoplasm as is typical of somatic isozymes of lactate dehydrogenase (Machado de Domenech et al., 1972; Blanco et al., 1975). Furthermore, LDH-X may be present on the sperm plasma membrane (Erickson et al., 1975). Mature sperm also express the male specific H-Y antigen (Wachtel, 1977), Ia determinants (Hämmerling et al., 1975; Vaiman et al., 1978), and probably histocompatibility antigens themselves (cf. Erickson, 1976). Knowledge of the temporal appearance of specific sperm surface molecules during spermatogenesis remains limited. Intracellular synthesis of LDH-X is initiated during the late primary pachytene stage and continues during the first half of spermiogenesis (Meistrich et al., 1977), but no data regarding its eventual expression on the cell surface are available. Fellous and co-workers (1976) have reported that Ia determinants are also synthesized concomitant with the appearance of pachytene primary spermatocytes. In contrast, the F9 teratocarcinoma antigen, shared by spermatogenic cells, is found on all developing male germ cells including spermatogonia (Gachelin *et al.,* 1976). Further investigation of the temporal expression of spermatogenic cell membrane antigens is obviously required before detailed hypotheses relating these components to functional roles during sperm development can be formulated.

Second, spermatozoa are also unusual in that their plasma membranes exhibit relatively little fluid mobility. Investigations of a variety of cell surface antigens (Koehler, 1975) as well as different lectin receptors (Millette, 1976) indicate that many, and perhaps the majority, of the surface constituents on mature mammalian sperm show little or no lateral translocation when exposed to external macromolecular ligands even in the presence of colchicine, vinblastine, or altered temperature. Although restricted areas of the sperm membrane, particularly the post-acrosomal region of the head, may contain some mobile glycoproteins (Nicolson and Yanagimachi, 1974), it is evident that the fluid mosaic model of the cell surface requires modification when applied to the male gamete. Questions arise, therefore, concerning the dynamic nature of the membranes of developing spermatogenic cells. Little has been done in this regard, but a recent report by Romrell and O'Rand (1978) indicates that cell surface isoantigens identified on rabbit seminiferous cells are able to cap and patch on the membranes of primary spermatocytes and early spermatids, but not on later stages of spermatogenic cells. In the mouse, however, plasma membrane antigens which first appear on primary pachytene spermatocytes do not cap to a significant extent on the surfaces of any spermatogenic cells assayed (Millette *vide infra*). Further biochemical and immunological analyses are required (1) to elucidate the restrictive mechanisms underlying the lack of mobility by the spermatozoan plasma membrane and (2) to detail the generation of these mechanisms at the molecular level during spermatogenesis.

As spermiogenesis proceeds, yet another alteration in sperm surfaces occurs. The plasma membrane of a mature spermatozoon is not uniform along the length of the cell, but exhibits extensive localized distributions of surface moieties including lectin receptors (Millette, 1976), serologically identified antigenic determinants (Koehler, 1975), and intramembranous particles presumably reflecting the placement of transmembrane proteins (Friend and Fawcett, 1974). The H-Y male specific antigen, for example, has been localized to the cell surface overlying the acrosomal cap of the sperm head (Koo et al., 1973). LDH-X, on the other hand, is found only on the post-acrosomal surfaces of spermatozoa (Erickson et al., 1975). The restricted topographical localizations of membrane components may be species-specific and probably reflect the function of these molecules during sperm capacitation or fertilization. Virtually nothing is known regarding the appearance during spermiogenesis of cell surface constituents which, on mature gametes, are restricted to particular regions of the cell surface. Data obtained for a variety of lectin receptors present on mouse spermatogenic cells and evidence relating to the selective partitioning of plasma membrane molecules just prior to spermiation suggest, however, that molecular rearrangements occur during spermiogenesis and not during sperm passage through the epididymis or vas deferens (Millette, *vide infra*).

II. Isolation of Mammalian Seminiferous Cells

The extreme cellular complexity of the mammalian testis has been the major hindrance to detailed analysis of spermatogenic cell surface antigens. Germ cells found in the mature seminiferous epithelium include type A spermatogonia, type B spermatogonia, preleptotene primary spermatocytes, leptotene primary spermatocytes, zygotene primary spermatocytes, pachytene primary spermatids, secondary spermatocytes, developing spermatids at many distinct morphological stages of differentiation, residual bodies, and testicular spermatozoa. Sertoli cells completely surround and support all of the differentiating spermatogenic cells. In addition, outside the seminiferous tubule the testicular interstitium is composed of Leydig cells and cells of the blood and lymphatic vessels. Perhaps 15% of the interstitial cell population is composed of macrophages. Finally, the prepuberal testis includes some precursor germ cells, which although not present in adult tissue are of great importance in understanding the generation of spermatozoa. In the mouse, for example, gonocytes are present only until day 4 after birth. These cells differentiate to yield primitive type A spermatogonia which may be found until day 6 when further cell division yields the first type A spermatogonia resembling those of the mature organ (Bellvé et al., 1977a,b). Such cellular diversity has required the development of cell isolation and cell separation procedures in order to achieve homogeneous cell populations at specific stages of differentiation.

Highly purified populations of mouse spermatogenic cells may be prepared from both the adult and prepuberal mouse testis using procedures developed by Bellvé and co-workers (Romrell et al., 1976; Bellvé *et al.*, 1977a,b). Briefly, after perfusion to remove blood cells testes are excised and incubated sequentially in collagenase and trypsin. Collagenase treatment produces lengths of intact seminiferous tubules and allows the removal of virtually all interstitial cells. Subsequent incubation of isolated seminiferous tubules with trypsin produces suspensions consisting primarily of single germ cells from the seminiferous epithelium. Purification of individual classes of mouse spermatogenic cells is then accomplished by sedimentation velocity at unit gravity using procedures modified from Miller and Phillips (1969).

Adult mouse testis provides populations of late pachytene primary spermatocytes, round spermatids, and residual bodies. Condensing spermatids may be enriched, but not highly purified unless centrifugal elutriation techniques are employed (Meistrich, 1977). In the adult mouse seminiferous epithelium pachytene spermatocytes, round spermatids, and condensing spermatids account for over 85% of the total cells present (Bellvé *et al.*, 1977b), making successful high-yield purification of earlier germ cells difficult. The use of prepuberal animals, however, has allowed the preparation of homogeneous populations of the following additional cell types: primitive type A spermatogonia, type A spermatogonia, type B spermatogonia, preleptotene primary spermatocytes, leptotene/zygotene spermatocytes, and early pachytene primary spermatocytes (Bellvé *et al.*, 1977a,b). In addition, prepuberal Sertoli cells may be obtained in high yield and purity from 6-day-old mice (Bellvé *et al.*, 1977a,b). Adult Sertoli cells are not obtained by the enzymatic dissociation procedure outlined above because of the continued presence after enzymatic treatment of occluding junctions between adjacent Sertoli cells which prevent the preparation of single cell suspensions. Adult Sertoli cells may be prepared by short-term *in vitro* culture techniques (Dorrington and Fritz, 1974; Steinberger *et al.*, 1975).

The relative purity the spermatogenic cell populations obtained by sedimentation velocity at unit gravity (Table II) indicates that at least nine distinct classes of seminiferous cells may be isolated from the mouse testis for immunological and biochemical studies. These populations include cells at all major stages of spermatogenic differentiation. Procedures for the further purification of the leptotene/zygotene primary spermatocyte population using fiber fractionation techniques (Rutishauer *et al.*, 1972) or alternative methods are now possible with the production of antisera directed specifically against Sertoli cell surface constituents (Tung and Fritz, 1978; Feig, Bellvé and Millette, unpublished observations). Depending upon particular experimental needs, the cell separation technique can readily provide a minimum of

TABLE II

Purity of Isolated Mouse Seminiferous Cells[a]

Cell type	Purity (%)[b]
Primitive type A spermatogonia	90
Type A spermatogonia	91
Type B spermatogonia[c]	76
Preleptotene spermatocytes	93
Leptotene/zygotene spermatocytes[d]	52
Pachytene spermatocytes	89
Round spermatids	87
Residual bodies	89
Sertoli cells (prepuberal)	99

[a] Data from Romrell et al. (1976) and Bellvé et al. (1977a).

[b] Data are expressed as percentage of total cells recovered in each population.

[c] The isolated type B spermatogonia may contain some intermediate spermatogonia which cannot always be identified with certainty.

[d] The population of leptotene and zygotene spermatocytes has as its major contaminant Sertoli cells (29%).

2×10^7 of any one cell type. This cell yield may often be increased 5-fold and is sufficient for detailed biochemical investigations.

A major experimental difficulty with the seminiferous cell separation procedure as described involves the exposure of cells to proteolytic enzymes which may remove or alter germ cell membrane components. However, sufficiently highly purified mammalian germ cell suspensions cannot be obtained without enzymatic treatment. Mechanical methods of testicular cell dispersion produce large numbers of multinucleated symplasts caused by the apparent coalition of the extensive intercellular bridges which connect developing spermatogenic cells at similar stages of differentiation (Romrell et al., 1976). For example, mechanical preparation of single cells from the mouse testis may produce up to 42% of the nuclei present as symplasts. By comparison, the sequential enzymatic treatment when applied to the adult mouse testis yields only 12–13% of the total nuclei as symplasts (cf. Romrell et al., 1976, Table I). These considerations are important for two reasons. First, increased percentages of symplasts reduce the theoretical yield of purified spermatogenic cells at any one stage of differentiation. Second, symplasts greatly reduce the resultant purity of isolated spermatogenic cell populations. Binucleated round spermatids, for example, have a volume similar to that of mid-pachytene primary spermatocytes in the mouse. As the unit gravity sedimentation procedure is most dependent upon cell volume (Miller and Phillips, 1969), a high number of spermatid symplasts results in isolated pachytene populations of low purity. All experiments detailed in this chapter, therefore, have been conducted using spermatogenic cells exposed to enzymatic treatment.

The effects of enzymes on the surface molecules of mouse spermatogenic cells are not known. There is substantial evidence, however, that major modification of some membrane antigens does not occur. NS-4, a cell surface antigen present on both cerebellar tissue and on mouse spermatozoa, is not susceptible to tryptic proteolysis (Schachner et al., 1975), and it has been reported that H2 and Ia determinants on mammalian spermatogenic cells are not adversely affected by collagenase (Erickson, 1976). Moreover, a variety of lectins bind to mouse germ cells isolated by the enzymatic procedures already described (Millette, 1976), suggesting that many surface glycoproteins on developing male germ cells are not removed by enzymes. Finally, in all experiments to be described below, antibodies prepared against isolated seminiferous cells following enzymatic dissociation have also been assayed by indirect immunofluorescence on mixed seminiferous cell suspensions obtained by purely mechanical methods. No differences were detected in antibody labeling of cells prepared in the presence or absence of collagenase and trypsin. These results suggest that although proteolytic cleavage of spermatogenic cell membrane components may occur, the particular antigenic determinants detected by those immunoglobulin preparations discussed here are not affected qualitatively by the cell separation procedure. Unfortunately, it is not currently possible to conduct short-term *in vitro* culture of isolated mammalian germ cells in order to allow the regeneration of cell surface constituents prior to experimental assay. *In vitro* conditions providing a satisfactory milieu for single cell cultures of germ cells have not yet been developed and the majority of cells die within hours (Steinberger, 1975).

Mouse spermatogenic cells isolated in high purity by unit gravity sedimentation have been used to investigate the following: (1) the temporal appearance of cell surface antigens during spermatogenesis, (2) the relative fluid mobility of plasma membrane components at various stages of spermatogenesis, (3) the selective partitioning of membrane antigens during late spermiogenesis, and (4) the isolation of purified plasma membranes from male germ cells.

III. Temporal Appearance of Surface Antigens during Spermatogenesis

Four antibody preparations were used to examine the expression of cell surface antigens on isolated populations of mouse spermatogenic cells. Xenogenic immunoglobulin preparations were made in rabbits and were directed against (1) purified adult pachytene spermatocytes (designated AP), (2) purified round spermatids (ARS), (3) a mixture of cells from the adult seminiferous epithelium (ASC), and (4) spermatozoa obtained from the vas deferens (AVDS). Analysis of unabsorbed IgG preparations by fluorescence microscopy indicated that none of these antibodies labeled mouse thymocytes, erythrocytes, or peripheral blood lymphocytes. Approximately 30% of mouse splenocytes were labeled by all four immunoglobulin samples (Millette and Bellvé, 1977). Complement-mediated cytotoxicity assays indicated that all four IgG fractions lysed 29–36% of mouse splenocytes, but killed no significant numbers of erythrocytes or other somatic cells tested (Table III). These results suggest that xenogenic antiserum directed against seminiferous cells does not recognize *a priori* cell surface components common to all somatic cells. The organ specificity of such antisera is facilitated by the blood–testis barrier at the level of the Sertoli cell tight junctions. The barrier is known to prevent the free interchange of macromolecules between the adluminal aspects of the seminiferous epithelium and the testicular interstitium. All cell populations used to raise the antisera listed above (AP, ARS, ASC, AVDS) are comprised of adluminal cells protected by the Sertoli cell tight junctions and thereby isolated from the general immune system.

To ensure immunological specificity, however, the four IgG fractions were absorbed exhaustively with somatic cells. Fluorescence microscopy revealed that all reactivity with mouse splenocytes was removed by a single absorption using 2×10^7 splenocytes per milligram isolated IgG. Additional absorptions were conducted with erythrocytes, thymocytes, and particulate fractions of kidney, liver, and 4-day-old mouse cerebellum. Following these procedures, none of the IgGs lysed any mouse somatic cells tested in significant numbers (Table III). Fluorescence microscopy confirmed these findings and also indicated that rabbit, guinea pig, and human spermatozoa were not labeled by absorbed AP, ARS, ASC, or AVDS.

The reactivity of absorbed IgG preparations on isolated populations of mouse spermatogenic cells was assayed by fluorescence microscopy, by complement-mediated cytotoxicity, and by quantitative tests using radioiodinated immunoglobulin. Purified cell populations assayed in-

TABLE III

Cytotoxic Effects of AP, ARS, ASC, and AVDS on Mouse Cells[a]

Cell type	Dead cells (%)			
	AP	ARS	ASC	AVDS
Unabsorbed IgG				
Splenocytes	33	29	36	26
Thymocytes	10	8	8	7
Erythrocytes	10	7	11	8
Absorbed IgG				
Splenocytes	4	4	3	4
Thymocytes	3	3	5	4
Erythrocytes	5	3	4	4
Primitive type A spermatogonia	<5	6	<5	<5
Type A spermatogonia	<5	14	<5	7
Type B spermatogonia	14	19	5	7
Preleptotene spermatocytes	<5	<5	<5	<5
Leptotene/zygotene spermatocytes	<5	<5	<5	<5
Pachytene spermatocytes	85	65	66	80
Round spermatids	55	90	76	80
Residual bodies	25	22	22	36

[a] Data represent the maximum percentage killed cells at an IgG concentration of 100 μg/ml. Results are expressed as the average of triplicate determinations, which differed <5%. Data from Millette and Bellvé, 1977.

cluded primitive type A spermatogonia, type A spermatogonia, type B spermatogonia, preleptotene spermatocytes, leptotene/zygotene spermatocytes, prepuberal pachytene spermatocytes, adult pachytene spermatocytes, round spermatids, and residual bodies.

All four of the antibodies revealed identical labeling patterns when tested on spermatogenic cells. Adult pachytene spermatocytes, round spermatids, and residual bodies were strongly labeled (Fig. 1). All earlier spermatogenic cells examined, from primitive type A spermatogonia to prepuberal pachytene primary spermatocytes, showed no immunofluorescence after treatment with either AP, ARS, ASC, or AVDS. None of these earlier germ cell classes were agglutinated at any IgG concentration, in contrast to adult pachytene spermatocytes, round spermatids, and residual bodies which were all agglutinated readily. These results indicate that cell membrane antigens specific to spermatogenic cells appear at the late pachytene stage of the first meiotic prophase in the mouse. The cell surface determinants are also expressed by male germ cells at subsequent stages of differentiation.

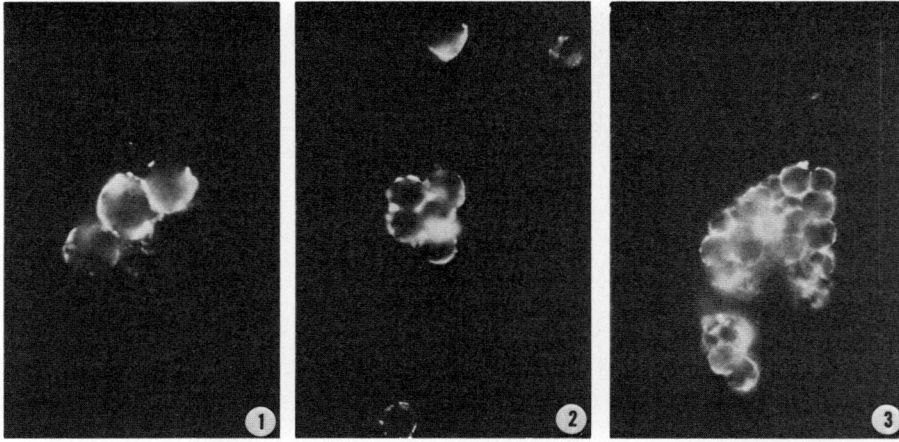

FIG. 1. Binding of rabbit antibody against adult mouse pachytene spermatocytes (AP) to purified mouse spermatogenic cells. Indirect immunofluorescence: (1) isolated pachytene spermatocytes (×646), (2) isolated round spermatids (×646), (3) isolated residual bodies (×646). No spermatogenic cells prior to late pachytene spermatocytes are labeled by AP. Identical results are obtained using antibodies ARS, ASC, and AVDS when assayed by indirect immunofluorescence.

Complement-mediated cytotoxicity was used to confirm the immunofluorescent results obtained with AP, ARS, ASC, and AVDS. As shown in Table III, primitive type A spermatogonia, type A spermatogonia, type B spermatogonia, preleptotene spermatocytes, and leptotene/zygotene spermatocytes were not lysed in significant numbers under any experimental conditions. In addition, pachytene primary spermatocytes obtained from prepuberal mice were not lysed by any IgG fraction (data not shown). Prepuberal mouse pachytene spermatocytes consist predominantly of cells in the first half of the prolonged pachytene stage of the first meiotic prophase and are significantly smaller and less developed than pachytene spermatocytes obtained from adult animals (Bellvé et al., 1977a).

Mouse spermatogenic cells at more advanced stages of differentiation, however, were lysed by AP, ARS, ASC, and AVDS (Table III). Although all antibody preparations exhibited weak cytotoxic titers, with little activity detected at immunoglobulin concentrations below 6.25 µg/ml, this may be attributed to the relative inefficiency of purified IgG in cytotoxic reactions. An optimal antibody levels, from 55 to 90% of pachytene spermatocytes and round spermatids, were lysed, in good agreement with the immunofluorescence results previously discussed. Residual bodies did not exhibit a marked capacity for cytotoxic lysis. This result remains unexplained (see Section V).

Finally, as a third measure of antibody binding to male germ cell surfaces, purified populations of intact mouse spermatogenic cells were incubated with radioiodinated preparations of AP, ARS, ASC, and AVDS. Saturating conditions were used in order to quantitate the number of antibody receptors per cell. As shown in Table IV, fewer than 1000 receptors per cell were detected on germ cells prior to late pachynema. Adult mouse pachytene spermatocytes, however, had approximately 2×10^5 receptors per cell while round spermatids, residual bodies, and vas deferens spermatozoa bound between 6×10^4 and 1.5×10^5 antibody molecules per cell depending upon the particular IgG fraction and germ cell type assayed.

Indirect immunofluorescence, cytotoxicity, and quantitative measurements all indicate, therefore, that new antigenic components are inserted into the plasma membrane of late pachytene primary spermatocytes and that these components are retained during the following stages of spermiogenesis. These membrane constituents are specific to spermatogenic cells and are not detected on spermatozoa of humans, guinea pigs, or rabbits. Immunofluorescent analysis of antibody binding to mature mouse spermatozoa suggests that AP, ARS, ASC, and AVDS have distinctive immunological specificities. It appears, therefore, that multiple cell surface antigens first appear during late pachytene of the first meiotic prophase (Millette and Bellvé, 1977).

Similar findings have been reported for both the rat (Tung and Fritz, 1978) and the rabbit (O'Rand and Romrell, 1977). Auto- and isoantisera raised in rabbits against whole rabbit semen were used to

TABLE IV

NUMBER OF ANTIBODY RECEPTORS ON ISOLATED GERM CELLS[a]

Cell type	Sites per cell ($\times 10^3$)			
	AP	ARS	ASC	AVDS
Primitive type A spermatogonia	<1	<1	<1	<1
Type A spermatogonia	<1	<1	<1	<1
Type B spermatogonia	<1	<1	<1	<1
Preleptotene spermatocytes	<1	<1	<1	<1
Leptotene/zygotene spermatocytes	<1	<1	<1	<1
Pachytene spermatocytes	274	180	241	183
Round spermatids	98	75	63	89
Residual bodies	103	86	99	106
Spermatozoa	62	61	90	147

[a] Data are expressed as the average of triplicate assays which differed by <10%. Background labeling (<5% of total binding) has been subtracted from these values. From Millette and Bellvé, 1977.

examine the temporal expression of cell surface antigens on purified populations of rabbit spermatogenic cells. Complement-dependent cytotoxicity studies indicated that developing germ cells bound antibody only after differentiation into pachytene primary spermatocytes. Sertoli cells, endothelial cells, Leydig cells, and erythrocytes were not lysed by the globulin fractions of the two antisera tested (O'Rand and Romrell, 1977). Tung and Fritz (1978) examined the same problem using homologous antisera prepared by the injection of pachytene spermatocytes, purified from Lewis/Wistar rats, into male animals of the identical strain. Heterologous rabbit antibody against rat pachytene cells was also prepared. Using cytotoxic measurements and immunohistochemical procedures, these investigators demonstrated that pachytene spermatocytes were the earliest cells in the spermatogenic lineage expressing cell membrane antigens recognized by the antisera. Antigens detected on rat spermatogenic cells were testis-specific and, moreover, did not bind to elongating spermatids or to maturing spermatozoa from the epididymis (Tung and Fritz, 1978). These findings are in good agreement with the results presented above regarding the expression of cell surface determinants during mouse spermatogenesis.

The appearance of new cell surface determinants on pachytene primary spermatocytes occurs concomitantly with extensive cell growth during the first meiotic prophase. Preleptotene primary spermatocytes are the smallest germ cells in the mouse having an average of diameter of 7.8 μm after isolation (Bellvé et al., 1977a) with a surface area of approximately 200 μm^2. As prophase proceeds, cell volume increases dramatically until mature pachytene primary spermatocytes attain an average diameter of 16 μm and a calculated surface area of 800 μm^2 (Romrell et al., 1976; Millette, 1976). These data imply that extensive amounts of plasma membrane must be synthesized during the first meiotic prophase. Such assembly would provide a ready mechanism for the insertion of new surface constituents not present on earlier spermatogenic cells. In addition, recent morphological data indicate that the plasma membranes of germ cells change sometime soon after the differentiation of type B spermatogonia into primary spermatocytes (Mollenhauer et al., 1977). Ultrastructural examination reveals a distinct asymmetry in the inner and outer leaflet organization of spermatocyte plasma membranes with an apparent reduction in the electron density of the outermost lamella. This alteration persists throughout spermatid development and is specific to the germinal elements of the seminiferous epithelium. The relationship of these gross morphological changes in the plasma membrane to the insertion of the antigenic determinants recognized by immunological techniques remains to be determined, but the available data strongly suggest that extensive cell surface reorganizations occur during late spermatogenesis in mammals.

Finally, the appearance of new surface antigens coincides temporally with the synthesis of new intracellular proteins. LDH-X, for example, first appears in pachytene primary spermatocytes (Hintz and Goldberg, 1977), and an elegant study by Meistrich and co-workers (Meistrich et al., 1977) demonstrates that LDH-X is first synthesized at this time. Similarly, a testis-specific form of cytochrome c, designated cytochrome c_t, is also absent from developing germ cells until the development of pachytene primary spermatocytes (Wheat et al., 1977), although it has not yet been determined that cytochrome c_t is first synthesized in these cells. The possibility exists, therefore, that coordinate gene activation during the first meiotic prophase is responsible for the simultaneous appearance of LDH-X, cytochrome c_t, and as yet uncharacterized cell surface markers (Wheat et al., 1977).

IV. Membrane Mobility of Spermatogenic Cells

Initial experiments have been conducted to determine the relative mobility of lectin and antibody receptors in the plasma membranes of late spermatogenic cells purified from the mouse seminiferous epithelium. Seminiferous cells obtained after collagenase and trypsin treatment from adult mouse testis were incubated at 0°C with either AP or ARS. After temperature shift to 33°C, testicular temperature in the mouse, or 37°C only, 18–25% of the cells exhibited distinct caps when assayed by indirect immunofluorescence. Purified populations of adult mouse germ cells were also assayed. Pachytene primary spermatocytes, round spermatids, and residual bodies were capped in similarly low percentage after labeling with either AP or ARS and subsequent temperature shift (Table V). The relative lack of observed membrane mobility in these cells may be the result of prior enzymatic treatment. Preliminary control experiments, however, indicate that mouse seminiferous cells prepared by mechanical methods also exhibit few capped cells after temperature shifts (Table V). Experiments conducted with concanavalin A and wheat germ agglutinin reveal that late mouse spermatogenic cells do not cap readily (data not shown). Patching of male germ cells is, however, detected in all experiments.

Using isoantisera prepared in female rabbits injected with whole rabbit semen, Romrell and O'Rand (1978) have also reported that membrane mobility decreases as spermatogenesis proceeds. These investigators detected capping by rabbit pachytene primary spermato-

TABLE V
Membrane Mobility of Mouse Spermatogenic Cells[a]

Cell type	Isolation treatment		Capping (%)		
			0°C	33°C	37°C
Seminiferous cell mixture	AP	Enzymes	11	24	25
	AP	No enzymes	0	19	23
	ARS	Enzymes	12	18	18
	ARS	No enzymes	7	21	21
Pachytene spermatocytes	AP	Enzymes	0	16	23
	ARS	Enzymes	5	15	16
Round spermatids	AP	Enzymes	0	8	10
	ARS	Enzymes	10	27	24
Residual bodies	AP	Enzymes	0	5	7
	ARS	Enzymes	0	5	8

[a] Cells were incubated in 10–100 μg/ml IgG for 15 minutes at 0°C and then observed continuously at 33°C or 37°C. Data recorded after 30 minutes; further incubations demonstrated no increase in capping percentage. Data are expressed as the average of four determinations. Control assays using mouse splenocytes and rabbit anti-mouse IgG yielded 97% caps after 30 minutes at 37°C.

cytes, but little such activity when residual bodies or mature spermatozoa were examined. Early and mid-condensing spermatids exhibited caps, but late spermatids did not. Although no controls were presented to investigate possible membrane damage caused by enzymatic treatment of cells prior to labeling, these results are in general agreement with the data discussed above regarding purified mouse spermatogenic cells.

It is obvious that additional experiments investigating the lateral mobility of spermatogenic cell membrane constituents are required. The answers to when, where, and how the developing germ cell surface restricts the movement of its components to achieve the apparently rigid phenotype of the mature spermatozoan surface may provide new insights into the regulation of spermatogenesis and fertilization, as well as the assembly and behavior of somatic cell membranes. During erythropoiesis, for example, the mobility of binding sites for both concanavalin A and *Ricinus communis* agglutinin undergoes marked alteration between the proerythroblast and the erythroblast stages of differentiation. As differentiation proceeds the membrane mobility of lectin receptors decreases (Chan and Oliver, 1976). Finally, changes in surface membrane mobility have also been implicated in the differentiation of embryonic cells (Robertson *et al.*, 1975). Thus, it appears that decreased mobility of cell surface moieties may be a general occurrence accompanying differentiative events in many cell types. The available techniques for the purification of mammalian germ cells suggest that spermatogenesis should be a primary system of investigation for detailed analysis of the regulation of membrane translocations.

V. Selective Partitioning of Membrane Antigens during Spermiogenesis

The molecular mechanisms responsible for the release of spermatozoa into the tubule lumen are not known. Spermiation begins with the breakdown of membrane junctional specializations between Sertoli cells and condensing spermatids (Ross, 1976). As the sperm loosens from the Sertoli cell, the spermatid cytoplasm becomes lobulated, but remains attached to the spermatid neck by a thin stalk (Fawcett and Phillips, 1969) which finally separates to release the spermatozoon. The cytoplasmic remnant or residual body is phagocytized by the Sertoli cell. Spermiation, therefore, results in the physical segregation of two segments of the spermatid plasma membrane.

It is possible that the segregation of cell surface components into the residual body membrane is nonrandom. For example, specific membrane receptors responsible for the adhesion of germ cell to Sertoli cell could be sequestered on the residual body to allow sperm release. Preliminary data demonstrating a greater density of concanavalin A sites on the surfaces of residual bodies in comparison with spermatids support the idea that individual membrane components are, in fact, partitioned selectively during late spermiogenesis (Millette, 1976). To examine this question directly a number of rabbit anti-mouse germ cell IgG preparations were assayed for binding to purified spermatogenic cells. Particular emphasis was placed on determining the relative binding of the antibodies to round spermatids, residual bodies, and maturing spermatozoa.

An IgG fraction directed against purified type B spermatogonia provided evidence for the selective partitioning of spermatogenic cell surface antigens. Binding of the rabbit anti-mouse type B spermatogonia IgG (ATBS), unabsorbed and absorbed, was first assayed by indirect fluorescence microscopy on mouse germ cells and somatic cells. Unabsorbed ATBS labeled all mouse somatic germ cells examined (Fig. 2). Mouse thymocytes and splenocytes were strongly labeled by ATBS in a patchy fashion. With the noted exception of spermatozoa, mouse germ cells were labeled diffusely. Spermatozoa from the testis, the caput epididymis, cauda epididymis, and vas deferens were all labeled

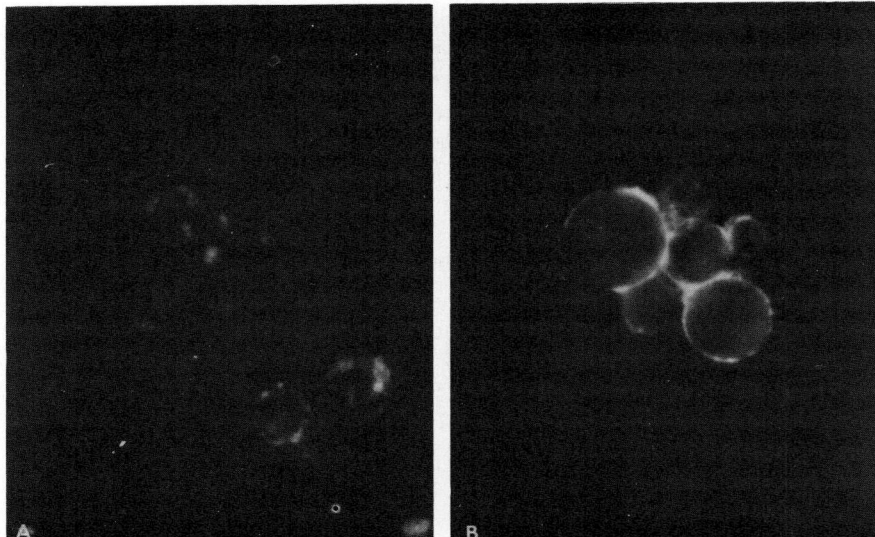

FIG. 2. Binding of unabsorbed rabbit antibody against purified mouse type B spermatogonia (ATBS) to mouse cells. (A) Splenocytes are labeled in a patchy manner (×723). (B) Mixed spermatogenic cells from the adult mouse seminiferous epithelium are labeled uniformly (×723). Cells shown include pachytene spermatocytes, round spermatids, and residual bodies.

in a distinctive pattern. The entire cell surface, except for lateral areas of the sperm head anterior and posterior to the post-acrosomal segment, was labeled uniformly (Fig. 3).

Immunofluorescence indicated that unabsorbed ATBS was not specific for spermatogenic cell surface antigens. Mouse thymocytes and splenocytes were labeled strongly. This was in contrast to the earlier results discussed for AP, ARS, ASC, and AVDS where unabsorbed IgG preparations labeled few somatic cells. The difference in immunological specificities may be explained by the respective cell populations used as immunogens; spermatogonia lie on the basal side of the Sertoli cell tight junctions which form the blood–testis barrier (Dym and Fawcett, 1970) and are not protected from the immune system as are pachytene spermatocytes, round spermatids, and spermatozoa. Spermatogonia, therefore, would not be expected to express cell surface components which could engender an autoimmune response.

Absorption of ATBS with mouse thymocytes or splenocytes removed all labeling of thymocytes, splenocytes, peripheral blood lymphocytes, erythrocytes, and Leydig cells as assayed by indirect immunofluores-

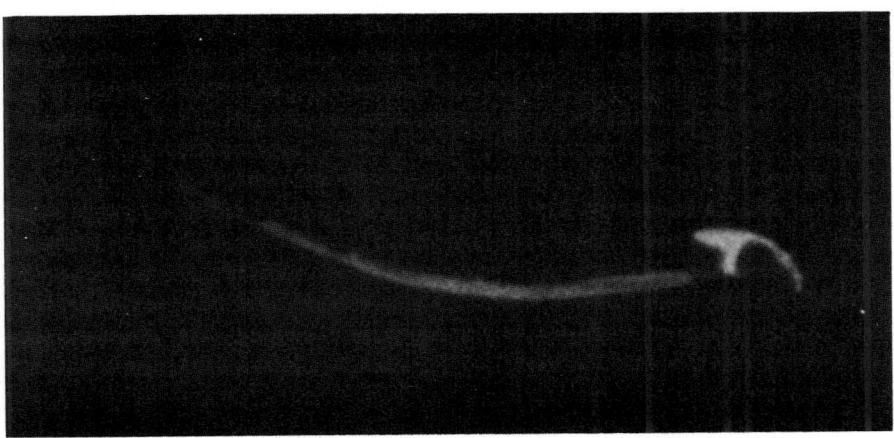

FIG. 3. Mouse vas deferens spermatozoon labeled by unabsorbed ATBS. Indirect immunofluorescence. The cell surface is uniformly labeled, except for lateral areas of the sperm head anterior and posterior to the post-acrosomal region. Absorbed ATBS does not bind to spermatozoa. ×1599.

cence. Mouse Sertoli cells, however, from both prepuberal and adult animals were still labeled strongly in a diffuse fashion. The significance of this binding activity is not yet clear. Experiments conducted using purified populations of mouse spermatogenic cells revealed that primitive type A spermatogonia, type A spermatogonia, type B spermatogonia, preleptotene primary spermatocytes, leptotene/zygotene primary spermatocytes, prepuberal pachytene spermatocytes, adult pachytene spermatocytes, round spermatids, and residual bodies were all labeled. In contrast, spermatozoa obtained from the testis, caput epididymis, cauda epididymis, or vas deferens showed no labeling by absorbed ATBS.

These data indicate that ATBS recognizes at least two classes of cell surface antigenic determinants. One class of antigens is shared by germ cells and somatic cells. These components are present on testicular, epididymal, and vas deferens spermatozoa and show a nonrandom topographical distribution on these cells. Other surface antigens are found only on Sertoli cells, on spermatogenic cells at early stages of differentiation, and on residual bodies. This class of antigens is not found on spermatozoa at any stage of maturation.

Complement-mediated cytotoxicity assays were used to obtain further information on the binding specificity of unabsorbed and absorbed antibody. All germ cell populations tested, in addition to prepuberal mouse Sertoli cells, were lysed by ATBS before and after absorption

(Table VI). No significant differences were noted between the two antibody preparations. With the exception of residual bodies, all testicular cell populations were killed in significant numbers (77–90%). Residual bodies, however, were not killed at levels greater than 50% by any concentration of either antibody preparation. The membrane of residual bodies has been found previously to be relatively resistant to complement-mediated lysis (cf. Table III).

Finally, radioiodinated ATBS, IgG, after absorption, was used to quantitate cell surface receptor sites on isolated mouse germ cells. Primitive type A spermatogonia bound the greatest number of absorbed antibody molecules, 2.7 million per cell, while round spermatids and residual bodies each had about 1 million surface receptor sites (Table VII). Estimations were made of the cell membrane density of antigenic receptors for absorbed antibody based upon the data shown in Table VII and the cell surface area calculated from the respective volumes of isolated spermatogenic cells (Bellvé et al., 1977b). These calculations indicate that the density of antibody receptors per cell increases approximately 2.5-fold during the mitotic proliferation of spermatogonia. In contrast, the density of surface receptors decreases approximately fourfold during the first meiotic prophase as the developing primary spermatocytes increase in size without concomitant cell division.

Vas deferens spermatozoa bound very low levels of absorbed ATBS. Only 0.003 million sites per cell were detectable (Table VII). Contrasted with the 1 million sites per cell seen on both round spermatids and residual bodies, these data are in good agreement with the immunofluorescent results already presented. Some cell surface antigens

TABLE VI

Cytotoxic Effect of ATBS on Mouse Seminiferous Cells[a]

Cell type	Unabsorbed IgG	Absorbed IgG
Primitive type A spermatogonia	86	85
Type A spermatogonia	87	77
Type B spermatogonia	86	88
Preleptotene spermatocytes	83	79
Leptotene/zygotene spermatocytes	77	81
Prepuberal pachytene spermatocytes	83	85
Adult pachytene spermatocytes	89	90
Round spermatids	90	84
Residual bodies	50	50
Prepuberal Sertoli cells	83	82

[a] Data represent the maximum percentage killed cells at an IgG concentration of 100 µg/ml. Results are expressed as the average of triplicate assays.

TABLE VII

NUMBER OF BINDING SITES FOR ABSORBED ATBS[a]

Cell type	Sites per cell ($\times 10^3$)
Primitive type A spermatogonia	2670
Type A spermatogonia	2150
Type B spermatogonia	1510
Preleptotene spermatocytes	2050
Leptotene/zygotene spermatocytes	2480
Prepuberal pachytene spermatocytes	1430
Adult pachytene spermatocytes	2230
Round spermatids	1030
Residual bodies	920
Vas deferens spermatozoa	3

[a] Data represent the average of triplicate assays differing by <10%. Background labeling was always <5% of the total binding and has been subtracted.

detected on mouse spermatogenic cells appear to be partitioned during spermiogenesis so that they are excluded from the spermatozoon.

Spermatozoa from the caput epididymis, cauda epididymis, and vas deferens were examined to quantitate the temporal disappearance of binding sites for absorbed ATBS. Only minor quantitative differences were detected between epididymal spermatozoa and vas deferens spermatozoa (Table VIII). Thus, the most dramatic alteration in the number of membrane receptors for ATBS, as depicted by the disparity in antigenic sites between round spermatids and vas deferens spermatozoa (Table VII), occurs before the cells enter the caput epididymis.

TABLE VIII

NUMBER OF ATBS RECEPTORS DURING SPERM MATURATION[a]

Cell type	Sites per cell ($\times 10^3$)	
	Unabsorbed IgG	Absorbed IgG
Caput spermatozoa[b]	8542	4
Cauda spermatozoa[c]	8710	4
Vas deferens spermatozoa	9720	3

[a] All cells were washed three times in enriched Krebs–Ringer buffer (EKRB) before labeling. Data represent the average of triplicate determinations, differing by <7%. Background labeling was <5% of total binding and has been subtracted.
[b] Sperm taken from the caput epididymis.
[c] Sperm taken from the cauda epididymis.

Furthermore, the immunofluorescent data obtained for testicular spermatozoa demonstrate that these membrane components assume a discrete topographical regionalization prior to spermiation.

In contrast to other spermatogenic cells, vas deferens spermatozoa bound less of the absorbed antibody and more unabsorbed ATBS (Table VII). It was important to test the possibility that sperm surface receptors for ATBS were peripheral as opposed to integral plasma membrane components. Fluid secretions of the epididymis or the vas deferens could be adsorbed onto the sperm surface, thereby either creating peripheral binding sites for unabsorbed ATBS or nonspecifically masking integral binding sites for the absorbed antibody. Therefore, vas deferens spermatozoa were first washed extensively in a variety of suspension media known to remove peripheral cell membrane constituents prior to the quantitation of antibody receptor sites.

The normal washing procedure resulted in values of 10 million receptor sites per cell for unabsorbed ATBS and only 0.003 million sites per cell for absorbed antibody. Treatment of vas deferens spermatozoa in five other media had only a minimal effect on the number of binding sites for unabsorbed ATBS (Table IX). The greatest decrease in surface binding sites obtained was 25% using 200 mM NaCl in Toyoda's medium (Toyoda et al., 1971). Although the small differences seen between incubation media were reproducible, these data strongly imply that the binding of unabsorbed ATBS to mouse vas deferens spermatozoa secretions was not due to components adsorbed from epididymal secretions.

TABLE IX

NUMBER OF ATBS RECEPTORS ON WASHED VAS DEFERENS SPERMATOZOA[a]

Wash solution	Sites per cell ($\times 10^3$)	
	Unabsorbed IgG	Absorbed IgG
EKRB[b]	9720	3
PBS	8535	3
0.15 M NaCl	8100	5
3 M KCl	7475	3
10 mM EDTA/NaCl	7370	4
200 mM NaCl/Toyoda	7355	4

[a] Cells were washed three times in the indicated medium before resuspension in EKRB and the addition of ^{125}I-labeled ATBS. Data are expressed as the average of triplicate assays which differed by <11%. Background labeling was <5% of total binding and has been subtracted.

[b] Abbreviations: EKRB = enriched Krebs–Ringer bicarbonate buffer, pH 7.25; PBS = phosphate-buffered saline, pH 7.4; Toyoda = from Toyoda et al., 1971.

The increased number of receptors may instead result from the elongation of the sperm flagellum during spermiation. Unabsorbed ATBS does bind to the plasma membrane on the sperm tail (Fig. 3).

Conversely, the alteration of incubation media had little or no effect on the binding of absorbed ATBS to mouse vas deferens spermatozoa (Table IX). In all intances, only 0.003–0.005 million receptor sites per cell were detected. These values are just above the sensitivity limit of the quantitative assay used for these experiments. The incubation media tested included physiological fluids [enriched Krebs–Ringer bicarbonate buffer (EKRB), phosphate-buffered saline (PBS), and 0.15 M NaCl], slightly hypertonic conditions known to remove decapacitation factors from mouse spermatozoa (200 mM NaCl/Toyoda), strongly hypertonic conditions (3 M KCl), and chelating agents (10 mM EDTA/NaCl). It does not appear, therefore, that receptor sites for absorbed ATBS are masked on spermatozoa by epididymal secretions. Instead, these receptors seem to be absent from the sperm surface entirely, having been partitioned selectively to the residual body membrane.

Several mechanisms could account for the apparent redistribution of plasma membrane components prior to spermiation. Receptors for ATBS may be masked nonspecifically by newly adsorbed peripheral surface molecules. Alternatively, the antigenic sites may be selectively removed from specific areas of the developing sperm plasma membrane by proteolysis or by internalization. A third and more plausible explanation is that these antigenic determinants are partitioned by lateral translocation in the plane of the membrane. The nonselective masking of ATBS receptors by adsorbed peripheral proteins appears unlikely for two reasons. First, the results of membrane partitioning are evident even on testicular spermatozoa which show no binding of absorbed ATBS. Testicular cells are not exposed to the membrane coating components present in epididymal secretions (Gordon *et al.*, 1975). Second, extensive washing of spermatozoa in media known to remove adsorbed peripheral membrane constituents and decapacitation factors failed to expose additional binding sites for the absorbed antibody (Table IX). The selective removal of the surface receptors for absorbed ATBS by hydrolytic processes or internalization cannot be ruled out completely. It should be noted, however, that the number of surface receptors detected on round spermatids is quantitatively similar to the number found on purified residual bodies (Table VII). It seems unlikely that residual bodies are able to synthesize plasma membrane receptors in order to replace quantitatively any molecules lost from round spermatids by proteolysis. There is no reported evidence for protein synthesis by residual bodies.

A selective and differential partitioning of membrane molecules is a more plausible explanation for the present observations. Selective lateral translocation of surface components has been demonstrated to occur during erythroid cell differentiation. Chan and Oliver (1976) have described the asymmetric distribution of concanavalin A receptors during the transition from proerythroblast to erythroblast in the chicken. Also, following enucleation during rabbit erythropoiesis, the residual membrane overlying the extruded nucleus is selectively enriched in both concanavalin A receptors and antigenic sites (Skutelsky and Danon, 1970). Finally, a discrete segregation of two intrinsic membrane glycoproteins and of acetylcholinesterase has been demonstrated during the vesiculation of erythrocytes (Lutz et al., 1977). The selective partitioning of plasma membrane components, therefore, is not limited to spermatogenic cell differentiation and may be of general importance for the membranes of diverse cell types.

VI. Isolation of Germ Cell Plasma Membranes

The physiological function of the different spermatogenic cell surface antigens that have been identified using serological techniques must yet remain speculative. To facilitate the biochemical characterization of these surface moieties, procedures have been developed for the purification of plasma membrane fractions from separated mouse spermatogenic cells. The availability of purified cell membranes should (1) allow the application of immunoprecipitation techniques to concentrate particular antigens, and (2) allow the production of monoclonal antibodies directed against specific membrane determinants using hybridoma technology (Barnstable et al., 1978).

Plasma membranes have been prepared from adult mouse seminiferous cells and from isolated populations of pachytene primary spermatocytes, round spermatids, and residual bodies using the methods of Atkinson and Summers (1971). After brief hypotonic shock, developing germ cells are gently homogenized and centrifuged to remove intact cells and nuclei. The supernatant material is then centrifuged using sucrose density gradients to yield purified plasma membrane.

A variety of experiments indicate that cell surface membranes are obtained in high purity. First, preincubation of seminiferous cells with either radioiodinated green pea lectin, wheat germ agglutinin, or *Ricinus communis* agglutinin, followed by density gradient fractionation, reveals that labeled lectin sediments predominantly to that position in the gradient containing membrane (Fig. 4). Previous experiments had

FIG. 4. Sucrose-density gradient fractionations of mouse spermatogenic cell membranes labeled with radioiodinated lectin. Linear gradients from 10 to 45% sucrose w/v in 0.01 M Tris, pH 8.0. (A) Cells were prelabeled with ^{125}I-labeled green pea lectin and

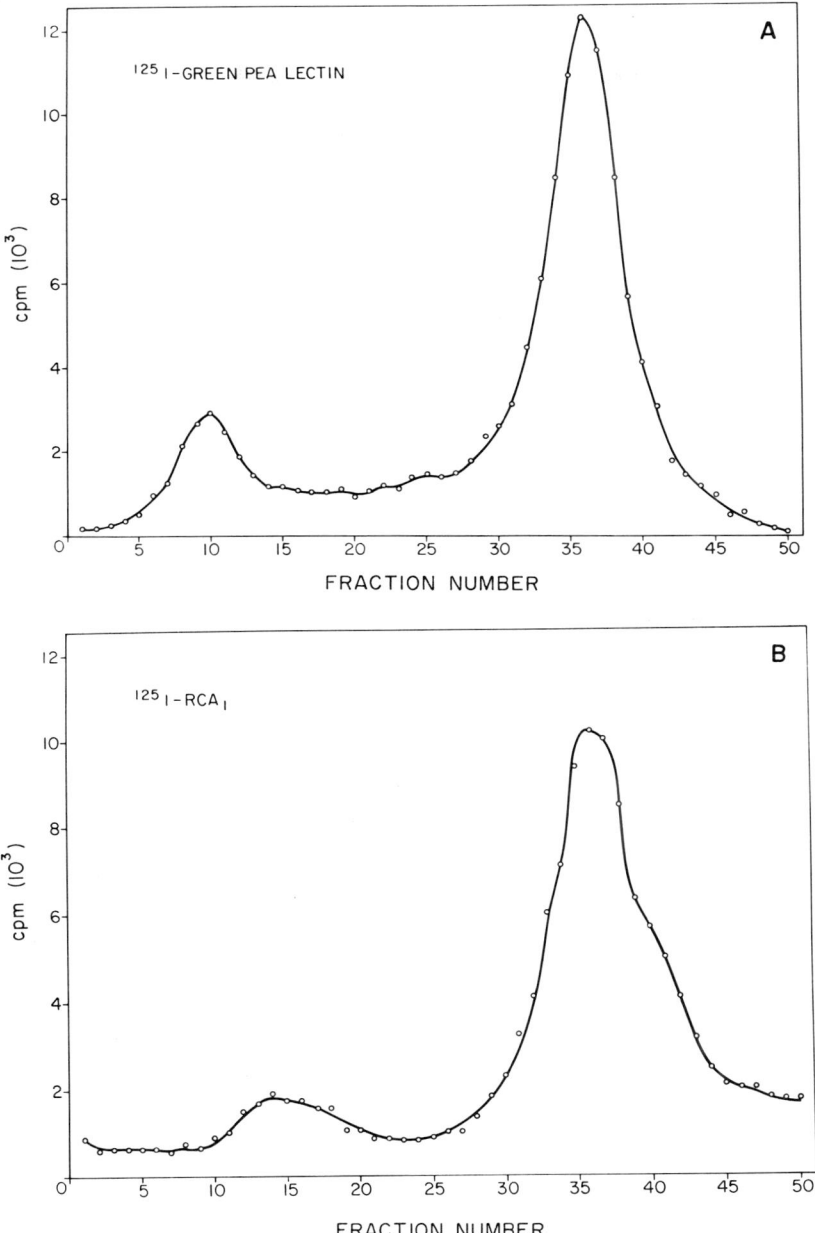

membranes then isolated by the method of Atkinson and Summers (1971). Unbound lectin was detected at the top of the gradient (tubes 6–13). Virtually all bound lectin was detected in a single peak centered on tube 36. Total gradient volume was 10 ml. Fraction size was 0.2 ml. (B) Similar results were obtained using adult mouse spermatogenic cells prelabeled with ^{125}I-labeled RCA_I.

demonstrated that all three lectins, under the conditions employed, labeled only the external cell membrane (C. F. Millette, unpublished observations). Second, ultrastructural examination of material obtained from density gradients reveals a collection of small membrane vesicles and sheets (Fig. 5) with little or no contamination by mitochondria, lysosomes, nuclei, or other intracellular organelles. Third, preliminary measurements of enzymatic activity isolated in the membrane fraction suggest little contamination by a number of cytoplasmic enzymes or by marker enzymes for mitochondria, lysosomes, and endoplasmic reticulum (C. F. Millette, unpublished observations). Unfortunately, this approach is currently limited by the fact that at present there is no known positive enzyme marker localized definitely to the outer membranes of mammalian spermatogenic cells. Possible candidates such as 5'-nucleotidase or Na^+–K^+-dependent ATPase have not yet proved reliable and these investigations are continuing. In this

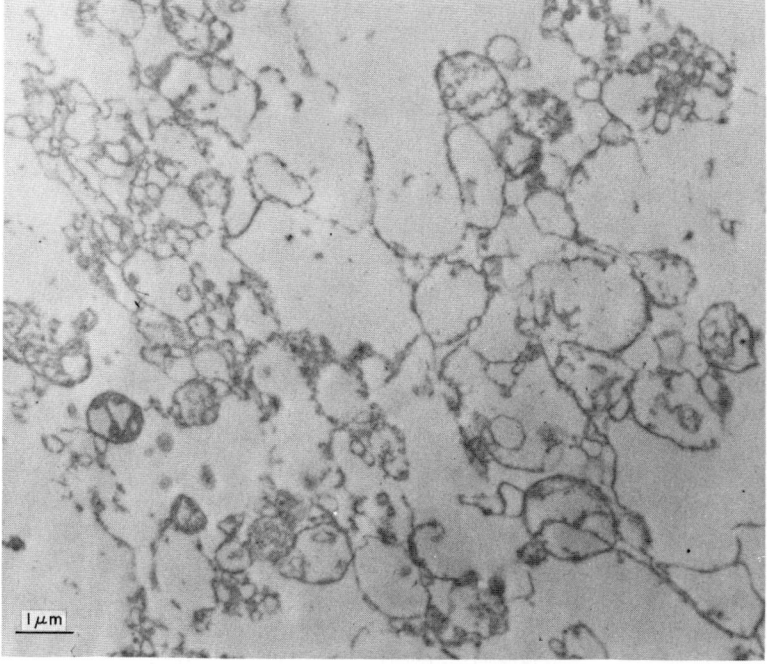

FIG. 5. Electron micrograph of membranes pelleted from sucrose density gradients. The material consists of vesicles and sheets of membrane with little visible contamination by intracellular organelles. ×5775.

regard the lectin-binding experiments already discussed have proved most helpful.

It appears likely, however, that surface membrane fractions have been obtained in high purity from mouse spermatogenic cells. The yield of membranes approximates 100–200 µg per experiment and is sufficient for biochemical and immunological study. These preparations should allow the eventual identification of individual cell surface antigens important for the regulation of mammalian spermatogenesis. This information, in turn, should suggest new approaches for the control of fertility and also provide new information concerning the assembly and functional roles of cell membranes during differentiation.

VII. Concluding Remarks

Immunological analysis of purified mouse spermatogenic cells indicates that the mammalian testis exhibits cell surface antigenic determinants which are specific to developing male gametes. Using immunofluorescence, complement-mediated cytotoxicity, and quantitative assays, a variety of antigens have been detected which first appear during late pachynema of the first meiotic prophase. These antigens are expressed by all late pachytene primary spermatocytes, round spermatids, and residual bodies in a uniform diffuse fashion on the surface membranes. Mature spermatozoa also express these surface components, but on these cells antigenic sites are detected only in restricted regions of the sperm plasmalemma. Earlier spermatogenic cells, including spermatogonia, preleptotene spermatocytes, leptotene spermatocytes, zygotene spermatocytes, and prepuberal pachytene spermatocytes do not express detectable antibody binding sites. These data provide evidence that mammalian spermatogenesis is a useful system for the molecular analysis of membrane constituents during cellular differentiation.

In addition, using antibody prepared against purified mouse type B spermatogonia, the selective partitioning of membrane antigens during spermiogenesis has been described. As spermatozoa mature morphologically by spermatid condensation and elongation, some cell surface constituents become sequestered on the residual body membrane. Primary spermatocytes, round spermatids, and residual bodies have 1–2 million antibody receptor sites. In contrast, testicular, epididymal, and vas deferens spermatozoa have only 0.003 million binding sites. Washing experiments suggest that adsorption of male reproductive tract secretions does not simply mask these particular antigenic determinants on spermatozoa. Instead, these antigens appear to be partitioned selectively and quantitatively to the residual body surface.

Finally, procedures have been developed for the isolation of mouse spermatogenic cell plasma membranes. These preparations will facilitate the biochemical characterization of cell surface determinants initially detected using immunological techniques.

REFERENCES

Atkinson, P. H., and Summers, D. F. (1971). *J. Biol. Chem.* **16,** 5162–5175.
Barnstable, C. J., Bodmer, W. F., Brown, G., Galfre, G., Milstein, C., Williams, A. F., and Ziegler, A. (1978). *Cell* **14,** 9–20.
Bellvé, A. R., Cavicchia, J. C., Millette, C. F., O'Brien, D. A., Bhatnagar, Y. M., and Dym, M. (1977a). *J. Cell Biol.* **74,** 68–85.
Bellvé, A. R., Millette, C. F., Bhatnagar, Y. M., and O'Brien, D. A. (1977b). *J. Histochem. Cytochem.* **25,** 480–494.
Blanco, A., Zinkham, W. H., and Walker, D. G. (1975). In "Isozymes, Vol. 3, Developmental Biology" (C. L. Markert, ed.), pp. 297–312. Academic Press, New York.
Byskov, A. G. (1975). *J. Reprod. Fertil.* **45,** 201–209.
Byskov, A. G., and Saxén, L. (1976). *Dev. Biol.* **52,** 193–200.
Chan, L.-N. L., and Oliver, J. M. (1976). *J. Cell Biol.* **69,** 647–658.
Clermont, Y., and Bustos-Obregon, E. (1968). *Am. J. Anat.* **122,** 237–248.
Connell, C. J. (1977). *J. Cell Biol.* **76,** 57–75.
Cooper, G. W., and Bedford, J. M. (1976). *J. Cell Biol.* **69,** 415–428.
Courot, M., Hochereau-de Riviers, M.-T., and Ortavant, R. (1970). In "The Testis" (A. D. Johnson, W. R. Gomes, and N. L. Vandemark, eds.), Vol. I, pp. 339–432. Academic Press, New York.
Dorrington, J. H., and Fritz, I. B. (1974). *Endocrinology* **94,** 395.
Dym, M., and Fawcett, D. W. (1970). *Biol. Reprod.* **3,** 308–326.
Edelman, G. M. (1976). *Science* **192,** 218–226.
Erickson, R. P. (1976). In "Immunobiology of Gametes" (M. Edidin and M. H. Johnson, eds.), pp. 85–114. Cambridge Univ. Press, London and New York.
Erickson, R. P., Friend, D. S., and Tennenbaum, D. (1975). *Exp. Cell Res.* **91,** 1–5.
Fawcett, D. W., and Phillips, D. M. (1969). *J. Reprod. Fertil.* Suppl. 6, 405–418.
Fawcett, D. W., Ito, S., and Slautterback, D. L. (1959). *J. Biophys. Biochem. Cytol.* **5,** 453–460.
Fellous, M., Erickson, R. P., Gachelin, G., Dubois, P., and Jacob, F. (1976). *Folia Biol.* **22,** 381–383.
Flickinger, C. J., and Fawcett, D. W. (1967). *Anat. Rec.* **158,** 207–222.
Friend, D. S., and Fawcett, D. W. (1974). *J. Cell Biol.* **63,** 641–664.
Gachelin, G., Fellous, M., Guenet, J.-L., and Jacob, F. (1976). *Dev. Biol.* **50,** 310–320.
Gachelin, G., Kemler, R., Kelley, F., and Jacob, F. (1977). *Dev. Biol.* **57,** 199–209.
Gilula, N. B., Fawcett, D. W., and Aoki, A. (1976). *Dev. Biol.* **50,** 142–168.
Gordon, M., Dandekar, P. V., and Bartoszewicz, W. (1975). *J. Ultrastruct. Res.* **50,** 199–207.
Hämmerling, G. J., Mauve, G., Goldberg, E., and McDevitt, H. O. (1975). *Immunogenetics* **1,** 428–437.
Hintz, M., and Goldberg, E. (1977). *Dev. Biol.* **57,** 375–384.
Huckins, C. (1978). *Anat. Rec.* **190,** 905–926.
Kaya, M., and Harrison, R. G. (1976). *J. Anat.* **121,** 279–290.
Koehler, J. K. (1975). *J. Cell Biol.* **67,** 647–659.

Koo, G. C., Stackpole, C. W., Boyse, E. A., Hämmerling, U., and Lardis, M. P. (1973). *Proc. Natl. Acad. Sci. U.S.A.* **70**, 1502–1505.
Lutz, H. U., Lomant, A. J., McMillan, P., and Wehrli, E. (1977). *J. Cell Biol.* **74**, 389–398.
Machado de Domenach, E., Domenach, C. E., Aoki, A., and Blanco, A. (1972). *Biol. Reprod.* **6**, 136–147.
Meistrich, M. L. (1977). *In* "Methods in Cell Biology" (D. M. Prescott, ed.), Vol. 15, pp. 15–54. Academic Press, New York.
Meistrich, M. L., Trostle, P. K., Frapart, M., and Erickson, R. P. (1977). *Dev. Biol.* **60**, 428–441.
Miller, R. G., and Phillips, R. A. (1969). *J. Cell. Physiol.* **73**, 197–201.
Millette, C. F. (1976). *In* "Immunobiology of Gametes" (M. Edidin and M. H. Johnson, eds.), pp. 51–71. Cambridge Univ. Press, London and New York.
Millette, C. F., and Bellvé, A. R. (1977). *J. Cell Biol.* **74**, 86–97.
Mollenhauer, H. H., Morré, D. J., and Hass, B. S. (1977). *J. Ultrastruct. Res.* **61**, 166–171.
Nicolson, G. L. (1976). *Biochim. Biophys. Acta* **458**, 1–72.
Nicolson, G. L., and Yanagimachi, R. (1974). *Science* **184**, 1294–1296.
O, W.-S., and Baker, T. G. (1976). *J. Reprod. Fertil.* **48**, 399–401.
Oakberg, E. F., and Huckins, C. (1976). *In* "Stem Cells of Renewing Cell Populations" (A. B. Cairnie, P. K. Lala, and D. G. Osmond, eds.), pp. 287–302. Academic Press, New York.
O'Rand, M. G., and Romrell, L. J. (1977). *Dev. Biol.* **55**, 346–358.
Phillips, D. M. (1974). "Spermiogenesis." Academic Press, New York.
Robertson, M., Neri, A., and Oppenheimer, S. B. (1975). *Science* **189**, 639–640.
Romrell, L. J., and O'Rand, M. G. (1978). *Dev. Biol.* **63**, 76–93.
Romrell, L. J., Bellvé, A. R., and Fawcett, D. W. (1976). *Dev. Biol.* **49**, 119–131.
Ross, M. H. (1974). *Anat. Rec.* **180**, 565–580.
Ross, M. H. (1976). *Anat. Rec.* **186**, 79–103.
Ross, M. H., and Dobler, J. (1975). *Anat. Rec.* **183**, 267–292.
Rutishauser, U., Millette, C. F., and Edelman, G. M. (1972). *Proc. Natl. Acad. Sci. U.S.A.* **69**, 1596–1600.
Russell, L. (1977a). *Am. J. Anat.* **148**, 301–312.
Russell, L. (1977b). *Am. J. Anat.* **148**, 313–328.
Schachner, M., Wortham, K. A., Carter, L. D., and Chaffee, J. K. (1975). *Dev. Biol.* **44**, 313–325.
Schleiermacher, E., and Schmidt, W. (1973). *Humangenetik* **19**, 75–85.
Skutelsky, E., and Danon, D. (1970). *J. Membr. Biol.* **2**, 173–179.
Steinberger, A. (1975). *In* "Methods of Enzymology" (J. G. Hardman and B. W. O'Malley, eds.), Vol. V, XXIX, Part D. Academic Press, New York.
Steinberger, A., Heindel, J. J., Lindsy, J. N., Elkington, J. S. H., Sanborn, B. M., and Steinberger, E. (1975). *Endocrinol. Res. Commun.* **2**, 261–272.
Toyoda, Y., Yokoyama, M., and Hosi, T. (1971). *Jpn. J. Anim. Reprod.* **16**, 147–157.
Tung, P. S., and Fritz, I. B. (1978). *Dev. Biol.* **64**, 297–315.
Vaiman, M., Fellous, M., Wiels, J., Renard, C., Lecointre, J., du Messil du Busson, F., and Dausset, J. (1978). *J. Immunogenet.* **5**, 135–142.
Wachtel, S. S. (1977). *Transplant. Rev.* **33**, 33–58.
Weeds, N. W. (1975). *Proc. Natl. Acad. Sci. U.S.A.* **72**, 4110–4114.
Wheat, T. E., Hintz, M., Goldberg, E., and Margoliash, E. (1977). *Differentiation* **9**, 37–41.

CHAPTER 2

IMMUNOPEROXIDASE LOCALIZATION OF BINDIN DURING THE ADHESION OF SPERM TO SEA URCHIN EGGS

Gary W. Moy and Victor D. Vacquier

MARINE BIOLOGY RESEARCH DIVISION
SCRIPPS INSTITUTION OF OCEANOGRAPHY
UNIVERSITY OF CALIFORNIA-SAN DIEGO
LA JOLLA, CALIFORNIA

I. Introduction	31
II. Preparation of Antigen and Antisera	33
III. Characterization of Antibindin by Immunodiffusion Assay	34
IV. Ultrastructural Immunohistochemical Localization of Bindin	36
A. Bindin Coats the Sperm Acrosome Process	36
B. Bindin in the Bond between Sperm and Egg	36
C. Bindin at the Site of Membrane Fusion	38
D. Controls on the Immunoperoxidase Procedure	40
E. Cross-Reactivity of Antibody to *S. purpuratus* Bindin with Sperm of Other Species	40
V. Concluding Remarks	42
References	44

I. Introduction

The plasma membrane of most animal eggs is covered by a glycoprotein layer of varying thickness. In marine invertebrates and amphibia this covering is termed the vitelline layer (VL), and in mammals it is known as the zona pellucida. Sperm adhere to these egg coverings prior to fusion with the egg. In both invertebrates (Metz, 1978) and mammals (Yanagimachi, 1977) the adhesion of sperm to eggs exhibits a high degree of species specificity. Loeb (1916) may have been the first to speculate that this specificity must result from the interaction of proteins on the surfaces of the gametes. Although during his time this was only an intuitive guess, today Loeb's idea would probably be accepted as dogma.

Sperm–egg adhesion during sea urchin fertilization seems to be an ideal model system for studying the molecular basis of a specific intercellular adhesion. The biological significance of sperm–egg interaction is well established, the gametes are homogeneous populations of single cells that can be obtained in large quantity, and the interaction of sperm and egg occurs with great synchrony in a time span of seconds. Also, in the past 2 years methods have been devised to isolate milligram quantities of the putative gamete surface components mediating sperm–egg adhesion.

The apex of the sea urchin spermatozoan contains a Golgi-derived acrosome granule approximately 0.3 μm in diameter composed of a uniformly electron-dense granular material (Dan, 1967). Immediately before or during contact of the sperm with the egg surface, exocytosis of the granule occurs (Colwin and Colwin, 1967; Dan, 1967; Summers et al., 1975). This reaction, known as the acrosome reaction, involves the fusion of the granule membrane with the overlying sperm plasma membrane, the externalization of the granule content, and the projection of the acrosome process by the polymerization of actin stored in the nuclear fossa (Jessen et al., 1973; Sanger and Sanger, 1975; Tilney et al., 1973). The protrusion of the acrosome process everts the former acrosome granule membrane which becomes the membrane covering of the rodlike acrosome process (Summers et al., 1975), approximately 1 μm in length in Strongylocentrotus purpuratus. The externalized acrosome content coats the membrane of the acrosome process as seen in electron micrographs from several laboratories (Collins, 1976; Dan et al., 1964; Decker et al., 1976; Sugiyama and Kato, 1977; Summers and Hylander, 1974). This extracellular coating is probably the first spermatozoan component to contact the egg surface and is presumed to be responsible for the species-specific adhesion of sperm to the vitelline layer of the egg (Summers et al., 1975; Summers and Hylander, 1976).

We have isolated the insoluble, membrane-free content of the sea urchin sperm acrosome granule and have found most of it to be a single, 30,500-dalton protein which, because of its suspected role in gamete binding, we refer to as "bindin" (Vacquier and Moy, 1977). Experiments with isolated bindin support the concept that it is the ligand bonding sperm to eggs. For example, bindin is a species-specific agglutin of unfertilized eggs (Glabe and Vacquier, 1977). The egg-agglutinating property of bindin is lost when it is mixed with glycopeptides produced by trypsinization of unfertilized eggs and trypsinized eggs are not agglutinated by bindin (Vacquier and Moy, 1977). Data indicate that bindin mediates gamete adhesion by interacting with glycoprotein "bindin receptors" on the outer surface of the egg vitelline layer (Glabe and Vacquier, 1978). Further substantiation of the role of bindin as a specific sperm adhesive necessitated localization of the protein on the surfaces of interacting gametes. As reported here, this was accomplished by use of peroxidase-conjugated swine anti-rabbit immunoglobulin reacted with rabbit antibody to electrophoretically purified bindin.

II. Preparation of Antigen and Antisera

Gametes of *S. purpuratus* were spawned by pouring 0.5 M KCl into opened body cavities. The acrosome reaction was induced by elevation of a fresh 0.1% sperm suspension to pH 9.2 with 1 M NH_4OH and fixation after 2 minutes in 5% formaldehyde in seawater. Spermatozoa bound to eggs were prepared by inseminating the eggs with dense suspensions of sperm and 15 seconds later fixing the preparation in seawater containing 5% formaldehyde or 2% glutaraldehyde (Vacquier and Payne, 1973).

Bindin was isolated from the sperm as previously described (Vacquier and Moy, 1977). For producing antibody, the final bindin pellet was dissolved in 5% sodium dodecyl sulfate (SDS) with 5% mercaptoethanol at a protein concentration of 2 mg/ml and 100 μg electrophoresed on 12% acrylamide gels containing 0.1% SDS in cylindrical tubes (5-mm diameter). The location of the bromphenol blue tracking dye was marked and the gels stored at $-20°C$. One gel was stained with Coomassie Blue to determine the position of the bindin band. Areas of frozen gels containing bindin were then excised, ground with equal volumes of phosphate-buffered saline (PBS) and Freund's complete adjuvant (Difco), and injected subcutaneously (100 μg bindin per inoculum) into New Zealand white rabbits at 7-day intervals for 1 month. Immune sera were collected 7 days after the final injection and immunoglobulins fractionated by ammonium sulfate precipitation; the sera were stored in PBS at $-20°C$ (Williams and Chase, 1967).

For reaction with antibody the gametes were washed extensively in PBS to remove the fixative and incubated for 4 hours at 23°C with a 1:800 dilution of rabbit antibindin (25 μg protein/ml). Cells were washed three times by settling or centrifugation and resuspension in PBS containing 0.1% bovine serum albumin (PBS-BSA). They were then incubated 4 hours with a 1:50 dilution of horseradish peroxidase (HRP) conjugated swine anti-rabbit immunoglobulin (BioRad, Richmond, California). The cells were then washed three times in PBS-BSA and incubated in a solution of 0.5 mg/ml 3,3-diaminobenzidine tetrahydrochloride (DAB), 0.5 M Tris-HCl, pH 7.6, and 0.01% H_2O_2. After 15 minutes the reaction was stopped by washing the cells in PBS. Positively reacting cells were readily visible under ×1000 magnification with the light microscope (Mazurkiewicz and Nakane, 1972).

For electron microscopy the reacted cells were fixed for 1 hour in 1% OsO_4 in seawater, dehydrated in ethanol and propylene oxide, and embedded in Epon. Sections were viewed unstained to achieve maximum contrast between the cell and the DAB precipitate.

Immunodiffusion assays in 1% agar were run to test the specificity of the antibindin. Whole sperm (4.6 mg/ml) and egg (1.2 mg/ml) protein was prepared by homogenization of fresh cells in PBS. Particulate bindin was partially solubilized by homogenization in PBS (0.1 mg protein/ml). Antibindin was absorbed for 4 hours with particulate bindin and centrifuged 20,000 g to remove the insoluble protein. Full-strength antibindin and preimmune sera (20 mg/ml) were used, 50 μl being added to each well and the plates incubated for 24 hours at 37°C.

III. Characterization of Antibindin by Immunodiffusion Assay

The insoluble bindin isolated in calcium-free seawater containing Triton X-100 and soybean trypsin inhibitor (Vacquier and Moy, 1977) is partially solubilized in PBS at 37°C, and as shown in Fig. 1, when diffused (well 1) against antibindin (center well) forms a single precipitin line. Bindin-absorbed antibindin (well 2) does not react with bindin (well 1). The reaction line between the absorbed antibindin (well 2) and antibindin (center well) results from soluble bindin in the absorbed sample. Reaction does not occur between the soluble bindin (well 2) and preimmune serum (well 3). Whole egg homogenates (well 4) also fail to react with antibindin. Antibindin absorbed with fixed, intact, unfertilized eggs (well 5) still reacts with a homogenate of whole sperm (well 6). Reaction also occurs between the whole sperm homogenate (well 6) and antibindin. These data indicate that the antibody used in these experiments is directed against a single sperm-borne protein which is bindin.

FIG. 1. Immunodiffusion plate of antibindin (center well) diffused against: (1) bindin solubilized in PBS at 37°C; (2) antibindin absorbed with an excess of bindin (note no precipitin line between 1 and 2); (3) preimmune serum (note no reaction between 2 and 3); (4) homogenate of whole eggs; (5) antibindin absorbed with egg surfaces (fixed, whole eggs); (6) homogenate of whole sperm in PBS. Abbreviations used in figures: AB, antibindin; AP, acrosome process; B, bindin; DAB, diaminobenzidine precipitate; EC, egg cytoplasm; ES, egg surface; M, mitochondrion; MV, microvillus; N, nucleus; NF, nuclear fossa; RAG, residual of acrosome granule; VL, vitelline layer.

FIG. 2. Bindin localized on the acrosome process and membrane of anterior tip of acrosome-reacted sperm. Anterior arrows mark area of deposition of DAB on sperm membrane. Arrow on tip of mitochondrion midpiece marks small patch of DAB probably caused by collision with the acrosome process of another sperm. ×23,750.

FIG. 3. The immunoperoxidase technique shows that bindin completely enshrouds the acrosome process. ×133,200.

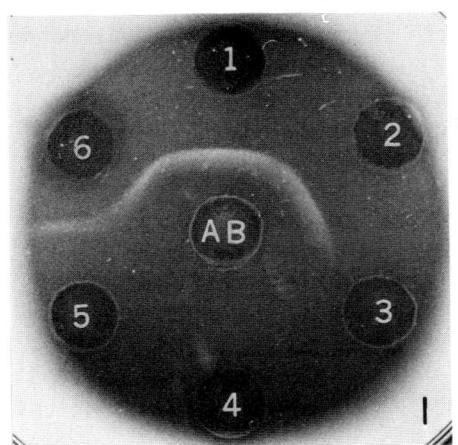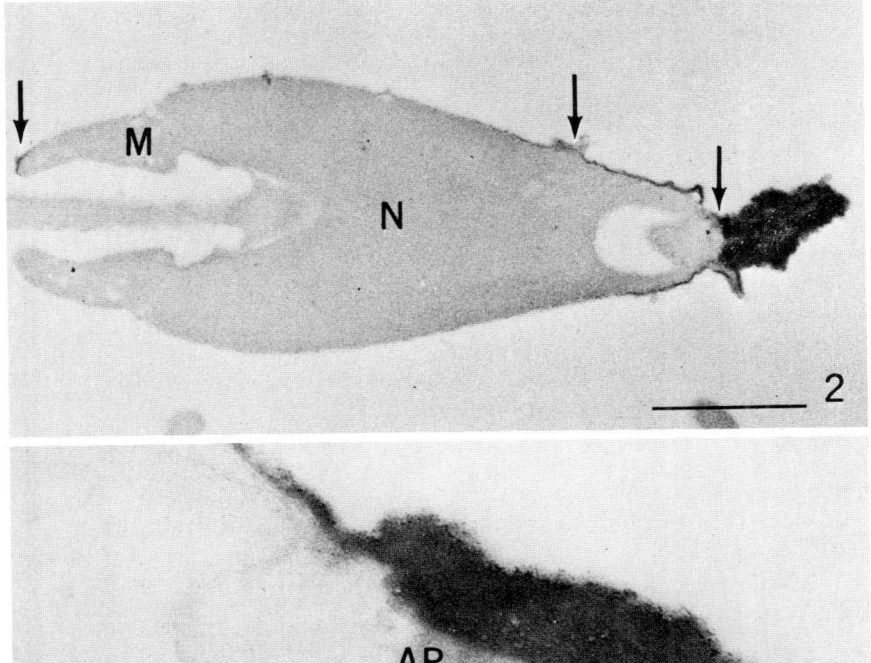

IV. Ultrastructural Immunohistochemical Localization of Bindin

A. BINDIN COATS THE SPERM ACROSOME PROCESS

When observed with the light microscope a dense layer of DAB precipitate appears in the anterior apex of each acrosome-reacted sperm. Electron microscopic examination shows the acrosome process of these sperm is enshrouded in a dense coating of DAB (Figs. 2 and 3). The bindin is also localized on the sperm plasma membrane for a distance of about 1–2 μm from the base of the acrosome process (area between arrows in Fig. 2). This membrane adjacent to the acrosome process may be partially derived from the acrosome granule membrane or, alternatively, the bindin may spread out on the anterior apex of the sperm during granule exocytosis. This distribution of bindin on acrosome-reacted sperm was essentially identical among all acrosome-reacted sperm we examined. The cell shown in Fig. 2 has a small deposit of DAB on the posterior tip of the mitochondrion (arrow). This condition was found only on the spermatozoan and was probably caused by contact with the acrosome process of another reacted sperm. The mitochondrion midpiece of sea urchin sperm is able to fuse with the acrosome process (Collins, 1976).

B. BINDIN IN THE BOND BETWEEN SPERM AND EGG

Fixed eggs with a maximum number of sperm bound to the surfaces (Vacquier and Payne, 1973) were reacted with the antibodies and DAB and then prepared for electron microscopy. Beginning with the sperm and progressing into the egg surface (Figs. 4–6), bindin is localized on that area of the spermatozoan membrane covering the nuclear fossa (Fig. 4) and along the base of the acrosome process, where a dense band of DAB precipitate is found. This band, or "collar," of DAB represents the residual bindin which is a remnant of the discharged acrosome granule. In Fig. 5, the acrosome process extends between two egg surface microvilli cut in cross section, whereas in Fig. 6 the side of the acrosome process is bonded to the side of an egg microvillus. In Figs. 5 and 6 it is especially apparent that bindin is present between the surfaces of the bonded gametes. A light coating of DAB precipitate was

---

FIGS. 4–6. Bindin localized in the sperm-to-egg bond. Bindin is on the sperm membrane to about the point of the arrow on Fig. 4. A collar of DAB at the base of the acrosome process marks the location of the bindin which is a residual or remnant of the acrosome granule. Bindin also appears on the surface of egg microvilli. This distribution of bindin on the surfaces of the acrosome process and the egg microvillus to which the process is attached is excellent supportive evidence that bindin functions in sperm–egg adhesion. Fig. 4: ×50,650. Fig. 5: ×87,700. Fig. 6: ×79,500.

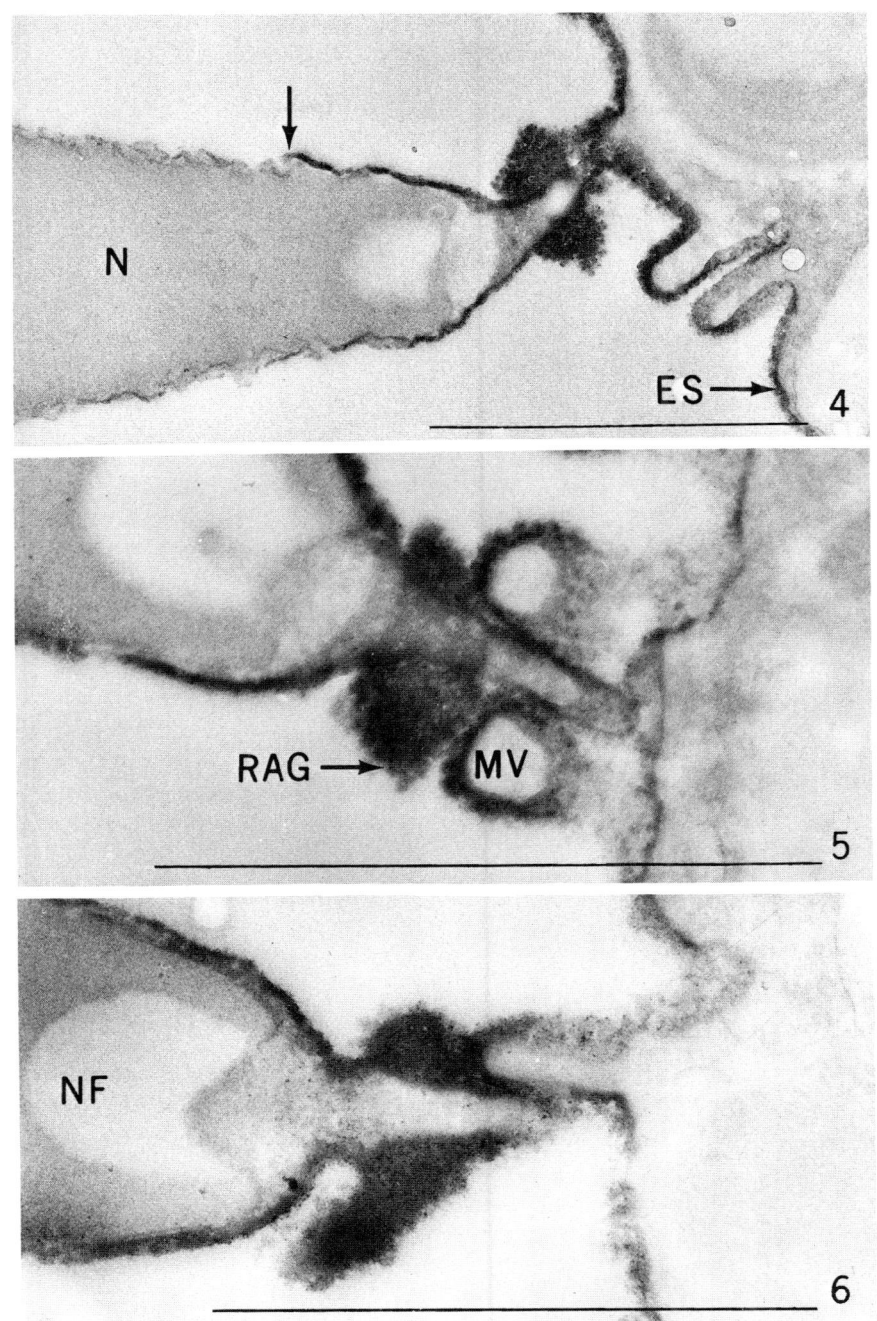

present on the entire egg surface and was especially dense on the microvilli adjacent to the adherent acrosome process (Figs. 4–6). Controls (see following discussion) show the surface of unfertilized eggs does not react with either antibindin or HRP-conjugated swine anti-rabbit sera used in these experiments. We believe that the thin coating of DAB on the surfaces of eggs to which sperm are bound is present because we made certain that between 1000 and 2000 sperm were bound to each egg in order to increase the probability of finding good sections through the sperm–egg bond (Vacquier and Payne, 1973). When exocytosis of the acrosome granule occurs against the egg surface, some of the bindin is deposited on the vitelline surface adjacent to the adhering sperm. Each attached sperm thus has a small "halo" of bindin on the egg surface around the point of attachment. In this study, so many sperm are attached to the egg that a thin coating of bindin is present on the entire egg surface.

C. Bindin at the Site of Membrane Fusion

During fertilization in invertebrates the membrane of the acrosome process fuses with the egg plasma membrane (Colwin and Colwin, 1967; Dan, 1967, Epel and Vacquier, 1978). Since the entire acrosome process is coated with bindin (Figs. 2 and 3), this protein must be present on the surfaces of fusing membranes and may even play a role in mediating the fusion process. Using the immunoperoxidase technique, we found that bindin is still present on the membrane surrounding the fusing sperm (Fig. 7). In this electron micrograph the remains of the core filaments of the acrosome process are visible in the egg cytoplasm. The DAB precipitate indicates that some bindin is present under the elevating vitelline layer. It is impossible to judge if the DAB deposit is present on only the sperm membrane or on a composite of both gamete membranes. The solubility properties of bindin indicate it to be a very hydrophobic protein with affinity for membrane environments. We are currently exploring the possibility that it is a fusigenic protein capable of enhancing the coalescence of biomembranes.

Fig. 7. Bindin in the area of gamete membrane fusion. The actin core filaments of the acrosome process are clearly visible in the egg cytoplasm. Bindin is present under the elevating vitelline layer and the residual acrosome granule is very prominent. ×78,800.

Fig. 8. Control. The antibindin activity is absorbed by bindin (compare Figs. 3 and 8). This micrograph shows the distinction between the DAB precipitate and the bindin. The coating of bindin on the acrosome process is similar to that described by other workers. ×88,800.

Fig. 9. Control. Preimmune serum does not react with sperm or egg surfaces. ×101,000.

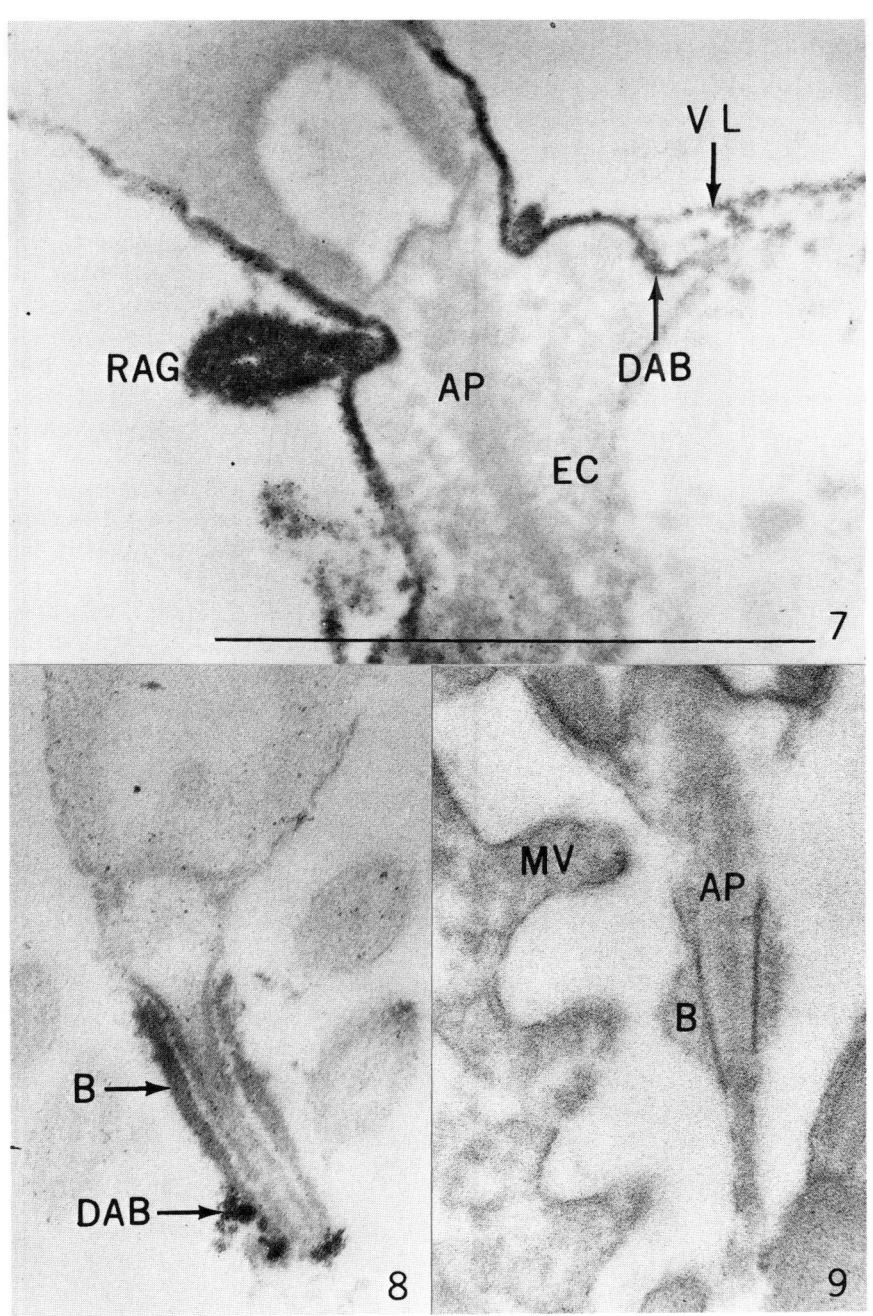

D. CONTROLS ON THE IMMUNOPEROXIDASE PROCEDURE

Omitting one reagent completely blocked the deposition of DAB on acrosome-reacted sperm and in the area of sperm–egg adhesion. For example, omission of rabbit antibindin or substitution with preimmune serum failed to produce a DAB reaction product as did omission of H_2O_2 or inhibition of the peroxidase with azide. Absorption of the antibindin with bindin (Fig. 1) removed almost all the antibody as shown in Fig. 8. It is important to note in this micrograph that the bindin coating along the acrosome process is readily distinguishable from the DAB precipitate (compare Figs. 3 and 8). This coating of bindin is similar in appearance to that of other published micrographs (Collins, 1976; Dan et al., 1964; Summers and Hylander, 1974; Decker et al., 1976). Absorption of antibindin serum with whole unfertilized eggs, total egg protein from egg homogenates, and BSA did not remove the antibindin activity. Incubation of gametes with preimmune serum followed by HRP conjugated swine and anti-rabbit sera and DAB were also negative for bindin as shown in Fig. 9.

Not all sperm elevated to pH 9.2 with NH_4OH undergo the acrosome reaction. Therefore, sperm with intact acrosomes are present in the same sample of acrosome-reacted cells shown in Figs. 2 and 3. Such acrosome-intact sperm are completely free of DAB precipitate (Fig. 10), showing that bindin is not present on the sperm surface before the acrosome reaction. This is important because in at least one sea urchin species, *Pseudocentrotus depressus,* sperm are capable of attaching to the egg surface before exocytosis of the acrosome granule (Aketa and Ohta, 1977). This suggests that a recognition–adhesion reaction must occur before the externalization of bindin and therefore other proteins in addition to bindin may be involved in gamete adhesion.

To show that the surface of the unfertilized egg does not react with antibindin, fixed eggs were washed in PBS containing BSA ore preimmune serum followed by incubation in antibindin, the HRP-conjugated serum, and the DAB reaction. Such eggs are negative for DAB precipitate as shown in Fig. 11 where the thin, shroudlike vitelline layer is clearly visible. We conclude from these micrographs that the immunoperoxidase procedure demonstrates that bindin is present solely in the acrosome granule of the sea urchin spermatozoan.

E. CROSS-REACTIVITY OF ANTIBODY TO *S. purpuratus* BINDIN WITH SPERM OF OTHER SPECIES

When invertebrate sperm are fixed in a mixture of 47% seawater (or PBS), 3% formaldehyde, and 50% glycerol, the plasma membranes over the acrosome granule and the granule membrane itself rupture,

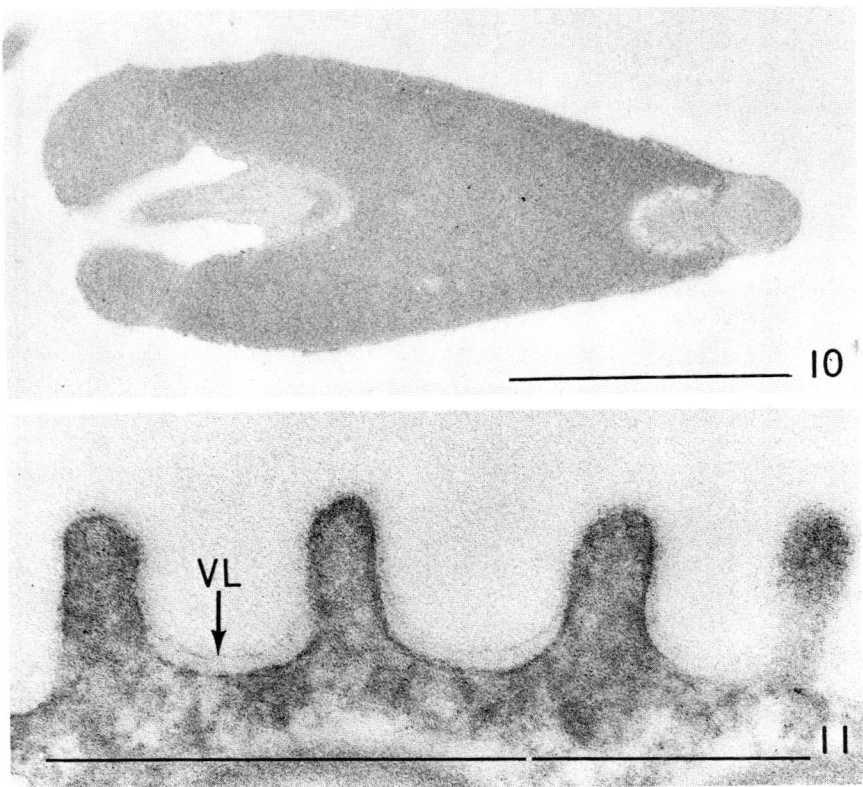

FIG. 10. Control. Rabbit antibindin and HRP-conjugated swine antirabbit sera do not react with sperm with intact acrosomes showing the bindin is only present in the acrosome vesicle. This sperm was in the same section as those shown in Figs. 2 and 3. ×37,150.

FIG. 11. Control. The surface of unfertilized eggs does not react with either the primary or the secondary serum. ×101,250.

externalizing the granule content. Colleagues from several countries generously supplied sperm samples fixed in this mixture which we reacted with *S. purpuratus* antibindin serum. Observation of phase microscopy under oil immersion at ×1000 was the method used to determine positive or negative cross-reactivity. We found that all species of echinoids (sea urchins and sand dollars) cross-reacted with the antibindin. Sperm from other classes of echinoderms or other phyla did not cross-react (Table I). Since *S. purpuratus* antibindin is not species-specific and reacts with all echinoid sperm, this particular bin-

TABLE I

Reactivity of *S. purpuratus* (Sea Urchin) Antibindin with Sperm of Various Species

Positive cross-reaction	Negative cross-reaction
Sea urchins	Starfish
Strongylocentrotus purpuratus (California)	*Asterias amurensis*
Strongylocentrotus franciscanus (California)	*Asterina pectinifera*
Lytechinus pictus (California)	*Pisaster bravispinus*
Tripneustes gratilla (Hawaii)	Annelids
Colobocentrotus atratus (Hawaii)	*Mercierella enigmatica*
Glyplocidaris crenulasis (Japan)	
Lytechinus variegatus (Florida)	Mollusca
Arbacia punctulata (Massachusetts)	*Cryptochiton stelleri* (chiton)
Hemicentrotus pulcherrimus (Japan)	*Acmae* sp. (limpet)
Pseudocentrotus depressus (Japan)	*Crassostrea gigas* (oyster)
Strongylocentrotus intermedius (Japan)	*Mytilus edulus* (mussel)
Arbacia lixula (Mediterranean)	*Macoma nasuta* (clam)
Paracentrotus lividus (Mediterranean)	*Clinocardium nutalli* (clam)
Sand dollars	*Prototheca staminea* (clam)
Dendraster excentricus (California)	*Pactinopectin yessoensis* (scallop)
Echinarachnius parma (Massachusetts)	Arthropods
	Pinnixa tubicola (crab)
	Urochordates
	Ciona intestinalis
	Amphibia
	Rana pipiens
	Mammals
	Guinea pig
	Hamster
	Rabbit
	Rat

din is immunologically conserved. The antibindin we produced must be directed against parts of the bindin molecule common to all Echinoidea. Somewhat surprisingly, closely related echinoderm bindin such as that of the starfish (asteroid, three species tested) does not react with sea urchin antibindin. The basic morphological differences between asteroid and echinoid spermatozoa are well known, however, and doubtless reflect the evolutionary separation of these two groups of echinoderms.

V. Concluding Remarks

Failure of antibindin to react with sperm in which the plasma membrane and acrosomal granule membrane remain intact (Fig. 10) supports the theory that the protein is not exposed until the acrosome reaction has occurred (Summers et al., 1975). The localization of bindin along the acrosome process and upon the microvillus of the egg to which the sperm is attached (Figs. 2–6) corroborates other experimental evidence that bindin mediates sperm–egg adhesion. This site of bindin in the sperm–egg bond closely resembles its distribution in the sand dollar, *Echinarachnius parma*, observed by Summers and Hylander (1974). In this species, bindin lay along the acrosome process and was also dispersed upon the vitelline layer in the vicinity of process attachment.

Although the presence of bindin on the acrosome process was to be expected from what is known about acrosome process generation, its presence upon the surface of egg microvilli to which sperm are bound was less certain. Perhaps the acrosomal granule undergoes an explosive and diffuse release upon contact with the surface of the egg, and bindin molecules may consequently adhere to "bindin receptors" on the vitelline layer close to the bound spermatozoa. Conversely, the DAB reaction product may dissipate from the site of initial reactivity and become bound to the vitelline layer neighboring the attached sperm.

The real importance of the bindin–bindin receptor system is that bindin is the first protein to be isolated in fairly pure form and milligram quantities that mediates a specific intercellular adhesion of a metazoan. Antiserum to this protein makes several experiments possible. For example, it could be used to study the timing of bindin synthesis during spermiogenesis by immunoperoxidase localization with either the light or electron microscope. It could also be used to precipitate the polyribosomes containing the mRNA for bindin. The cDNA probe to the bindin mRNA could then be prepared and the number of copies of the bindin gene determined.

An extensive experimental attack on the chemistry of bindin and its glycoprotein receptor would seem crucial to our future understanding of the molecular basis of gamete recognition and adhesion. How does bindin adhere to the membrane of the acrosome process? Is the hydrophobic portion of the molecule embedded in the hydrophobic region of the lipid bilayer, or is bindin strictly a peripheral coating on the outer surface? How do bindin monomers associate with each other to form the compact matrix of the acrosome granule? One ultimate goal would be to determine the structure of bindin by X-ray diffraction analysis and to demonstrate how the receptor oligosaccharide sequence interacts with bindin. If the structures of bindin and its receptor glycopeptide were known for several species, we would undoubtedly understand the basis of species-specific gamete interaction during fertilization.

ACKNOWLEDGMENTS

Special thanks are given to Dr. Daniel S. Friend for his interest and help in this project. We are also grateful to the following colleagues for providing samples of sperm: Drs. G. Anderson, J. Dan. Y. Hiramoto, N. Holland, H. Kanatani, R. Kane, S. Katsura, T. Kishimoto, K. Kleene, R. Lallier, C. Lambert, A. Lopo, S. Meizel, K. Okuno, K. Osanai, M. Roberson, L. Suer, T. Yanaka, Y. Uno, H. Van Veldhuizen, and I. Yasumasu. This work was supported by NIH Grant HD 12896A.

REFERENCES

Aketa, K., and Ohta, T. (1977). *Dev. Biol.* **61**, 366–372.
Collins, F. (1976). *Dev. Biol.* **49**, 381–394.
Colwin, L. H., and Colwin, A. L. (1967). *In* "Fertilization" (C. B. Metz and A. Monroy, eds.), Vol. 1, pp. 295–367. Academic Press, New York.
Dan, J. C. (1967). *In* "Fertilization" (C. B. Metz and A. Monroy, eds.), Vol. 1, pp. 236–293. Academic Press, New York.
Dan, J. C., Ohori, Y., and Kushida, H. (1964). *J. Ultrastruct. Res.* **11**, 508–524.
Decker, G. L., Joseph, D. B., and Lennarz, W. J. (1976). *Dev. Biol.* **53**, 115–125.
Epel, D., and Vacquier, V. D. (1978). *In* "Cell Surface Reviews" (G. Poste and G. L. Nicolson, eds.), Vol. 5, pp. 1–63. Elsevier North-Holland, Amsterdam.
Glabe, C. G., and Vacquier, V. D. (1977). *Nature (London)* **26**, 836–383.
Glabe, C. G., and Vacquier, V. D. (1978). *Proc. Natl. Acad. Sci. U.S.A.* **75**, 881–885.
Jessen, H., Behne, O., Wingstrand, K. G., and Rostgaard, J. (1973). *Exp. Cell Res.* **80**, 47–54.
Loeb, J. (1916). "The Organism as a Whole," pp. 71–94. Putnam, New York.
Mazurkiewicz, J. E., and Nakane, P. K. (1972). *J. Histochem. Cytochem.* **20**, 969–974.
Metz, C. B. (1978). *Curr. Top. Dev. Biol.* **12**, 107–147.
Sanger, J. W., and Sanger, J. M. (1975). *J. Exp. Zool.* **193**, 441–447.
Sugiyama, M., and Kato, K. (1977). *Dev. Growth Differ.* **19**, 23–29.
Summers, R. G., and Hylander, B. L. (1974). *Cell Tissue Res.* **150**, 343–368.
Summers, R. G., and Hylander, B. L. (1976). *Exp. Cell Res.* **100**, 190–194.
Summers, R. G., Hylander, B. L., Colwin, L. H., and Colwin, A. L. (1975). *Am. Zool.* **15**, 523–551.
Tilney, L. G., Hatano, S., Ishikawa, H., and Mooseker, M. S. (1973). *J. Cell Biol.* **59**, 109–126.
Vacquier, V. D., and Moy, G. W. (1977). *Proc. Natl. Acad. Sci. U.S.A.* **74**, 2456–2460.
Vacquier, V. D., and Payne, J. E. (1973). *Exp. Cell Res.* **82**, 227–235.
Williams, C. A., and Chase, M. W. (1967). "Methods in Immunology and Immunochemistry." Academic Press, New York.
Yanagimachi, R. (1977). *In* "Immunobiology of Gametes" (M. Edidin and M. H. Johnson, eds.), pp. 255–296. Cambridge Univ. Press, London and New York.

CHAPTER 3

CELL SURFACE DIFFERENTIATIONS DURING EARLY EMBRYONIC DEVELOPMENT

Alberto Monroy and Floriana Rosati

STAZIONE ZOOLOGICA
NAPOLI, ITALY

I. Introduction	45
II. The Problem of Polarity	46
III. The Segregation of Cell Lines	49
A. Some Speculations on the Phylogenesis of Cell Lines	50
B. An Example of Segregation of Cell Lines: The Ascidian Embryo	50
C. The Immunological Approach to the Study of the Segregation of Cell Lines	52
D. Differential Expression of Surface Properties and of Antigens in the Various Cell Lines in the Embryo	54
IV. The Sperm–Egg Interaction in Fertilization	59
V. Concluding Remarks	65
References	67

I. Introduction

Attempts at exploiting immunological concepts and techniques for the analysis of developmental problems go back to the discovery of organ-specific antigens in the adult animal. Indeed, the apparent simplicity and the high specificity of antigen–antibody recognition and interaction appeared to offer a unique opportunity of answering some of the major questions of development. Two approaches have been followed: (1) an analogical approach whereby use has been made of immunological concepts to interpret developmental events; (2) an analytical approach whereby immunology is used as a tool to ask questions such as those related to the appearance and localization of specific antigen molecules and their role in the morphological and functional shaping of the embryo. The analogical approach, unless used as a preliminary step to formulating working hypotheses, does not usually lead far beyond a mental exercise. Indeed, it "may do more harm than good. This mistake would be interesting methodologically as showing that one should not use concepts taken from a level of *unnecessarily* high complexity" (Needham, 1942).

In preparing this chapter our aim has been not so much that of presenting results of work on developing systems in which use has been made of immunological concepts and/or techniques, but rather that of trying to direct the attention of immunologists to two of the major problems of development: the acquisition of polarity by the egg and the segregation of cell lines. We believe that in both cases the cell surface plays a major, if not *the* major, role. Although our belief may be considered an act of faith, we will present the data which we consider evidence, even though circumstantial, in support of our view. No attempt has been made to approach the problem of the acquisition of polarity from the immunological point of view. There are, however, quite a few old and some new immunological data relevant to the problem of the segregation of cell lines. Indeed, it appears that immunology may make a major contribution to this field. Finally, we discuss fertilization, not only because this is the problem in which we are chiefly interested, but because the problem of the sperm–egg interaction is the most exquisite process (and, in fact, the most important one from the point of view of evolution) of cell recognition and interaction. Should immunology prove to be one of the most powerful tools to approach this fundamental problem we owe it to the intuition of F. R. Lillie (1919).

II. The Problem of Polarity

The earliest and the most fundamental property to appear in any developing organism, be it animal or plant, is the acquisition of a differential organization of the egg components which results in the expression of different properties at the two opposite ends of the polarity axis. To our knowledge, in all animal eggs it is the so-called animal–vegetal (a–v) polar axis that is acquired first, the dorsoventral axis appearing later in the course of development. The a–v polarity together with the structural organization of the egg which is related to it conditions the whole development of the embryo (for a detailed discussion, see Monroy and Moscona, 1979).

The question is, what is the molecular basis of polarity? Does it depend on a molecular lattice of the egg cytoplasm which arises during oogenesis or after fertilization; or on the organization of the egg surface, or on a combination of both? There is a large body of evidence that the eggs of many animals are already polarized at the time of fertilization, i.e., that they exhibit different axial properties whereby an animal and a vegetal pole can be recognized. However, the evidence that these "properties" are linked to a differential organization of the egg plasma membrane, or, to use a more noncommittal expression, of the egg surface, is tenuous. In the egg of *Xenopus laevis,* the overall organization of the egg surface, as revealed by scanning electron microscopy, differs at the animal pole from that at the vegetal pole (Monroy and Baccetti, 1975). Even more interesting is the case of another amphibian, *Discoglossus pictus,* in which not only is sperm entry restricted to the animal pole (as in most amphibians), but the small area—the animal dimple—where sperm–egg fusion occurs is characterized by the presence of many small microvilli anchored to microfilaments (Campanella, 1975) and by the fact that this is the only area of the whole egg surface which specifically binds the lectin from *Fucus,* fucose-binding protein (FBP) (Denis-Donini and Campanella, 1977). The lectin-binding ability of this area is lost after fertilization; in fact, the dimple itself disappears, and the organization of the surface of the animal territory then becomes essentially identical to that of the rest of the egg.

Although great caution is required in interpreting the results of lectin binding experiments, they nevertheless suggest a specialization of the egg surface relative to its binding to the spermatozoon. In this case the observation that the site of the sperm–egg interaction has an affinity, for the FBP acquires special importance in view of the recent observations (Rosati and De Santis, personal communication) showing that in the ascidian egg, fucose plays a key role in the binding of spermatozoa to the sperm receptors located on the chorion of the egg. These data are compatible with the view of a regionalization of the egg surface. Attempts to find structural specializations in the area of the grey crescent have failed (Monroy and Baccetti, 1975; see also Brachet, 1977). An answer to these questions may arise from the use of monoclonal antibodies (Köhler and Milstein, 1975; Milstein and Lennox, Volume 14, this series).

The most important question for developmental biologists is how these differences arise. One widely accepted notion is that the relationship of the oocyte to the ovarian wall causes a specific alteration of the oocyte surface (and possibly of the oocyte cytoplasm as well) which results in the a–v axis.

An interesting example is that of the oocyte of the snail, *Limnea* (Raven, 1970, 1976), in which the area of the oocyte surface that is in contact with the basal membrane of the ovarian acinus becomes the vegetal pole of the egg. This is compatible with the notion that this interaction results in a specific organization of the oocyte surface which is different from that at its free pole. Furthermore, this area of the oocyte is surrounded by six (in *L. stagnalis*) (Fig. 1) or nine (in *L. peraegra*) follicle cells which give rise to an equal number of so-called "subcortical accumulations" (SCA) made up of finely granular, densely packed subcortical material in the equatorial region of the oocyte. The

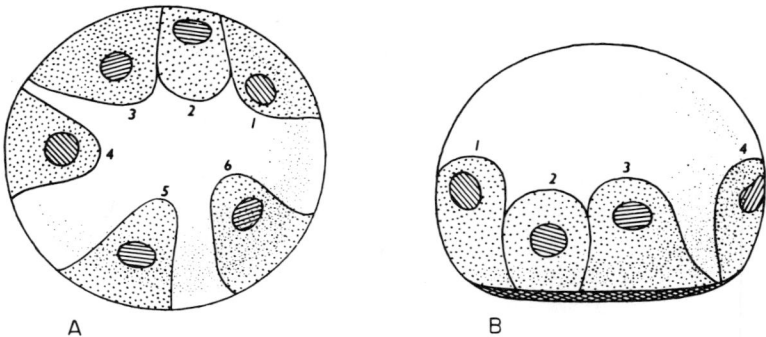

FIG. 1. Arrangement of the six inner follicle cells around the oocyte of *Limnea stagnalis*. (A) View from the inner pole of the oocyte. (B) View from the side. The crosshatched area in B indicates the contact with the ovarian wall. From Raven (1970).

pattern of symmetry of the SCA coincides with the future dorsoventral axis of the embryo (Fig. 2). Since, in this spirally cleaving egg, the direction of the cleavage plans, whether dextral or sinistral, is maternally determined (Boycott *et al.*, 1931), it was an interesting question

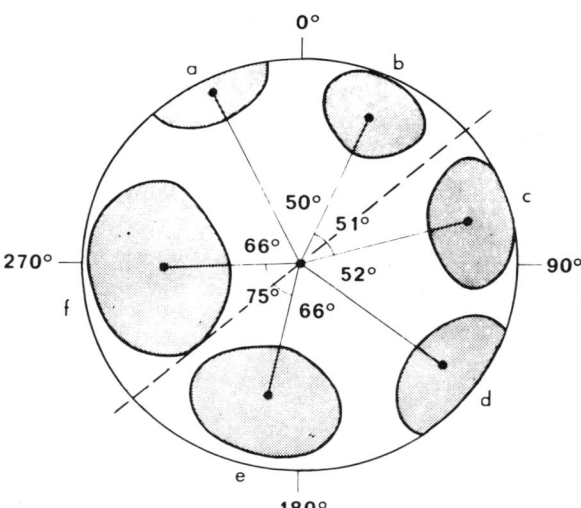

FIG. 2. Position of the six (a–f) subcortical accumulations (SCA) in the egg of *Limnea stagnalis* as viewed in a horizontal projection of the vegetal hemisphere of the egg. The vegetal pole is at the center. The SCA (stippled) are arranged in a subequatorial ring. The broken line indicates the approximate plane of symmetry of the SCA pattern. From Raven (1970).

whether or not the pattern of the SCA was different in the genetically dextral and sinistral broods. The answer was that it is not (Ubbels *et al.*, 1969). Hence, from these observations it may be concluded that, while the a–v polarity is likely to be epigenetically controlled by the relationships that the oocyte happens to establish with the ovarian wall, the orientation of the cleavage plans is under genetic control. This is an important conclusion as it suggests that the a–v polarity may depend on a specific molecular organization of the egg surface; on the other hand, the direction of the cleavage plans is likely to result from the specific organization of the egg cytoplasm.

Recent studies on the processes underlying the attachment of cells in culture to the substrate (Culp *et al.*, 1978) may help in understanding how the contact with the ovarian wall influences the organization of the oocyte surface *and* of the oocyte cytoplasm. The contact with the ovarian wall may impose a constraint on the membrane components which, in its turn, may result in a specific organization of the subsurface cytoskeletal components (e.g., microfilaments), thus eventually giving rise to a molecular lattice. The recently available methods for the study of microfilaments may help to answer this question.

III. The Segregation of Cell Lines

In all multicellular organisms, the cleavage of the egg gives rise to cells which differ from one another and which, through successive cell divisions, will eventually give rise to homogeneous cell populations (cell lines), each endowed with its own specific developmental program. (For the historical aspects of the research on cell lines, the reader is referred to the recent article by Maienschein, 1978.) This not only implies a process of sorting out of molecules (either preexisting in the egg before fertilization or being synthesized in the course of development) into the various blastomeres, but also a process of cells recognizing one another and coordinating their movements, their rate of cleavage, their metabolic activities, and the like.

It is almost a truism that these events will have their counterpart in a specific organization of the cell plasma membrane. The question is: To what extent does the molecular organization of the membrane of the blastomeres control the segregation process? We assume that the expression of the genes in the different blastomeres in the course of cleavage is controlled by cytoplasmic factors and *hence* also by the molecular organization of the plasma membrane that has differentiated during oogenesis (such as the properties related to polarity), or as a result either of the expression of previously unexpressed genes in different blastomeres, or of the interaction with other blastomeres.

A. Some Speculations on the Phylogenesis of Cell Lines

We have recently suggested (Monroy and Rosati, 1979) that one of the major events—if not *the* major event—connected with the appearance of multicellular organisms is the segregation of the somatic from the germ cell line. We have postulated that the dichotomy between the two cell lines involves the following:

1. In the somatic cell line, the genes, which in the unicellular organisms code for the surface structures responsible for the recognition of and interaction between cells of the two gametic types, are silenced. The evidence for this is indirect. Although the matter has never been investigated with this question in mind, the formation of mouse chimeras (Tarkowski, 1961; Mintz, 1962; see also review by Herbert and Graham, 1974) shows that genetically male and female embryonic cells do not distinguish each other as different. Also, hybrid histotypic aggregates can be formed in culture from such species as far apart as chick and mouse (Moscona, 1957; Moscona and Moscona, 1965). (However, the possibility should be taken into consideration that *in vitro* conditions may alter the organization of the cell surface in such a way that some of its properties such as the species specificity are lost while the tissue specificity is retained.) These observations are compatible with the view that the structures discriminating between male and female are not expressed at the surface of these cells.

2. A largely derepressed genome is retained by the cells of the germ line. This is inferred from the fact that in the oocyte, the complexity of the transcripts is several-fold greater than in the somatic cells (see, e.g., Galau *et al.*, 1976). Although to our knowledge there is no such direct evidence in the case of the male germ cells, it has been shown that at least in *Drosophila,* spermatocytes exhibit lampbrush chromosomes comparable to those of the oocyte (Hess, 1968).

B. An Example of Segregation of Cell Lines: The Ascidian Embryo

The smaller the number of cells, the greater the accuracy with which cell lineage can be traced in the embryo. From this point of view, an almost ideal organism is the ascidian egg in which the major cell lines are completely segregated from one another at the 64-cell stage. Figure 3 shows the lineage of each of the two animal anterior (a 4.2), vegetal–anterior (A 4.1) and vegetal–posterior (B 4.1) blastomeres from the 8-cell to the 64-cell stage. (The development of the animal posterior blastomere (b 4.2) has been omitted as it gives rise only to ectoderm of the tail, see Table I.) The data illustrated in Fig. 3 are

3. CELL SURFACE DIFFERENTIATIONS

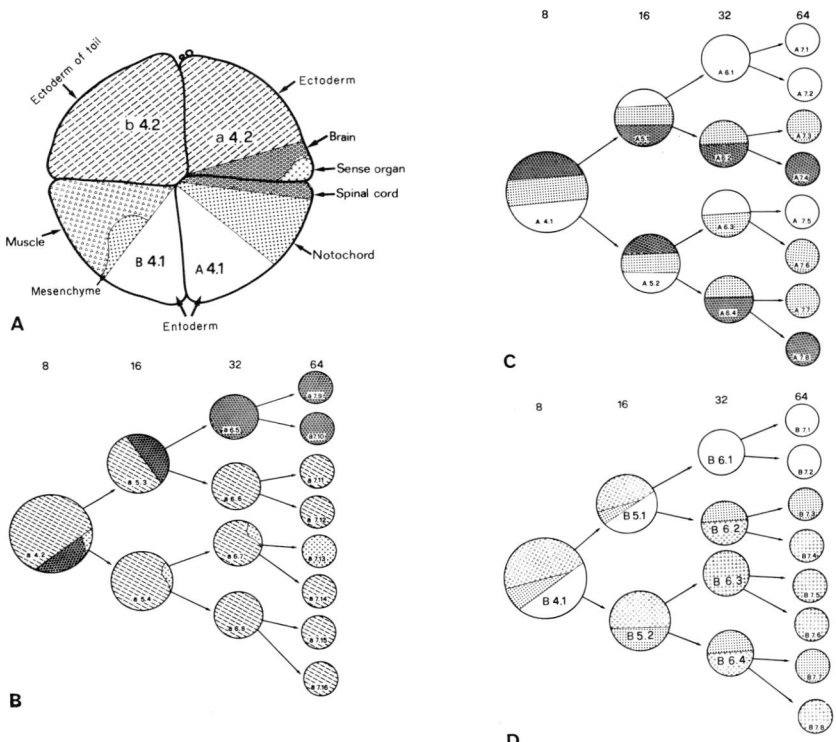

FIG. 3. Segregation of the major cell lines from the 8- (A) through the 64-cell stage of the blastomeres a 4.2 (B), A 4.1 (C), and B 4.1 (D) in the ascidian embryo.

TABLE I

SEGREGATION OF THE MAJOR CELL LINES IN THE ASCIDIAN EMBRYO FROM THE 8-CELL STAGE TO THE 64-CELL STAGE

Blastomeres	Derivatives (No.)	Total at the 64-cell stage	
a 4.2	Nerve cells (2)	Nerve cells	8
	Ectoderm cells (5)	Sense organ	2
	Sense organ cell (1)	Ectoderm	26
A 4.1	Entoderm cells (3)	Entoderm	10
	Nerve cells (2)	Notochord	6
	Notochordal cells (3)	Mesenchyme	4
B 4.1	Entoderm cells (2)	Muscle	8
	Mesenchymal cells (2)		
	Muscle cells (4)		
b 4.2	Ectoderm cells (8)		

summarized in Table I in which the total number of cells in each cell line at the 64-cell stage is also given; indeed, the 64-cell stage is the stage at which the segregation of all the cell lines is completed. Gastrulation begins in the 128-cell embryo.

Figure 3 shows that the sorting out of the individual lineage cells is not only a stepwise process, but that even within the same cell line some of the cells become "clean" lineage cells at the 32-cell stage while others shall have to wait yet another cleavage. Consider, for example, blastomere a 5.3 which at the 32-cell stage segregates into a 6.5 and a 6.6, which are already *pure lineage cells for the nervous system and the ectoderm,* respectively. On the other hand, the lineage cells of the nervous system deriving from blastomere A 4.1 have to wait until the 64-cell stage to segregate from the notochord lineage cells. Even more striking is the case of the lineage cells of the tail muscle which all derive from the B 4.1 blastomere. Here, the B 6.3 blastomere has already become a muscle lineage cell at the 32-cell stage, whereas the segregation of the two additional muscle lineage cells—B 7.4 and B7.8—occurs at the 64-cell stage. The factors underlying these differences so far escape us.

In an attempt to detect changes in the organization of cell membrane that might be related to the process of segregation of cell lines, we have studied the binding of several lectins to the surface of the blastomeres during the course of cleavage. A remarkable result was obtained with the lectin from *Dolichos biflorum* (which has a binding affinity for the nonreducing α-linked N-acetyl-D-galactosamine) on the segregation of the muscle cell line in the embryo of *Ascidia malaca* (Ortolani *et al.*, 1977). The lectin binds rather uniformly to the surface of the fertilized egg. Concurrently with the cytoplasmic redistribution that follows fertilization, the binding of the lectin concentrates at the vegetal pole in the area of the yellow crescent and then follows the segregation of the lineage cells of the tail musculature (Fig. 4). Since the binding of the lectin is restricted to the cell surface, this finding is, to our knowledge, the first evidence of a cell membrane specialization preceding by several cell divisions the overt differentiation of the cell. Indeed, the specific alteration of the cell surface is restricted to the area of the B 4.1 blastomeres which contain the material that is going to be segregated into the tail muscle cells.

C. The Immunological Approach to the Study of the Segregation of Cell Lines

To our knowledge, Ruth M. Clayton (1953) should be credited for the first immunological attempts to answer the problem of the molecu-

3. CELL SURFACE DIFFERENTIATIONS   53

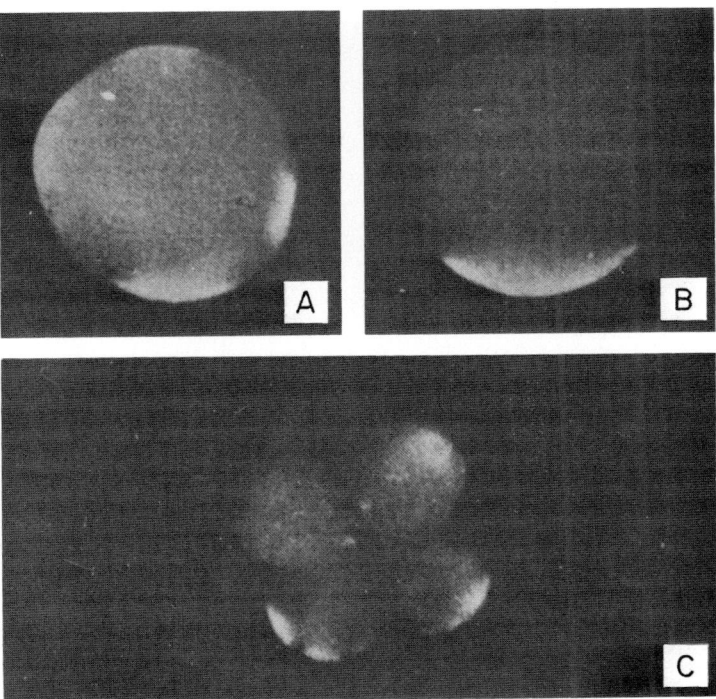

Fig. 4. (A) Patchy fluorescence of the unfertilized egg of *Ascidia malaca* stained with fluorescein-conjugated *Dolichos* lectin. (B) Fertilized egg after the ejection of the two polar bodies. The staining is now concentrated at the vegetal pole. (C) The posterior quartet (b 4.2 and B 4.1) of an 8-cell embryo. The staining is in the vegetal part of the B 4.1 blastomeres. The staining on the upper region of the right animal blastomere is an artifact due to the damage of the blastomere during separation. From Ortolani *et al.* (1977).

lar basis of the differences between the presumptive territories of an embryo, i.e., of the segregation of cell lines. As a matter of fact, previous immunological work had mainly aimed at identifying the appearance of new antigens in the course of development as a general criterion of differentiation, i.e., the appearance in the embryo *as a whole* of molecular species not to be found in the unfertilized egg (see, for example, the work of Perlman and Gustafson, 1948, on the sea urchin embryo). Another interesting line of work has focused mainly on the development in the embryo of the immune system as a whole (for review, see Ebert, 1958). In a painstaking work, Clayton was able to show that in the newt embryo, the ectoderm, the mesoderm (archenteron roof),

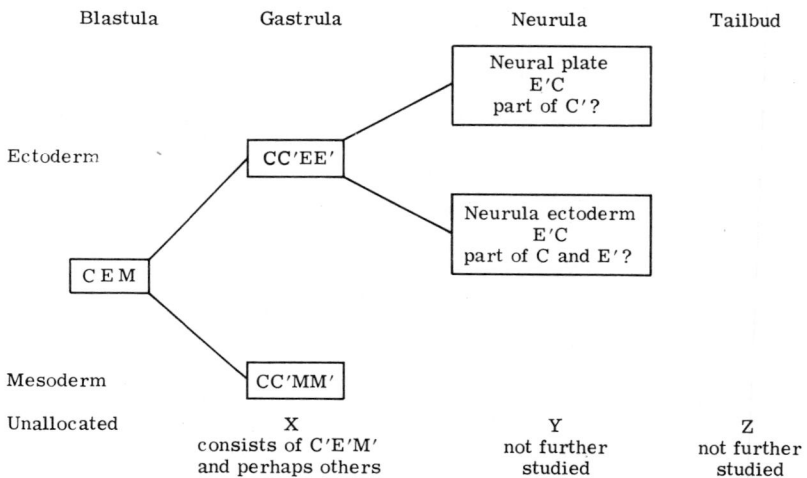

FIG. 5. Three main groups of antigens can be identified in the blastula: C,E,M. C antigen is common to ectoderm and mesoderm of the early gastrula. Part of this antigen is present in the blastula, while part of it (C') arises in the gastrula and segregates to the neural ectoderm and possibly also to the neural plate. E is the ectoderm antigens which segregate to the neuronal ectoderm, while a part arises after gastrulation (E') segregates into the neural plate. M is the mesoderm antigens which are specific for the mesoderm (archenteron roof), a part of which reappear (M') after gastrulation. Clayton was also able to distinguish in the ectoderm of the early gastrula, antigens distinct for the neural plate (P group) and for the neural ectoderm (N group) (not shown). From Clayton (1953).

and the neural plate contain specific antigens and that "an antigenic fraction specific to a given tissue may be detected at a stage earlier than that at which this tissue first appears." Also important was the finding that "two antigen fractions later characteristic of distinct tissues may coexist in their common precursor." Her main conclusions are diagramatically represented in Fig. 5.

D. DIFFERENTIAL EXPRESSION OF SURFACE PROPERTIES AND OF ANTIGENS IN THE VARIOUS CELL LINES IN THE EMBRYO

We do not intend here to review the large recent literature on the immunological analysis of the surface organization of individual cell types in the course of their differentiation, a subject that will be dealt with in other contributions of this series (see also Brackenbury et al., 1977; Jacob, 1978; McClay and Gooding, 1978; Gachelin, 1978; Dewey et al., 1978). In fact, we will confine ourselves to the discussion of some observations which are more pertinent to the problem of the segregation of cell lines.

Since the genome of the germ cells is largely derepressed, their surface may be expected to be a mosaic of a much greater variety of antigens inserted into the cell membrane during oogenesis and spermatogenesis than that of any somatic cell. And, in fact, they may share a number of antigens with highly specialized somatic cells. For example, acetylcholine receptors have been found at the surface of both the *Xenopus* (Kusano *et al.*, 1977) and mouse (Eusebi *et al.*, 1979) oocytes. Similarly, the nervous system (NS) antigens 6 and 7 are both present at the surface of mouse spermatozoa (Chaffe and Schachner, 1978a,b). On the other hand, they may be expected to contain specific molecules involved in the process of sperm–egg interaction at their surface. Fertilization will be discussed in a later section of this chapter; suffice it here to point out that one expects these molecules to disappear or be inactivated following fertilization. Interesting data have emerged from immunological studies using the mouse embryo.

The F9 antigen, a surface antigen coded for by a gene at the T locus, is expressed only at the surface of the spermatozoa and at the surface of the egg after fertilization, while it is absent at the surface of all somatic cells (Artzt *et al.*, 1973). The antigen then increases during cleavage up to the morula stage and is detectable both on the trophectoderm and on the inner cell mass (ICM) (Artzt *et al.*, 1973) and then slowly decreases. Interestingly, after implantation it becomes again detectable at the surface of the primordial gonocytes (Gachelin *et al.*, 1976). It has been suggested (Gachelin *et al.*, 1976) that these are the only cells at whose surface it remains present. On the other hand, these findings are also compatible with the interpretation that there is no time of disappearance of the antigen, but rather blocking of its exposure (see Weiss and Klavins, 1977). In the case of the acetylcholine receptors, they are no longer detectable even in the 2-cell embryo (Eusebi *et al.*, 1979). The cell surface antigen, PCC4, which is common to the multipotent embryonal carcinoma cells and spermatozoa, is first detectable at the surface of the ICM, while it is absent in the trophoblast (Gachelin *et al.*, 1977).

Other interesting antigens are two antigens expressed by a mouse testicular teratoma and which are also differentially expressed at the surface of the cells of mouse embryo (Gooding *et al.*, 1976) as shown in Table II.

In this case, the behavior of the two antigens is interestingly different. While both antigens are present at the surface of the unfertilized

TABLE II

EXPRESSION OF TESTICULAR TERATOMA HETEROANTIGENS ON EGGS AND EARLY MOUSE EMBRYO[a]

Stage	Antigen I	Antigen II
Unfertilized egg	+	+/−
4- to 16-cell embryo	+	−
Early blastocyst		
Polar trophoblast	+	−
Mural trophoblast	Not determined	+
Late blastula		
Trophoblast	−	+
Postimplantation embryo		
ICM	+	−
Trophoblast	−	−

[a] From Gooding et al. (1976).

egg, antigen I remains detectable at the cell surface throughout the early blastocyst stage when it specifically segregates to the cells of the ICM. By contrast, antigen II, although its expression at the surface of the unfertilized egg is weak, disappears altogether during cleavage only to become transiently detectable again at the surface of the trophoblast in the late blastula, while in the postimplantation embryo it is only present in the ICM.

Another interesting observation concerns the expression of the large external transformation-sensitive protein (LETS) (Zetter and Martin, 1978) during the early development of the mouse embryo. The LETS protein is a major cell surface component in many types of cells. By an indirect immunofluorescence assay it was shown that this protein cannot be detected at the surface of the blastomeres during cleavage or in the morula. In the blastocyst, both young (3-day) and late (4-day), the trophoblast cells were constantly negative. However, when the trophoblast was lysed by immunosurgery, thus exposing the ICM, it was seen that the ICM cells were only slightly positive in the young blastocyst but their surface became strongly positive in the late blastocyst. Furthermore, when the ICM was cultured for 2 days, thus allowing the differentiation of the entoderm, only the ectoderm cells expressed the LETS protein at their surface while the entoderm cells did not. These observations may thus suggest that fertilization sets in motion a process of reorganization of the oocytes surface which results in (1) the disappearance of molecules involved in the process of maturation and/or fertilization; (2) their differential sorting out into certain cell lines while in others they are no longer expressed; (3) the temporary repression of their expression until some later stage when they are expressed again in a specific cell line.

We like to think of the blastomeres as originally having a large variety of determinant molecules exposed at their surface. At each step of the sorting-out process, the molecular population at the surface of each cell becomes progressively restricted, eventually ending up with a relatively small population of specific molecules which permits a limited number of interactions with the cells of the same type or of a compatible type.

This interpretation is somewhat similar to the suggestion recently advanced by Changeux and Dauchin (1976) on the differential stabilization of the surface receptors during neuronal differentiation under the influence of stimuli from different neurons. Very little is known about the changes of the surface organization of the blastomeres during cleavage. The selective localization of the *Dolichos* lectin-binding sites at the surface of the muscle lineage cells in *Ascidia* (Ortolani et al., 1977) has already been quoted (Section III,B). Another observation worth mentioning is that of the differential mobility of the concanavalin A (Con A)-binding sites of the surface of the micromeres of the sea urchin embryo with respect to the macro- and mesomeres (Roberson et al., 1975).

It has also been recently found that in the mouse embryo during the compaction process that takes place at the 8-cell stage, the Con A-induced agglutination of the blastomeres drops dramatically. This suggests a change in the organization of the membrane of the blastomeres and, specifically, a change of its fluidity, during compaction (Rector and Granholm, 1978).

Hence, as a first approximation, we suggest that the process of segregation of cell lines consists of two main processes: (1) the progressive *differential restriction* of the developmental program in each line; (2) the *differential expression* of genes in the different lines. Both processes are expressed in a specific molecular organization of the surface of the cells belonging to a cell line and of their derivatives. Although in the early stages of development the first process is likely to be the prevailing process whereas in later stages the situation appears to be reversed in favor of the second, the two processes are not to be thought of as being mutually exclusive and may indeed proceed concurrently.

A differential sorting out of antigens at the cell surface into different blastomeres may not necessarily imply an irreversible commitment of the blastomeres, and it may be anticipated that the situation will be different in different animals depending on whether they belong to the so-called regulative or determinative type. For example, when an 8-cell mouse embryo was dissociated into individual blastomeres which were then attached to the surface of another 8- or 16-cell intact embryo, the outside cells differentiated into trophoblast cells (Herbert and Graham, 1974). This result is in accord with the experiments showing that in the mouse embryo up to the 8-cell stage at least, the blastomeres are still totipotent; in fact, mixing of the cells from two 32-cell embryos results in a normal blastocyst (Tarkowski, 1961; Mintz, 1962).

On the other hand, in the ascidian embryo, substitution of the 4 b.1 blastomeres (presumptive ectoderm of the tail) for the 4 a.1 (presumptive cephalic ectoderm + brain and sense organs) results in an embryo lacking the nervous sytem (Reverberi and Minganti, 1946). In this case, already at the 8-cell stage, the two animal anterior and the two animal posterior blastomeres appear to be irreversibly committed to differentiation along a well-defined pathway. This means that the 4 b.1 blastomeres are no longer able to respond to the stimuli from the underlying A 4.1 blastomeres which are responsible for the evocation of the nervous system from the ectoderm (without this interaction, the 4 a.1 blastomere gives rise to an ectodermal vesicle). This implies an irreversible restriction of the developmental program of each blastomere, and hence the irreversible silencing of certain genes. Whether this process is accompanied by a molecular reorganization of the surface of the blastomeres is not at present known. However, from the cited observation on the muscle cell line in the ascidian embryo (Ortolani *et al.,* 1977), this appears very likely. Experiments to test this hypothesis are presently in progress in our laboratory.

In our view the differential irreversible gene repression is not the only event at the gene level that controls differentiation (Caplan and Ordahl, 1978). We agree that gene silencing is probably the most important event underlying the segregation of cell lines during cleavage in all the embryos in which experiments have been conducted. A turning point appears to be the blastula stage. Even before the advent of molecular biology, the study of hybrid embryos (the pioneering studies in this field were carried out by F. Baltzer, 1910, in the sea urchin) had shown that in the nonviable hybrids, development is arrested prior to gastrulation. As early as 1948, Perlmann and Gustafson had shown that in the mesenchyme blastula of the sea urchin, an antigen appears that is different from those present in the unfertilized egg and in all subsequent stages of development until the mesenchyme blastula stage. This is not the place to review fully the literature on the subject (more detailed information can be found in Monroy and Moscona,

1979). We wish to confine ourselves to the information pertinent to the differentiation of the cell surface especially in relation to immunological analysis.

Interesting observations have recently been made by McClay *et al.* (1977; McClay, this volume) on the viable sea urchin hybrids of *Tripneustes esculentus* (T) and *Lytechinus variegatus* (L). Control experiments showed that cells disaggregated from the two species are incapable of forming hybrid aggregates. Cells from pregastrula hybrid crosses T♀ × L♂ aggregate only with cells of the maternal species; on the other hand, when prepared from gastrulas or later stages they proved to be able to aggregate also with cells of the paternal species. This result strongly suggests that around the time of gastrulation, paternal determinants are exposed at the surface of the cells. Antisera were then raised against cell membranes of *Lytechinus* gastrulas, and the indirect immunofluorescence test showed that the cells of the hybrid embryos begin to express L-specific antigens at gastrulation. Furthermore, Fab fragments of the antiserum specifically inhibited aggregation with L-cells. Hence, these experiments are compatible with the notion that the molecular organization of the cell surface up to the blastula stage is (largely if not exclusively) under the control of the maternal genome. This is in accordance with the widely accepted idea that development to the blastula stage is largely controlled by the transcripts of the maternal genome stockpiled in the oocyte during oogenesis.

The previously mentioned differential expression of the LETS protein in the ectoderm and entoderm of the mouse blastocyst is also compatible with a differential gene expression in the two cell lines. In addition, antisera against the ectoplacental cone of the mouse do not react either with the unfertilized or fertilized 1-cell egg. The antigens begin to be detectable at the blastomere surface of the 8-cell embryo and the reaction is strong in the morula; it then becomes confined to the throphectoderm of the blastocyst. In the postimplantation embryos, the labeling is confined to the trophectoderm derivatives, while it is absent in the embryonic ectoderm and entoderm which are derivatives of the ICM (Searle and Jenkison, 1978).

IV. The Sperm–Egg Interaction in Fertilization

In this section the question of the specificity of sperm–egg interaction will be briefly discussed. Fertilization is the major event for the continuity of the species. The two gametes involved in the process must be able to recognize one another as a preliminary to their merging into a single cell, the zygote. It is thus imperative for the gametes to be endowed with surface properties which must be different in the different species.

At the molecular level, this implies that specific molecules must be present at the surface of both gametes on whose specific interaction the success of fertilization depends (for a discussion on the phylogenesis of the surface structures involved in mating, see Monroy and Rosati, 1979).

The concept that "specificity" in the mating reaction is due to specific molecules at the surface of the gametes or secreted by them was first formulated by F. R. Lillie (1913). It originated from the observation of the species specificity of the agglutination of the spermatozoa of a variety of animals when mixed with the supernatant from an egg suspension. According to Lillie, fertilization depends on the interaction between a substance called "fertilizin" at the surface of the egg (and which slowly diffuses from the egg when left standing in the suspension medium) with an "antifertilizin" present at the surface of the spermatozoon. By analogy with the antibody-induced cell agglutination, Lillie (1913–1919) also suggested an analogy between fertilization and the antigen–antibody reaction. Although currently the fertilizin theory has only historical interest, because of the part it has played in the studies on fertilization, it is worth spending a few words to explain how Lillie conceived of its role in fertilization. This is the best done through the scheme of Fig. 6.

The prevailing current idea is that the specificity of fertilization depends on a direct cell-to-cell interaction rather than on factors in solution. Nevertheless, the great contribution of Lillie's theory was that of having stimulated the immunological approach to the study of fertilization. This approach has proved to be most suitable to obtain information about the organization of the gametes surface and about the components involved in the recognition process. Indeed, work along these lines must take into account the peculiarities of the interaction between sperm and egg and related problems.

The egg is covered by a glycoprotein coat (called vitelline coat or zona pellucida or chorion) which in its turn is surrounded either by a jelly envelope or by a layer of follicle cells. As for the spermatozoon, the organization of its apical region is such that under appropriate conditions it breaks open to expose the acrosomal contents (acrosome reaction). The inner (adnuclear) part of the acrosomal membrane is thus exposed and in some cases everted (this has been shown to be due to the rapid polymerization of actin filaments, Tilney et al., 1973). Hence, the peculiarity of the sperm–egg interaction as compared to that between

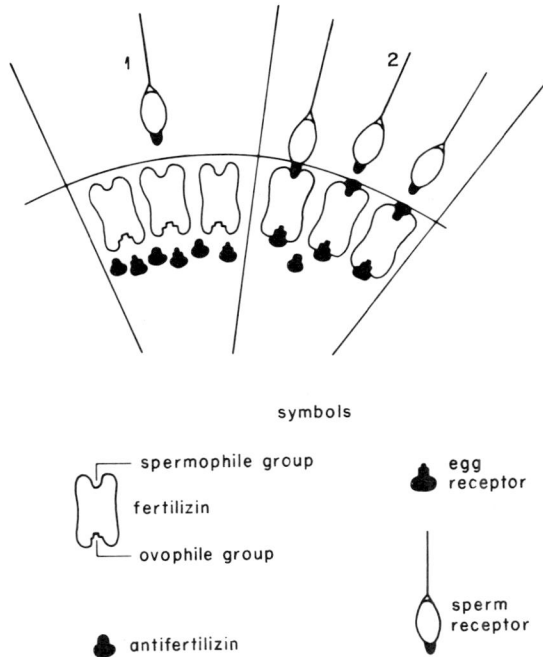

Fig. 6. Illustration of F. R. Lillie's theory of fertilization. In the unfertilized egg, fertilizin molecules are present in the cortical layer of the egg. Each one of them bears an ovophile and a spermophile group (sector 1). Upon fertilization (sector 2), the sperm receptors of the spermatozoon combine with the spermophile groups of a group of fertilizin molecules (for the sake of simplicity, one sperm receptor per spermatozoon is indicated in the diagram). This reaction activates the fertilizin molecules and the activation propagates all over the egg surface. As a result of such activation, molecules of antifertilizin rapidly block all the spermophile groups of the fertilizin molecules, thus preventing reaction with the other spermatozoa. This is Lillie's hypothesis concerning the mechanism of prevention of polyspermy. Another result of the activation of the fertilizin molecules is the reaction of all ovophile groups each with one egg receptor. This is the hypothetical mechanism of the activation of the egg. From Monroy (1965); adapted from Lillie (1914).

somatic cells is that, while the latter interaction involves their plasma membrane, the interaction between gametes requires a complex series of steps which are preliminary to (and, in fact, indispensable for) the final event of gamete fusion and which involve the various surfaces exposed by the gametes.

The preliminary steps of interaction in the groups more extensively studied may be summarized as follows: (1) interaction of the spermatozoon with the outermost egg envelope; (2) firm attachment (binding) of the spermatozoon to the inner glycoprotein coat (vitelline coat); (3) penetration of this envelope and fusion with the egg plasma membrane. The work of the last few years suggests that the binding of the spermatozoon to the inner glycoprotein coat (the vitelline coat) is the key event of these preliminary steps. This structure appears to be endowed with the property to recognize and to bind spermatozoa species-specifically.

Regarding the spermatozoon, although the events leading to the acrosome reaction are fairly well known, the question as to whether or not the acrosome reaction is triggered by the interaction with the outermost egg envelope(s) or with the vitelline coat is still open. It is also not quite clear which component of the spermatozoon is involved in the binding reaction.

To our knowledge, the case of the starfish is the only one (Metz, 1945, 1957) in which the jelly coat can be considered to be the specific trigger of the acrosome reaction. And yet, even in this case, it is not at all clear if other components of the vitelline coat participate in the binding reaction. On the other hand, in the groups in which the binding reaction has been studied in greater depth, such as the sea urchin and the mammals (Hartmann et al., 1972; Hartmann and Hutchinson, 1976), the timing of the acrosome reaction with respect to the following steps, and the specificity of the triggering reaction, are still unclear (see Gwatkin, 1976; Metz, 1978).

For example, in the case of the sea urchin, two alternative possibilities can be considered. The former possibility is that the acrosome reaction occurs upon contact of the spermatozoon with the jelly coat (Summer and Hylander, 1975, 1976); this does not necessarily need to be a specific interaction. Hence, the binding of the spermatozoon to the vitelline envelope would require the prior opening of the acrosome (see also Metz, 1978). According to this view, the acrosomal component responsible for the binding of the spermatozoon to the egg is a protein, "bindin," which is released from the acrosome vesicle upon its dehiscence (Vacquier and Moy, 1977). On the side of the egg, a glycoprotein fraction has been isolated from the vitelline coat which exhibits species specific affinity for bindin (Glabe and Vacquier, 1978). On the other hand, peroxidase-conjugated antibodies to bindin have localized bindin on the acrosomal process and on the vitelline coat adjacent to the point of the acrosomal process attachment (Moy et al., 1977; Moy and Vacquier, this volume).

The second alternative is that the spermatozoon reaches the vitelline coat in an unreacted condition, and hence binding is mediated by the sperm plasma membrane (Aketa and Ohta, 1977). This finding is consistent with the observation (Decker *et al.*, 1976) that under physiological pH, jelly coat solutions of *Arbacia* fail to induce an acrosome reaction of the homologous spermatozoa. Even in this case, a species-specific component has been obtained from the vitelline coat of sea urchin eggs (Aketa *et al.*, 1968; Schmell *et al.*, 1977) which, when combined with *unreacted* spermatozoa, inhibits their fertilizing ability. These different results might be explained by assuming that the whole series of events, i.e., the attachment of the spermatozoon and the triggering of the acrosomal reaction which results in the release of bindin, are mediated by the same egg surface component (Glabe and Vacquier, 1977). In this case the jelly coat should be thought of as a nonspecific barrier which reduces the number of spermatozoa reaching the vitelline coat (as already suggested by Hagström, 1956). On the other hand, it should be borne in mind that it is very difficult to remove the jelly coat of the sea urchin egg completely and cleanly (Kidd, 1978).

The ascidians have proved to be a new interesting model with which to study the sperm–egg interaction. Indeed, using ascidians, it is possible to separate the events of the binding of the spermatozoon to the egg from those of the acrosome reaction and of the following steps of sperm penetration through the vitelline coat (called the *chorion*) and its final incorporation into the egg (Rosati and De Santis, 1978; Rosati *et al.*, 1978; De Santis *et al.*, 1979).

The major breakthrough in this work has been the discovery that glycerol treatment of the eggs results in the shedding of the follicle cells and cytolysis of the egg and of the test cells, while the chorion remains apparently undamaged. When returned to seawater, these eggs retain the ability to bind spermatozoa species-specifically; however, none of the attached spermatozoa undergo the acrosome reaction (Rosati and De Santis, 1978). On the other hand, when the spermatozoa attach to the chorion of eggs from which the follicle cells have been mechanically removed (by shaking) (Fig. 7), *some* do undergo the acrosome reaction (Rosati and De Santis, 1978). From these "egg ghosts," clean chorions can be prepared, and these are currently being used both for chemical analyses and for the preparation of antibodies.

Before discussing the mechanism of the sperm–egg recognition and binding, it is worth discussing the physiological role and the composition of the chorion. It was known that the chorion is the main barrier to self-fertilization (Morgan, 1923) and to heterologous fertilization (Minganti, 1948). This appears to depend on the fact that the specific recognition and binding of the spermatozoa take place on the chorion (Rosati and De Santis, 1978). Hence, the chorion must be endowed with highly specific receptors located at its outer surface.

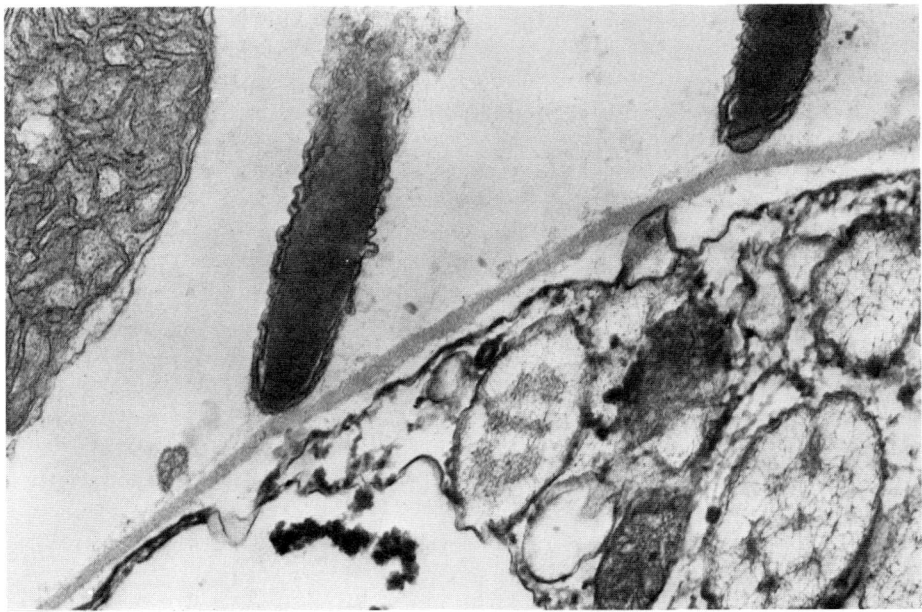

Fig. 7. Two spermatozoa of *Ciona intestinalis* attached to the chorion of an egg mechanically deprived of its follicle cells by shaking. Fixation 15 minutes after addition of sperm. ×37,400.

A major effort is being made in our laboratory toward the chemical and antigenic identification of the receptors both in the chorion and at the surface of the spermatozoon. In a first series of experiments (Rosati et al., 1978), it was shown that the surface of the chorion and the surface of the spermatozoa share binding sites for Con A and for FBP from *Lotus tetragouolobus,* thus suggesting that mannose, glucose, and fucose residues are exposed at the surface of the chorion and the spermatozoon. Chemical analyses in progress suggest that the chorion is a glycoprotein and that the major carbohydrate component is fucose, followed by glucose; mannose is present in much smaller amounts. Further experiments (Rosati and De Santis, in preparation) have shown that both sperm attachment to the chorion of glycerol-treated eggs and fertilization of living eggs is interfered with by an excess of fucose. The reversibility of the fucose action and the stereospecific requirement of L-fucose (D-*fucose has proved to be totally ineffective*) encourage the view that fucose plays a key role in the sperm–egg interaction and binding and, in fact, that fucose is the essential carbohydrate component of the receptor. Glucose and mannose interfere only slightly with binding and fertilization; hence, they may possibly play a role as "adaptors" which aid the molecular fitting of the receptors located on the chorion and the surface of the spermatozoon. Preliminary experiments in which IgG or Fab fragments from antichorion and antisperm sera have been used suggest that the chorion and the sperm plasma membrane have at least some antigens in common.

The mechanism whereby the spermatozoon binds to the chorion have also been studied (De Santis *et al.*, 1979). "Tufts" made up of very thin fibrils have been detected both at the outer and the inner surface of the chorion. The fibrils stain strongly with Ruthenium Red (RR) and bind ferritin-conjugated Con A. When chorions isolated from glycerol-treated eggs are challenged with spermatozoa, attachment occurs only at the outer surface of the chorion (Fig. 8). The connection between the spermatozoa and the chorion is established through very thin fibrils extending from the plasma membrane of the tip of the spermatozoon to the chorion (Fig. 8). The fibrils are also RR-positive and bind ferritin-conjugated Con A (Fig. 9). It was then suggested that the tufts at the outer surface of the chorion are the sperm receptors or, at least, that the sperm receptors are part of the tufts. The question then arises as to whether or not every tuft at the outer surface of the chorion is a sperm receptor. A similar situation exists in the sea urchin egg in which sperm attachment appears to occur at the tip of microvillar protrusions of the vitelline envelope, though not every protrusion is a site of sperm attachment (Tegner and Epel, 1976; see also Epel, 1978). Since there is no obvious morphological difference between the microvillar protrusions, one possibility is that binding of more spermatozoa (1300–1800 bound spermatozoa per egg, Vacquier and Payne, 1973) is limited by the movement of the attached ones (Epel, 1978). Our observations show that morphological identity, even though at the cytochemical level, is still too crude a criterion to rule out functional diversity. The best example is that though the branched filaments at the inner surface of the chorion react positively to RR and bind the ferritin-conjugated Con A, they fail to bind spermatozoa. Hence, whether or not a tuft or a microvillar protrusion can function as a sperm receptor must depend on its molecular organization at a finer level than can be detected by present morphological methods.

V. Concluding Remarks

In conclusion, the vitelline coat appears to be the egg component responsible for sperm recognition and binding; the concentration of sperm receptors in this structure may result in a kind of amplification. Although further investigation is warranted, studies carried out thus

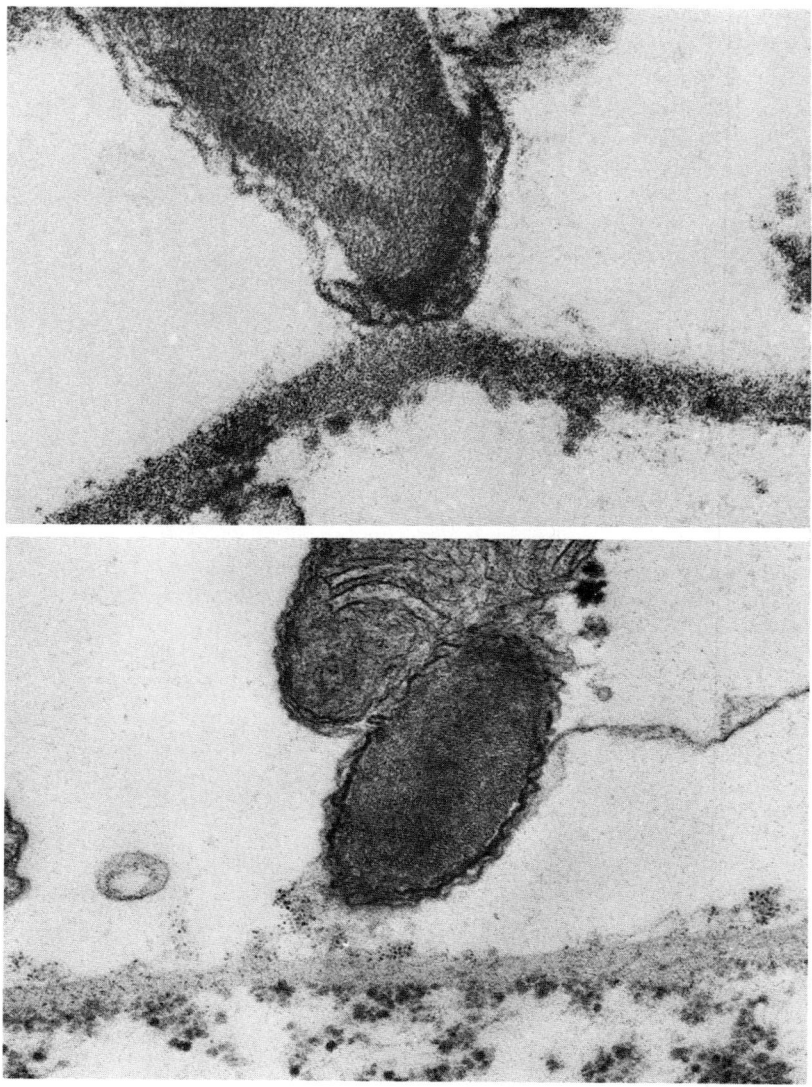

Fig. 8. A spermatozoon of *Ciona intestinalis* attached to an isolated chorion. Thin fibrils connect the plasma membrane of the tip of the head of the spermatozoon to the outer surface of the chorion. The acrosome vesicle is still intact. ×119,350.

Fig. 9. A spermatozoon bound to the chorion of a glycerol-treated egg; staining with ferritin-conjugated Con A. The tufts of fibrils at the outer surface of the chorion and the fibrils connecting the outer surface of the chorion to the plasma membrane of the spermatozoon selectively bind the ferritin-conjugated Con A. ×63,140.

far suggest that the material of the vitelline coat is largely manufactured by the cells of the ovarian follicle, i.e., by somatic cells. The participation of the oocyte in building up the vitelline coat is not very clear. Therefore, we seem to be faced with a kind of paradox: namely, that the oocyte delegates the synthesis of a structure which is essential for the specificity of fertilization to the somatic cells. However, the possibility should be considered that, while the component molecules of the coat are synthesized by the follicle cells, their stereospecific arrangement at the oocyte surface is dictated by the molecular organization of the oolemma.

Finally, in considering the process of fertilization in a teleological perspective, one is struck by the immense disproportion of the sperm–egg ratio. Although the disproportion may be more apparent than real, since the vast majority of spermatozoa may fail to reach their target either because of the very large dilution they undergo in the milieu in animals with external fertilization or because they are destroyed in the female genital tract, still the sperm–egg ratio is vastly in favor of the spermatozoa. Does this mean that a large percentage of the spermatozoa are somehow "defective" and hence unable to participate in fertilization? This interpretation is compatible with some recent claims (Timourian et al., 1972) that in the sea urchin only about 2% of the spermatozoa are able to participate in fertilization. However, an alternative interpretation that we would like to suggest stems from an analogy with the lymphocytes.

According to the widely accepted theory of clonal selection, each lymphocyte has the genetic information to express only one class of immunoglobulins. We consider the possibility that spermatogenesis results in the production of *clones* of spermatozoa, each carrying one or a very few specific receptors at their surface. According to this hypothesis, recognition and attachment of the spermatozoa to the sperm receptors on the egg depends on the general species-specific organization of the surface of the two gametes, while "fusion" requires a highly specific molecular matching between *receptors* at the gamete surface. Interestingly, a similar analogy has been made by Granit (1977) to interpret the immense diversification and specialization of the cells of the nervous system.

REFERENCES

Aketa, K., and Ohta, T. (1977). *Dev. Biol.* **61**, 366–372.
Aketa, K., Tsuzuki, H., and Onitake, K. (1968). *Exp. Cell Res.* **50**, 676–79.
Artzt, K., Dubois, P., Bennett, D., Condamine, H., Babinet, C., and Jacob, F. (1973). *Proc. Natl. Acad. Sci. U.S.A.* **70**, 2988–2992.
Baltzer, F. (1910). *Arch. Zellforsch.* **5**, 497–621.

Boycott, R. E., Diver, C., Garstang, S. L., and Turner, F. M. (1931). *Philos. Trans. R. Soc. B* **219**, 51–131.
Brachet, J. (1977). *In* "Current Topics in Developmental Biology (A. A. Moscona and A. Monroy, eds.), Vol. 11, pp. 133–186. Academic Press, New York.
Brackenbury, R., Thiery, J. P., Rutishauser, U., and Edelman, G. M. (1977). *J. Biol. Chem.* **252**, 6835–6840.
Campanella, C. (1975). *Biol. Reprod.* **12**, 439–447.
Caplan, A. I., and Ordhal, C. P. (1978). *Science* **201**, 120–130.
Chaffee, J. K., and Schachner, M. (1978a). *Dev. Biol.* **62**, 173–184.
Chaffee, J. K., and Schachner, M. (1978b). *Dev. Biol.* **62**, 185–192.
Changeux, J. P., and Dauchin, A. (1976). *Nature (London)* **264**, 705–712.
Clayton, R. M. (1953). *J. Embryol. Exp. Morphol.* **1**, 25–42.
Culp, L. A., Buniel, J. F., and Rosen, J. J. (1978). *In* "Cell Surface Carbohydrate Chemistry" (R. E. Harmon, ed.), pp. 205–224. Academic Press, New York.
Decker, G. L., Joseph, D. B., and Lennon, W. J. (1976). *Dev. Biol.* **53**, 115.
Denis-Donini, S., and Campanella, C. (1977). *Dev. Biol.* **61**, 140–152.
De Santis, R., Jamunno, G., and Rosati, F. (1979). *Dev. Biol.* (in press).
Dewey, M. J., Filler, R., and Mintz, B. (1978). *Dev. Biol.* **65**, 171–182.
Ebert, J. D. (1958). *In* "The Chemical Basis of Development" (W. D. McElroy and B. Glass, eds.), pp. 526–545. Johns Hopkins Press, Baltimore, Maryland.
Epel, D. (1978). *Dev. Biol.* **12**, 185–246.
Eusebi, F., Mangia, F., and Alfei, L. (1979). *Nature (London)* **277**, 651–653.
Gachelin, G. (1978). *Biochim. Biophys. Acta* **516**, 27–60.
Gachelin, G., Fellous, M., Guenet, J. L., and Jacob, F. (1976). *Dev. Biol.* **50**, 310–320.
Gachelin, G., Kemler, R., Kelly, F., and Jacob, F. (1977). *Dev. Biol.* **57**, 199–209.
Galau, G. A., Klein, W. H., Davis, M. M., Wold, B. J., Britten, R. J., and Davidson, E. H. (1976). *Cell* **7**, 487–505.
Glabe, C. G., and Vacquier, V. D. (1978). *Proc. Natl. Acad. Sci. U.S.A.* **75**, 881–885.
Gooding, L. R., Hsu, Y. C., and Edidin, M. (1976). *Dev. Biol.* **49**, 479–486.
Granit, R. (1977). "The Purposive Brain." MIT Press, Cambridge, Massachusetts.
Gwatkin, R. B. L. (1976). *In* "The Cell Surface in Animal Embryogenesis and Development" (G. Poste and G. L. Nicholson, eds.), Vol. 1, pp. 1–54.
Hagström, B. E. (1956). *Exp. Cell Res.* **11**, 160–168.
Hartman, J. F., and Hutchinson, C. F. (1976). *J. Cell. Physiol.* **88**, 219–226.
Hartman, J. F., Gwatkin, R. B. L., and Hutchinson, C. F. (1972). *Proc. Natl. Acad. Sci. U.S.A.* **69**, 2767–2769.
Herbert, M. C., and Graham, C. F. (1974). *In* "Current Topics in Developmental Biology" (A. A. Moscona and A. Monroy, eds.), Vol. 8, pp. 151–178. Academic Press, New York.
Hess, O. (1968). *Mol. Gen. Genet.* **103**, 58–71.
Jacob, F. (1978). *Proc. R. Soc. London B* **201**, 249–270.
Kidd, P. (1978). *J. Ultrastruct. Res.* **64**, 204–215.
Köhler, G., and Milstein, C. (1975). *Nature* **256**, 495–497.
Kusano, K., Miledi, R., and Stinnakre, J. (1977). *Nature (London)* **270**, 739–741.
Lillie, F. R. (1913). *Science* **38**, 524–528.
Lillie, F. R. (1914). *J. Exp. Zool.* **16**, 523–590.
Lillie, F. R. (1919). "Problems of Fertilization." Univ. of Chicago Press, Chicago, Illinois.
Maienschein, J. (1978). *J. Hist. Biol.* **11**, 129–158.
McClay, D., and Goodin, L. R. (1978). *Nature (London)* **274**, 367–368.
McClay, D. R., Chambers, A. F., and Warren, R. H. (1977). *Dev. Biol.* **56**, 343–355.

Metz, C. B. (1945). *Biol. Bull.* **89**, 84–94.
Metz, C. B. (1957). *In* "Physiological Triggers," (Bullock, eds.), pp. 17–45.
Metz, C. B. (1978). *In* "Current Topics in Developmental Biology" Academic Press, New York. (A. Moscona and A. Monroy, eds.), vol. 12, pp. 106–140.
Minganti, A. (1948). *Nature (London)* **161**, 643–644.
Mintz, B. (1962). *Am. Zool.* **2**, 541.
Monroy, A. (1965). "Chemistry and Physiology of Fertilization." Holt, New York.
Monroy, A., and Baccetti, B. (1975). *J. Ultrastruct. Res.* **50**, 131–142.
Monroy, A., and Moscona, A. A. (1979). "Introductory Concepts in Developmental Biology." Univ. of Chicago Press, Chicago, Illinois.
Monroy A., and Rosati, F. (1979). *Nature (London)* **278**, 165–166.
Morgan, T. H. (1923). *Proc. Natl. Acad. Sci. U.S.A.* **9**, 170–173.
Moscona, A. A. (1957). *Proc. Natl. Acad. Sci. U.S.A.* **43**, 184–194.
Moscona, A. A., and Moscona, M. (1965). *Dev. Biol.* **11**, 402–423.
Moy, G. N., Friend, D. S., and Vacquier, V. D. (1977). *J. Cell Biol.* **75**, 61a.
Needham, J. (1942). "Biochemistry and Morphogenesis." Cambridge Univ. Press, London and New York.
Ortolani, G., O'Dell, D. S., and Monroy, A. (1977). *Exp. Cell Res.* **106**, 402–404.
Perlmann, P., and Gustafson, T. (1948). *Experientia* **4**, 481–482.
Raven, C. P. (1970). *Int. Rev. Cytol* **28**, 1–44.
Raven, C. P. (1976). *Am. Zool.* **16**, 395–403.
Rector, J. T., and Granholm, N. H. (1978). *J. Exp. Zool.* **203**, 497–502.
Reverberi, G., and Minganti, A. (1946). *Pubbl. Staz. Zool. (Naples)* **20**, 135–151.
Roberson, M., Neri, A., and Oppenheimer, S. B. (1975). *Science* **189**, 639–640.
Rosati, F., and De Santis, R. (1978). *Exp. Cell Res.* **112**, 111–119.
Rosati, F., De Santis, R., and Monroy, A. (1978). *Exp. Cell Res.* **116**, 419–427.
Schmell, E., Earles, B. J., Brenx, C., and Lemar, W. G. (1977). *J. Cell Biol.* **72**, 35–46.
Searle, R. F., and Jenkinson, E. J. (1978). *J. Embryol. Exp. Morphol.* **43**, 147–156.
Summers, R. G., and Hylander, B. L. (1975). *Exp. Cell Res.* **96**, 63–68.
Summers, R. G., and Hylander, B. L. (1976). *Exp. Cell Res.* **100**, 190–194.
Tarkowsky, A. K. (1961). *Nature (London)* **190**, 857–860.
Tegner, M. J., and Epel, D. (1976). *J. Exp. Zool.* **197**, 31–58.
Tilney, L. G., Hatano, S., Ishikawa, H., and Mooseker, M. S. (1973). *J. Cell Biol.* **59**, 109–126.
Timourian, H., Hubert, C. E., and Stuart, R. N. (1972). *J. Reprod. Fertil.* **29**, 381–385.
Ubbels, G. A., Bezern, J. J., and Raven, C. (1969). *J. Embryol. Exp. Morphol.* **21**, 445–466.
Vacquier, V. D., and Payne, J. E. (1973). *Exp. Cell Res.* **82**, 227–235.
Vacquier, V. D., and Moy, G. W. (1977). *Proc. Natl. Acad. Sci. U.S.A.* **74**, 2456–2460.
Weiss, M. J., and Klavins, J. V. (1977). *In* "Membranes and Cellular Functions" (G. A. Jamieson and D. M. Robinson, eds.), Vol. 4, pp. 72–97. Butterworth, London.
Zetter, B. R., and Martin, G. R. (1978). *Proc. Natl. Acad. Sci. U.S.A.* **75**, 2324–2328.

CHAPTER 4

IMMUNOFLUORESCENT ANALYSIS OF CHROMATIN STRUCTURE IN RELATION TO GENE ACTIVITY: A SPECULATIVE ESSAY

Lee M. Silver

SECTION ON DEVELOPMENTAL GENETICS
SLOAN-KETTERING INSTITUTE
NEW YORK, NEW YORK

and Sarah C. R. Elgin

HARVARD UNIVERSITY
THE BIOLOGICAL LABORATORIES
CAMBRIDGE, MASSACHUSETTS

I. Introduction	71
II. The Technique of Immunofluorescence Antibody Staining	72
III. Patterns of Chromosomal Protein Distribution in Relation to Gene Activity...	76
A. Staining in Proportion to the DNA Distribution	78
B. Staining of Heterochromatin or Euchromatin Only	78
C. Staining of a Set of Developmentally Active Loci	80
D. Staining of the Immediately Active Loci	84
IV. Concluding Remarks	86
References..	87

I. Introduction

A unique advantage of working with antibodies as specific reagents to identify certain macromolecules is that they allow the correlation of biochemical and cytological data. In studies of chromosome structure and function this property can be exploited best using *Drosophila*. The polytene chromosomes of the third instar larval salivary glands allow the visualization (at the level of the light microscope) of particular genetic loci from the well-defined and highly reproducible banding pattern, which is correlated with the extensive genetic map. The technique of *in situ* hybridization has been used to localize the complements of particular DNA and RNA sequences to particular polytene chromosome bands (Pardue and Gall, 1975). Using antisera against specific chromosomal proteins, one can similarly assess the pattern of distribution of these components in relation to the established banding pattern.

It is of particular interest to consider the distribution pattern of chromosomal proteins in relation to patterns of gene activity. Sites of intense transcriptional activity in the polytene chromosomes can be identified in the form of puffs (decondensed, swollen chromosomal regions). Puffs appear and regress in a highly coordinate and regular manner during normal development of the third instar larva. Specific changes in the normal puffing pattern can be induced by external stimuli such as heat shock. Placing the larva (or the isolated glands) in buffer at 37°C for even 10 minutes drastically reduces normal gene activity and causes the specific induction of puffs at nine chromosomal loci (Ritossa, 1962; Ashburner, 1970; Spradling *et al.*, 1975; McKenzie *et al.*, 1975; Mirault *et al.*, 1978).

In the last few years our laboratory and others have developed immunofluorescence procedures to analyze the *in situ* distribution patterns of specific chromosomal proteins on *Drosophila* polytene chromosomes (Silver and Elgin, 1976; Alfageme *et al.*, 1976). In this chapter we will not discuss in detail all the results that have been obtained with this technique, but rather focus on a set of results indicating distribution patterns of nonhistone chromosomal (NHC) proteins which may reflect structural differences in chromatin related to the control of gene expression.

II. The Technique of Immunofluorescence Antibody Staining

In preparing polytene chromosomes for microscopic examination, the conventional procedure is to squash the dissected salivary gland between the slide and coverslip in a solution of 45% acetic acid. The slide can then be frozen and the coverslip removed. Following postfixation in 3.7% formaldehyde–phosphate-buffered saline and washing in Tris-buffered saline, the chromosomes can be "stained" by an indirect immunofluorescence procedure. The chromosomes are incubated first with the test antiserum, washed extensively in Tris-buffered saline, incubated with fluorescein-conjugated antibodies directed against IgG molecules of the type used initially, washed again, and mounted for viewing by phase contrast and ultraviolet (UV) illumination. This procedure, using "conventional fixation" in 45% acetic acid, works well in many cases (for an example, see Fig. 1). However, it has been demonstrated with isolated chromatin that 45% acetic acid can extract the bulk of the histone and a selected portion of the NHC proteins (Silver and Elgin, 1976). A number of fixation techniques have been devised to overcome this problem. In the work described from our laboratory, preparations designated as "formaldehyde-fixed" were obtained by preincubating the intact *Drosophila* salivary gland first in buffered

0.5% Nonidet P40 for 10 minutes and then in buffered 2% formaldehyde for 30 minutes prior to squashing in acetic acid and staining as above. Such a fixation procedure prevents the extraction of chromosomal proteins by 45% acetic acid. However, the converse problem can arise. In certain cases it appears that the fixation is so extensive as to either denature or render inaccessible certain antigens. For example, following this formaldehyde fixation procedure, the densest chromomeres fail to stain with antihistone 3 serum (anti-H3), although with less extensive fixation protocols one observes staining with anti-H3 serum roughly in proportion to the distribution of chromosome mass, as would be anticipated from biochemical data (Silver, 1978). Clearly it is of advantage to examine the staining pattern obtained with each antiserum using a variety of fixation conditions before venturing an interpretation of the protein distribution pattern. A detailed description of the fixation techniques we have used and a discussion of the advantages and disadvantages of each has been presented elsewhere (Silver and Elgin, 1978a). Others have developed analogous but somewhat different formaldehyde fixation techniques for this purpose (Alfageme et al., 1976; Plagens et al., 1976). Hill and Watt (1977) have proposed to circumvent the problems caused by acetic acid by using hand-dissected polytene chromosomes (fixed with low concentrations of formaldehyde) rather than conventional squashes.

Antisera against specific *Drosophila* chromosomal proteins can be readily obtained. Since one frequently uses sodium dodecyl sulfate (SDS) gel electrophoresis to separate and purify the chromosomal proteins, investigators have often followed the technique of Tjian et al. (1974) in using the protein in the isolated gel band without extraction as the immunogen. Sera have also been prepared directly against isolated enzymes of chromosomal interest (e.g., RNA polymerase) (Greenleaf and Bautz, 1975). The core histones have highly conserved amino acid sequences; these proteins tend to be relatively poor immunogens. Thus antihistone sera could conceivably contain significant activity against NHC proteins which were low-level contaminants in the immunizing histone fraction. Consequently, it is generally advantageous to use antihistone sera prepared using a heterologous source of histones, in our case those from calf thymus. Such sera show extensive cross-reaction with the histones of *Drosophila* (Bustin et al., 1977), but should not react with the less conserved NHC proteins.

In general, indirect immunofluorescence staining of polytene chromosomes using a serum directed against a chromosomal protein as a primary reagent will cause prominent staining of the chromosomes and a very low-level staining of the cytoplasmic debris (Fig. 1). Con-

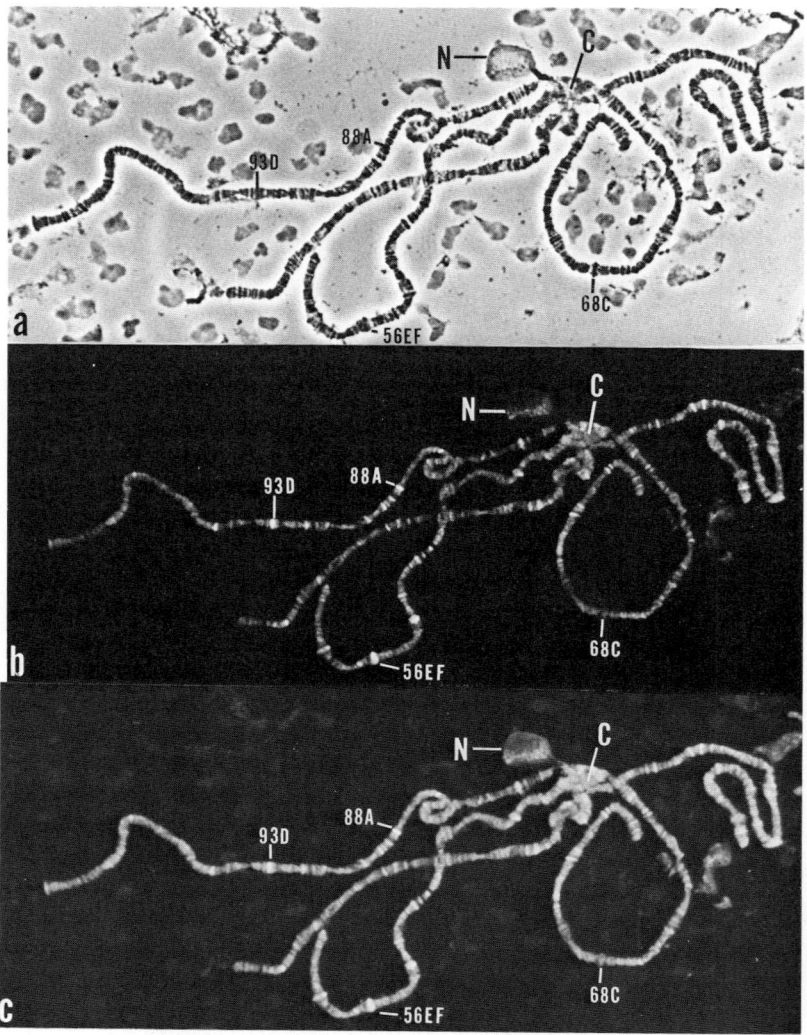

Fig. 1. Polytene chromosome spread prepared by the conventional fixation technique and stained using an antiserum against *Drosophila* embryo total nonhistone chromosomal proteins. N is the nucleolus, C the chromocenter. (a) Phase-contrast view. (b) High-contrast fluorescence view. (c) Moderate-contrast fluorescence view. Adapted from Silver and Elgin (1978a).

versely, staining using a preimmunization serum or a serum directed against a cytoplasmic protein will produce a general low- or high-level fluorescence, with the chromosomes stained at or below the level of the cytoplasmic debris (Silver, 1978). (It should be noted that the evaluation of staining intensities is usually a relative and not an absolute process. It is a courtesy to the reader to use micrographs including cytoplasmic debris for reference when possible.) In a few cases exceptions to the above generalities have been observed. In two cases (one rabbit and one mouse) from approximately 20 animals we have examined, the preimmunization or control serum caused staining of the chromosomes (Silver, 1978; S. M. Abmayr and S. C. R. Elgin, unpublished observations). In one case we have also failed to observe chromosome staining using an antiserum against a known chromosomal component, the H3–H4 tetramer from calf thymus. This result was obtained using chromosomes prepared by both fixation techniques, even though the serum reacted well with soluble *Drosophila* chromatin as shown by complement fixation (Silver et al., 1978). Presumably the result is a consequence of problems of denaturation or inaccessibility of this particular antigen.

With use of all other antisera against chromosomal proteins that we have tested, one observes a general widespread chromosomal staining at a level significantly above that of background cytoplasmic debris. When selective staining patterns have been observed, they are always superimposed upon this general chromosomal staining. By manipulating photographic variables, it is possible to emphasize either the selective staining pattern or the general chromosomal fluorescence, as shown in Fig. 1. The intensity of the general staining obtained with many antisera is dependent upon the fixation conditions and in some cases is greatly enhanced on chromosomes prepared by the more extensive formaldehyde fixation procedure. In many cases where specific anti-NHC protein sera or antihistone sera have been used to stain conventional or formaldehyde-fixed polytene chromosomes, this general staining is present primarily in the interband (phase contrast light) regions (Silver, 1978). At least in the case of antisera against the core histones, the dominant interband staining pattern observed on formaldehyde-fixed chromosomes must be an artifact of the formaldehyde fixation, since the unstained bands contain 95% of the polytene chromosome DNA and, on the basis of biochemical evidence, must therefore contain 95% of the core histones. Indeed, staining of all bands more intensely than interbands is observed with the use of the same antihistone sera on polytene chromosomes which have not been exposed to formaldehyde (Silver and Elgin, 1978a).

In addition, background staining of chromosomes (at a level below that of the cytoplasmic debris) is observed with all preimmunization and control antisera; this staining is also localized at interbands. The basis for this staining is not known, but the observation raises the possibility that patterns of interband staining obtained with specific antisera might similarly be artifactual. The lack of fluorescent staining of a particular chromosomal locus cannot be accepted as firm evidence for the lack of antigens at that site. The accessibility to an antibody probe of one antigen at a particular locus under a particular set of fixation conditions does *not* guarantee the accessibility of a second antigen at the same locus under the same set of conditions (Silver and Elgin, 1978a; Silver, 1978).

In light of these results, it is clear that several precautionary procedures and control observations should be used routinely to establish the specificity of the staining reaction: (1) the preimmunization serum of the animal to be used must be tested; (2) the antiserum should be tested using several chromosome fixation techniques (at least conventional squashes and formaldehyde-fixed chromosomes); (3) the fluorescence of the chromosomes should be significantly greater than that of the cytoplasmic debris; (4) the fluorescence staining pattern of the chromosomes should be reproducible among all nuclei within a single squash. All of these criteria have been met in the experiments from our laboratory discussed below. One obviously places greatest confidence in specific staining patterns obtained using the formaldehyde fixation technique. In several cases the specific staining patterns obtained with sera against NHC proteins have been found to be invariant for the different fixation techniques.

III. Patterns of Chromosomal Protein Distribution in Relation to Gene Activity

During the last few years we have learned a considerable amount about chromatin structure. The basic chromatin fiber for all DNA sequences is now believed to consist of a chain of repeating subunits, the nu bodies or nucleosomes, each made up of 165–240 base pairs of DNA wrapped around a core of eight of the small histones (probably two each of H2A, H2B, H3, and H4). Histone H1 and the NHC proteins apparently interact with the outer surface of the DNA–histone bead. (See Elgin and Weintraub, 1975, and Felsenfeld, 1978, for reviews of the supporting evidence.) Obviously, the basic chromatin fiber must be folded or organized into higher order structures culminating ultimately in the metaphase chromosome (e.g., Laemmli *et al.*, 1978). While it appears that both active and inactive chromatins possess to some degree the fundamental structure dictated by association with histones (Lacy and Axel, 1975; Kuo *et al.*, 1976; Reeves, 1976), distinct structural differences at a secondary level have been detected. Weintraub and Groudine (1976) first reported that the active globin gene in chick reticulocyte nuclei was preferentially susceptible to DNase I digestion. This differential sensitivity of active loci has been confirmed using several systems (Garel and Axel, 1976; Panet and Ceder, 1977; Flint and Weintraub, 1977; Levy and Dixon, 1977; Bellard *et al.*, 1978). The altered state of the active gene is not dependent on the actual transcription process in that (1) sequences can remain DNase I sensitive after transcription has stopped (Weintraub and Groudine, 1976; Palmiter *et al.*, 1978); (2) sequences can be DNase I sensitive in cell types that are the developmental precursors of the cell that produces the mRNA (Miller *et al.*, 1978); and (3) the degree of DNase I sensitivity is not correlated with the apparent frequency of transcription (Garel *et al.*, 1977). The type of structural alteration leading to DNase I sensitivity has not yet been clearly identified. It is interesting that in a recent analysis of the heat shock loci by nuclease digestion we have observed, on gene activation, a disruption of higher order structures involving 2–20 kb of DNA as well as a smearing of the nu body pattern (C. E. C. Wu, Y. C. Wong, and S. C. R. Elgin, in preparation).

One can infer a two-tiered (at a minimum) mechanism of the control of gene expression. Prior to the events that actually initiate gene transcription, a structural transition occurs in the chromatin. It is logical to suggest that some of the NHC proteins may play important determinative and structural roles in this process. Accordingly, it is of interest to look for examples of the following types of NHC protein distribution patterns:

1. Staining of all chromosomal loci in proportion to DNA distribution. Such a protein, like the histones, would be a component of the basic machinery needed to fold and organize the DNA.

2. Staining of the heterochromatin only, indicating a protein exclusively required to form a chromatin structure which does not allow gene activation; and conversely, staining of the euchromatin only.

3. Staining of a set of loci, each of which is active at some point during a given developmental period. This pattern might be defined by two different types of proteins. In one case, heat shock loci would also be stained (but only after induction) in addition to the developmental pattern. In the other case, induced heat shock loci would not become stained. Proteins associated with both the developmentally active and the heat shock induced loci could define a condition which is necessary

but not sufficient for gene activity. Those which fail to become associated with the active heat shock loci are clearly not necessary for the active gene structure, but might be of developmental significance.

4. Staining only of the set of loci which is actively being transcribed at the time of testing ("immediately active" loci). This pattern should be indicative of a protein involved in the transcription process per se.

In the last few years evidence has been obtained suggesting that there are different NHC proteins showing most of these types of distribution patterns. The results obtained support the hypothesis of a multitiered or combinatorial model of gene regulation. We shall discuss these cases in consecutive order.

A. Staining in Proportion to the DNA Distribution

One clear case of this type has been observed. An antiserum [anti-(5.2/21)] was produced against an acidic (pI 5.2), low-molecular-weight (21,000) *Drosophila* NHC protein that had been isolated by sequential isoelectric focusing and SDS gel electrophoresis (Silver and Elgin, 1978b). The specificity of the antiserum for the immunizing protein was confirmed by radioimmunostaining of isoelectric focusing gels and SDS gels on which total NHC proteins had been fractionated.

When anti-(5.2/21) serum is used in the staining of conventionally fixed polytene chromosomes, one obtains a pattern of staining which correlates positively with DNA density; for example, the fluorescent pattern, at a first approximation, corresponds with the phase contrast banding pattern (Fig. 2). The chromocenter is prominently stained. The puffs–regions of decondensed chromatin—are correspondingly stained at a very low level. Protein (5.2/21) could be an NHC protein which is bound to chromatin *in vivo* without DNA sequence specificity. One can infer from rough quantitative estimates of protein yield that only one (5.2/21) molecule is present for every 10–100 nucleosomes. Hence, it is possible that (5.2/21) could be involved in a general role in the packaging of nucleosomes into higher order structures.

B. Staining of Heterochromatin or Euchromatin Only

Recently H. Will and E. Bautz have observed a pattern of heterochromatin staining using a serum prepared against a *Drosophila* NHC protein fraction prepared by chromatography on hydroxylapatite (H. Will and E. Bautz, personal communication, 1978). The NHC protein D1 isolated by Alfageme *et al.* (1976) appears to be present and accessible to antibody probes at a very limited number of loci, all of which correspond with (A–T)-rich regions of the genome.

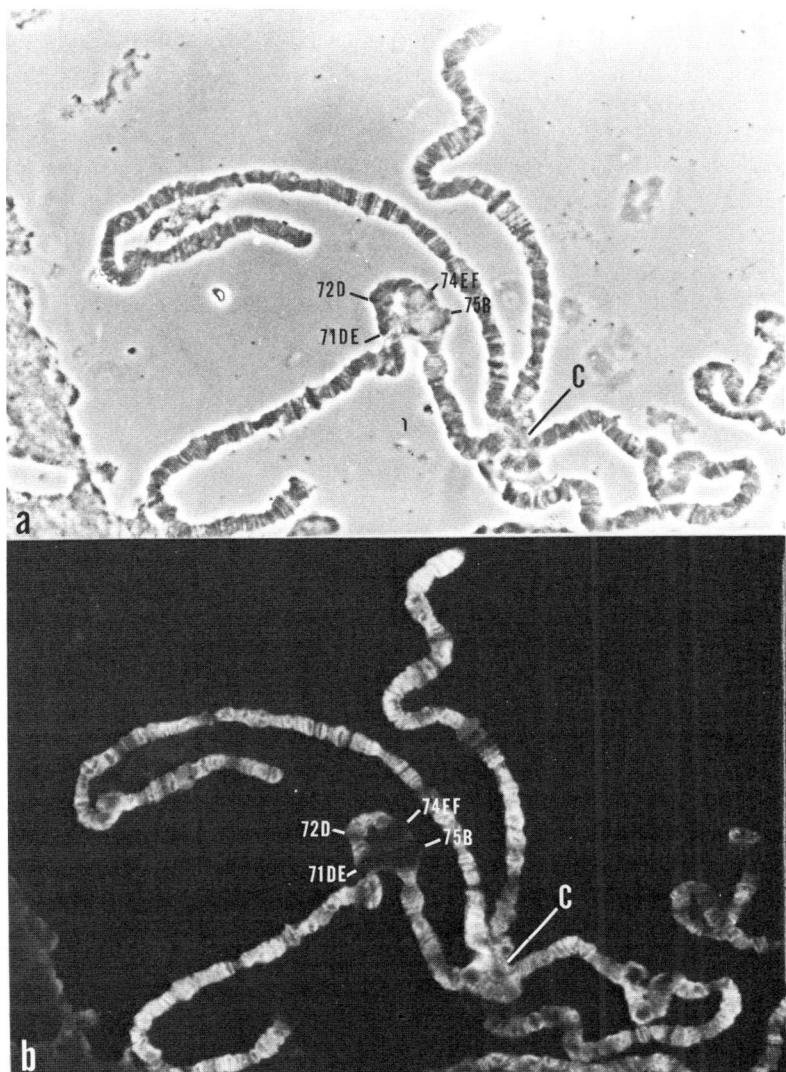

FIG. 2. Polytene chromosome spread from heat-shocked larva prepared by conventional fixation technique and stained using an antiserum prepared against a *Drosophila* NHC protein of isoelectric point 5.2, molecular weight 21,000. C is the chromocenter; 71DE, 72D, 74EF, and 75B are developmentally induced puffs. (a) Phase-contrast view. (b) Fluorescence view. See Silver and Elgin (1978b) for details of this study.

While there are several antisera against NHC proteins which do not stain the chromocenter (Silver, 1978), it is difficult to evaluate this data because of the relative reduction in number of gene copies, and hence assay sensitivity, for this region in the polytene chromosomes (Gall, 1973). Potential candidates for proteins associated with all euchromatin but no heterochromatin probably would have to be tested by immunofluorescent staining of unfixed metaphase chromosomes. No systematic work of this type has been done.

C. Staining of a Set of Developmentally Active Loci

During late third-instar larval development, the relatively quiescent salivary gland chromosomes carry out a highly controlled program of specific puff formation and regression. This tissue-specific developmental program of gene activity has been mapped by Ashburner (1972) and is identical in all wild type members of the species. A correlation has been observed between the complete set of loci active at any time during salivary gland development and the prominent fluorescence staining patterns obtained with two independently derived antisera. One, anti-ρ serum, was produced against a molecular-weight subfraction of the total NHC proteins (Silver and Elgin, 1977); the other, anti-band 2 serum, was produced against a molecular-weight subfraction of 60,000 from the proteins released by mild DNase I digestion of *Drosophila* nuclei (Mayfield et al., 1978). Using the anti-ρ serum, a 90% correlation was obtained, in a careful analysis of chromosome 3, between those loci prominently and consistently stained and those listed by Ashburner (1972) as puffing at some time in the third instar or prepupal periods (Silver and Elgin, 1977). The band 2 pattern (shown in Fig. 3) is similar although not completely identical (Mayfield et al., 1978).

Fluorescent staining of essentially the complete set of developmentally active loci was observed with the use of either antiserum on chromosomes obtained from larvae at all stages of development; thus, staining was correlated with the potential for gene activity and the history of gene activity as well as gene activity per se. In contrast to these results, the pattern of staining obtained with anti-RNA polymerase II serum was observed to correlate with immediately active loci only (see below); the RNA polymerase staining pattern was different on polytene chromosomes obtained from larvae at different stages of development.

As previously mentioned, it is possible to alter the normal pattern of polytene chromosome gene activity by subjecting the larva to a temperature of 37°C (larvae are usually grown at 25°C). The heat shock

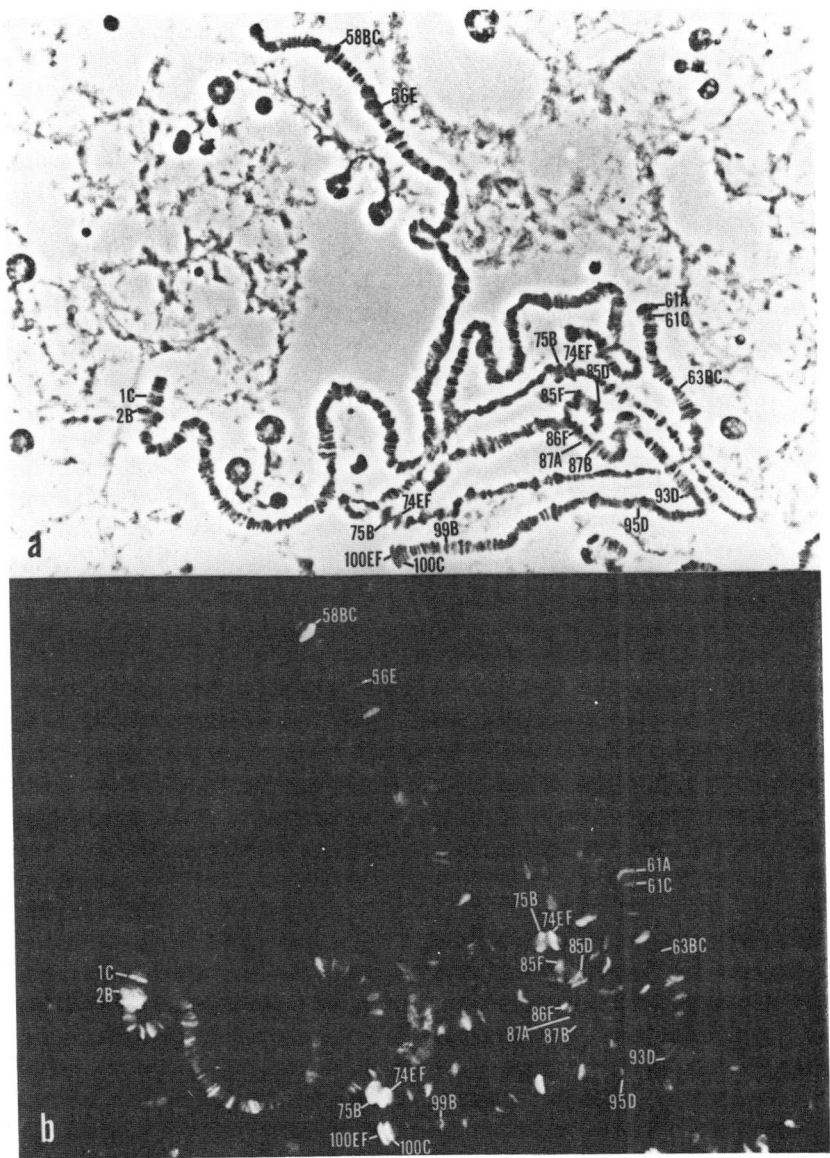

Fig. 3. Polytene chromosome spread prepared using the formaldehyde fixation technique and stained using an antiserum raised against a 60,000-dalton fraction from *Drosophila* chromosomal proteins released following limited digestion of nuclei with DNase I. (a) Phase-contrast view. (b) Fluorescence view. From Mayfield *et al.* (1978). Copyright © by The MIT Press.

response of the salivary glands involves a drastic reduction in transcriptional activity at normal developmentally active puffs concurrent with the induction of intense transcriptional activity at nine specific loci, several of which are never active during any stage of normal larval development. The fluorescent staining pattern obtained using either anti-ρ or anti-band 2 serum on formaldehyde-fixed chromosomes from larvae that have been heat-shocked is shown in Fig. 4. The staining pattern obtained following heat shock includes those sites which stained prior to heat shock stimulation and in addition the heat shock loci, which previously were not prominently stained. The results suggest that a different chromatin structure, as indicated by positive staining, is associated with gene activity as indicated by puffing. The retention of staining by the now inactive developmental sites is reminiscent of the retention of DNase I sensitivity by previously active loci (see previous discussion). One can suggest that the chromatin state defined by positive ρ or band 2 staining is necessary but not sufficient for gene activity. The set of loci so defined correlates well with the set of loci known to be active in this given developmental period (Silver and Elgin, 1977; Mayfield et al., 1978). It seems likely, although not established, that this set of loci will also correlate well with the set of loci that is DNase I sensitive in this cell type.

It is interesting to note that under conditions of conventional fixation one does not observe chromosome staining using anti-band 2 serum; apparently the antigen is extracted by the acetic acid. Using the anti-ρ serum under these conditions one observes staining of the developmentally active loci against a high background of general chromosome staining, but does not observe staining of the induced heat shock loci (Fig. 5). Staining is observed at *93D*, a heat shock puff also induced during normal third instar larval development, but no staining is observed at loci *87A* and *87B-C1*, which are uniquely active in the heat shock state for this tissue. This could indicate different binding coefficients for the same antigen at the two different types of loci. Alternatively, one could suggest that the anti-ρ serum at a minimum reacts with two antigens. One is acetic acid-extractable and analogous to the band 2 antigen, while the second is not acid-extractable and shows the above staining pattern. Such an antigen as this must not be essential for transcriptional activity per se, but may be of importance in developmentally defining a program of gene activity. It is interesting that in several experiments examining the staining patterns using antisera directed against NHC proteins of unknown function, one observes staining of reproducible defined sets of loci for which there is no known physical or biological correlation (Silver and Elgin, 1977; Silver and

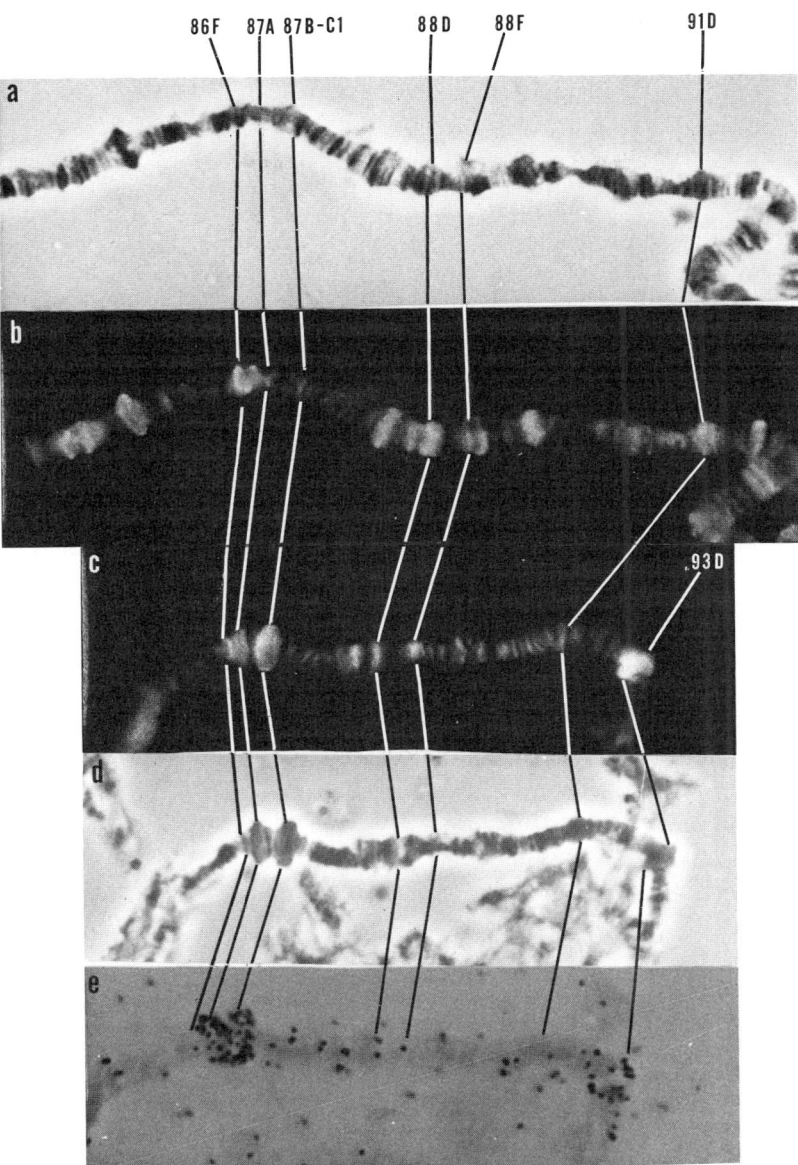

FIG. 4. Staining pattern of a central region of chromosome *3R* using anti-ρ serum on chromosome spreads prepared by formaldehyde fixation. (a,b) Phase-contrast and fluorescent views of a chromosome from a larva maintained at 25°C. (c–e) Fluorescent, phase-contrast, and autoradiographic (using [^{3}H]uridine) views of a chromosome from a larva maintained at 37°C for 20 minutes. *87A, 87B-C1,* and *93D* are loci activated by heat shock; *86F, 88D, 88F,* and *91D* are loci active in this developmental period. See Silver and Elgin (1977) for details of this analysis.

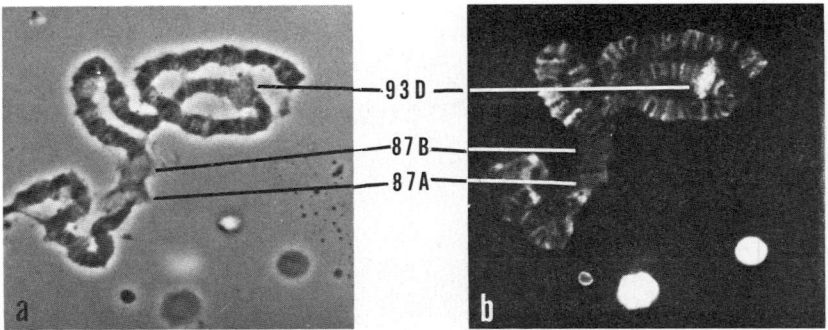

Fig. 5. Staining of a conventionally fixed chromosome arm 3R from a heat shocked larva using anti-ρ serum. (a) Phase-contrast view. (b) Fluorescence view. From Silver (1978).

Elgin, 1978b). Perhaps these sets correlate with other patterns (for other tissues or times) of gene expression. One caveat should be noted; only in a few of the above cases has it been demonstrated that the staining patterns observed are the consequence of single antigens.

D. STAINING OF THE IMMEDIATELY ACTIVE LOCI

Antibodies against RNA polymerase II of *Drosophila* have been prepared by Greenleaf and Bautz (1975) and kindly supplied to us as well as used in their own work. Using this antiserum one obtains prominent staining at all puffs; following heat shock, staining is lost at the developmentally active loci but is prominent at the newly induced loci (Plagens *et al.*, 1976; Jamrich *et al.*, 1977a; Elgin *et al.*, 1978). Thus, the pattern of staining is similar to the pattern of uridine incorporation and corresponds to the pattern of prominent RNA synthesis (Fig. 6). [Careful examination of Figs. 4 and 6 reveals low-level staining at the heat shock loci using both ρ and RNA polymerase antisera in the control cases. It is known that the heat shock response can be induced by other environmental stress (Ashburner, 1972). The response detected is thus most likely a background level associated with dissection and gland incubation. Puffs are not visible.]

In addition to the prominent puff staining, secondary level interband staining is also observed using antibodies directed against RNA polymerase II. As a consequence, Jamrich *et al.* (1977b, 1978) have suggested that all interbands are active genes. The possibility that such a staining pattern is an artifact has been discussed in Section II. In addition, it should be noted that studies with the *lac* repressor (Lin

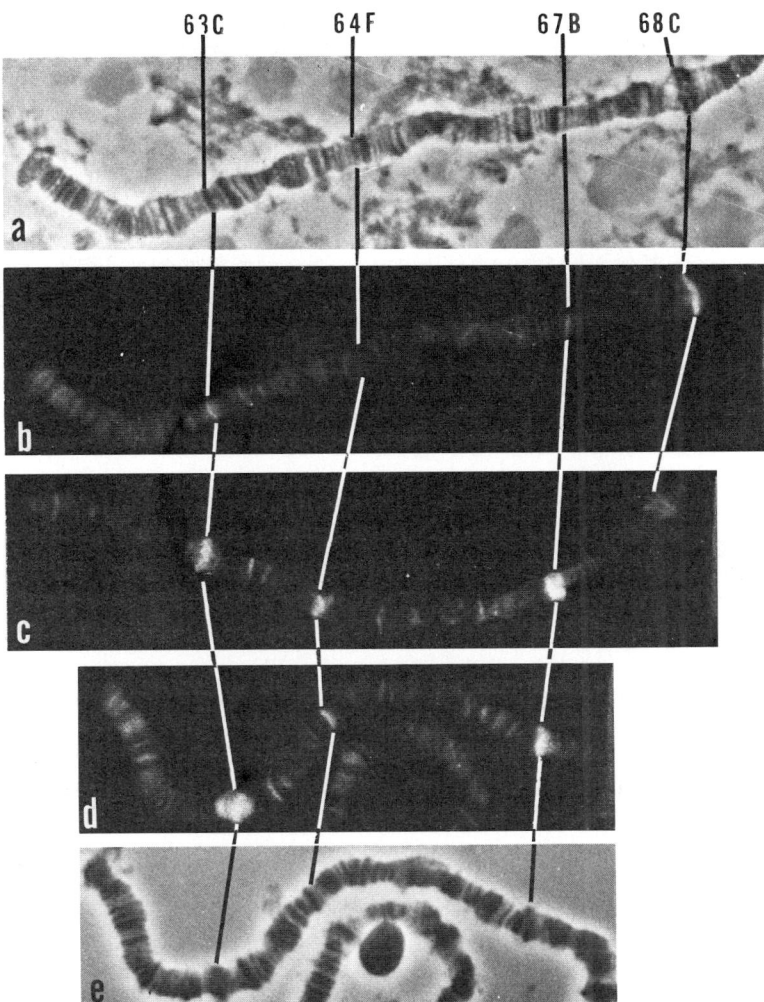

FIG. 6. Staining of a region of chromosome arm *3L* using an antiserum directed against *Drosophila* RNA polymerase II (provided by Dr. Arno Greenleaf, Duke University). (a,b) Phase-contrast and fluorescence views of a chromosome from a larva maintained at 25°C. (c) Fluorescence view of a chromosome from a larva maintained at 37°C for 20 minutes. (d,e) Phase-contrast and fluorescence views of a chromosome from a second larva maintained at 37°C for 20 minutes. Chromosomes used in (a), (b), and (c) were prepared by the conventional squashing technique; the chromosome used in (d) and (e) was prepared by formaldehyde fixation. *68C* is a locus puffed early in the third instar period; *63C, 64F,* and *67B* are induced by heat shock. *67B* can normally be active at later stages of development (Ashburner, 1972).

and Riggs, 1972, 1975) and with *E. coli* RNA polymerase (Hinkle and Chamberlin, 1972) have demonstrated the existence of two classes of DNA binding sites. Class A sites are specific sequences of DNA to which regulatory proteins bind with high affinity. Class B sites include all other DNA sequences. The existence of class B binding sites *in vivo* may be essential for the normal functioning of these proteins, although actual transcription is restricted to class A sites (Lin and Riggs, 1975). The existence of an analogous situation in eukaryotic nuclei has been proposed by Yamamoto and Alberts (1975) for the particular case of steroid hormone–receptor complexes.

It is possible that eukaryotic RNA polymerases partake in a similar dual binding regime. The selective high level fluorescent staining pattern would represent the stage-specific set of class A binding sites, while the interband staining pattern would represent those class B binding sites which were accessible, after fixation, to an antibody probe.

Jamrich *et al.* (1978) have indicated that the level of interband staining by anti-RNA polymerase II serum is reduced as a consequence of the heat shock response. We have not observed this to be the case (see Fig. 6). We have found that the intensity of interband staining generally increases as a function of more extensive fixation conditions, with some experimental variation. Such a result would be expected if interband staining were a consequence of low affinity binding of RNA polymerase II to chromatin.

IV. Concluding Remarks

The results obtained to date support the notion of a hierarchy of chromatin states or structures related to gene expression. The use of immunological probes to determine the distribution patterns of chromosomal proteins will be an important tool in testing this hypothesis further. There remain technical ambiguities in any reliance on cytological techniques alone. This work would be strengthened if one could demonstrate an appropriate fractionation of the DNA sequences based on interactions of the chromatin complex with the appropriate antibodies (by immunoprecipitation or affinity chromatography). Such experiments are being pursued.

The evidence for at least three types of NHC proteins bearing different relationships to active and inactive genes is relatively convincing: (1) those widely distributed in chromatin, such as (5.2/21) (Silver and Elgin, 1978b); (2) those associated with the developmentally active loci and heat shock-activated loci of the third instar salivary gland, that is, those which appear to be necessary but not sufficient for gene activity (Silver and Elgin, 1977; Mayfield et al., 1978); and (3) those associated primarily with active loci, such as RNA polymerase II (Jamrich et al., 1977a,b; Elgin et al., 1978). More evidence is needed to substantiate the other classes suggested. Further, it should be noted that in several cases the actual proteins involved remain to be identified, and in all cases the structural roles we have suggested remain to be substantiated. It is intended that this discussion will indicate an interesting approach to the problem and illustrate the potential power of antibody techniques in an analysis of the relation of chromatin structure to function.

ACKNOWLEDGMENTS

We thank Sue Abmayr for preparation of Fig. 2. During the writing of this chapter our work was supported in part by a grant GM20779 from the National Institutes of Health. S.C.R.E. is a recipient of an NIH Research Career Development Award; L.M.S. is a recipient of a Biomedical Research Fellowship from the Population Council.

REFERENCES

Alfageme, C. R., Rudkin, G. T., and Cohen, L. H. (1976). *Proc. Natl. Acad. Sci. U.S.A.* **73,** 2038–2042.
Ashburner, M. (1970). Chromosoma **31,** 356–376.
Ashburner, M. (1972). *In* "Developmental Studies on Giant Chromosomes" (W. Beerman, ed.), pp. 101–151. Springer-Verlag, Berlin and New York.
Bellard, M., Gannon, F., and Chambon, P. (1978). *Cold Spring Harbor Symp. Quant. Biol.* **42,** 779–792.
Bustin, M., Reeder, R. H., and McKnight, S. L. (1977). *J. Biol. Chem.* **252,** 3099–3101.
Elgin, S. C. R., and Weintraub, H. (1975). *Annu. Rev. Biochem.* **44,** 725–774.
Elgin, S. C. R., Serunian, L. A., and Silver, L. M. (1978). *Cold Spring Harbor Symp. Quant. Biol.* **42,** 839–850.
Felsenfeld, G. (1978). *Nature (London)* **271,** 115–122.
Flint, S. J., and Weintraub, H. M. (1977). *Cell* **12,** 783–794.
Gall, J. G. (1973). *In* "Molecular Cytogenetics" (B. A. Hamkalo and J. Papaconstantinou, eds.), pp. 59–74. Plenum, New York.
Garel, A., and Aexl, R. (1976). *Proc. Natl. Acad. Sci. U.S.A.* **73,** 3966–3970.
Garel, A., Zolan, M., and Axel, R. (1977). *Proc. Natl. Acad. Sci. U.S.A.* **74,** 4867–4871.
Greenleaf, A. L., and Bautz, E. K. F. (1975). *Eur. J. Biochem.* **60,** 169–179.
Hill, R. J., and Watt, F. (1977). *Chromosoma* **63,** 57–78.
Hinkle, D. C., and Chamberlin, M. J. (1972). *J. Mol. Biol.* **70,** 157–185.
Jamrich, M., Haars, R., Wulf, E., and Bautz, F. A. (1977a). *Chromosoma* **64,** 319–326.
Jamrich, M., Greenleaf, A. L., and Bautz, E. K. F. (1977b). *Proc. Natl. Acad. Sci. U.S.A.* **74,** 2079–2083.
Jamrich, M., Greenleaf, A. L., Bautz, F. A., and Bautz, E. K. F. (1978). *Cold Spring Harbor Symp. Quant. Biol.* **42,** 389–398.
Kuo, M. T., Sahasrabuddhe, C. G., and Saunders, G. F. (1976). *Proc. Natl. Acad. Sci. U.S.A.* **73,** 1572–1575.
Lacy, E., and Axel, R. (1975). *Proc. Natl. Acad. Sci. U.S.A.* **72,** 3978–3982.
Laemmli, U. K., Cheng, S. M., Adolph, K. W., Paulson, J. R., Brown, J. A., and Baumbach, W. R. (1978). *Cold Spring Harbor Symp. Quant. Biol.* **42,** 351–360.

Levy, W. B., and Dixon, G. H. (1977). *Nucl. Acids Res.* **4,** 883–898.
Lin, S. Y., and Riggs, A. D. (1972). *J. Mol. Biol.* **72,** 671–690.
Lin, S. Y., and Riggs, A. D. (1975). *Cell* **4,** 107–111.
Mayfield, J. E., Serunian, L. A., Silver, L. M., and Elgin, S. C. R. (1978). *Cell* **14,** 539–544.
McKenzie, S. L., Henikoff, S., and Meselson, M. (1975). *Proc. Natl. Acad. Sci. U.S.A.* **72,** 1117–1121.
Miller, D. M., Turner, P., Nienhuis, A. W., Axelrod, D. E., and Gopalakrishnan, T. V. (1978). *Cell* **14,** 511–521.
Mirault, M. E., Goldschmidt-Clermont, M., Moran, L., Arrigo, A. P., and Tissieres, A. (1978). *Cold Spring Harbor Symp. Quant. Biol.* **42,** 819–828.
Palmiter, R. D., Mulvihill, E. R., McKnight, G. S., and Senear, A. W. (1978). *Cold Spring Harbor Symp. Quant. Biol.* **42,** 639–648.
Panet, A., and Cedar, H. (1977). *Cell* **11,** 933–940.
Pardue, M. L., and Gall, J. G. (1975). *Methods Cell Biol.* **X,** 1–16.
Plagens, U., Greenleaf, A. L., and Bautz, E. K. F. (1976). *Chromosoma* **59,** 157–165.
Reeves, R. (1976). *Science* **194,** 529–532.
Ritossa, F. (1962). *Experientia* **18,** 571–573.
Silver, L. M. (1978). Ph.D. Thesis, Biophysics, Harvard University.
Silver, L. M., and Elgin, S. C. R. (1976). *Proc. Natl. Acad. Sci. U.S.A.* **73,** 423–427.
Silver, L. M., and Elgin, S. C. R. (1977). *Cell* **11,** 971–983.
Silver, L. M., and Elgin, S. C. R. (1978a). *In* "The Cell Nucleus" (H. Busch, ed.), Vol. 5, pp. 215–262. Academic Press, New York.
Silver, L. M., and Elgin, S. C. R. (1978b). *Chromosoma* **68,** 101–114.
Silver, L. M., Feldman, L., Stollar, B. D., and Elgin, S. C. R. (1978). *Exp. Cell Res.* **114,** 479–484.
Spradling, A., Penman, S., and Pardue, M. L. (1975). *Cell* **4,** 395–404.
Tjian, R., Stinchcomb, D., and Losick, R. (1974). *J. Biol. Chem.* **250,** 8824–8828.
Weintraub, H., and Groudine, M. (1976). *Science* **193,** 848–856.
Yamamoto, K. R., and Alberts, B. (1975). *Cell* **4,** 301–310.

CHAPTER 5

IMMUNOGENETICS OF MAMMALIAN CELL SURFACE MACROMOLECULES

Carol Jones and Theodore T. Puck

ELEANOR ROOSEVELT INSTITUTE FOR CANCER RESEARCH
FLORENCE R. SABIN LABORATORIES FOR GENETIC AND DEVELOPMENTAL MEDICINE
DEPARTMENT OF BIOCHEMISTRY, BIOPHYSICS AND GENETICS
UNIVERSITY OF COLORADO MEDICAL CENTER
DENVER, COLORADO

I.	Introduction	89
II.	Human–Chinese Hamster Ovary Cell Hybrids	90
III.	Immunologic Specificity of Human and Chinese Hamster Ovary Cell Surface Structures	92
IV.	Human Antigens (A_L, B_L, C_L, etc.) Expressed on Human–Chinese Hamster Somatic Cell Hybrids	93
V.	Analysis in Depth of the A_L Antigenic System	95
	A. Theoretical Analysis Using Tissues with Some but Not All the Antigenic Activity	95
	B. The Use of Rabbit Antiserum (Human Red Blood Cell) and the Human Red Blood Cell to Resolve the A_L System into Two Separate Antigens, a_1 and a_2	96
	C. Demonstration of a_3	98
	D. Use of the Variant Clones for Regional Gene Mapping	99
	E. Complementation Analysis	102
	F. Biochemical Characterization of the a_1 Cell Surface Antigen	104
	G. Direct Detection of Antigens on Tissue Cells by Antibody Titration and by Use of Immunofluorescence and the Horseradish Peroxidase Reaction	106
VI.	Use of A_L System to Detect Environmental Mutagens	107
VII.	Cell Surface Antigens of Wide and of Limited Distribution among Different Tissues	112
VIII.	Discussion	113
IX.	Concluding Remarks	114
	References	115

I. Introduction

Studies of many different kinds indicate that cell surface molecules play a crucial role in the control of growth, differentiation, development, and communication in mammalian cells. However, the genetics of surface membrane structures is meager and unsatisfactory. Methods for elucidation of the genetic and genetic biochemical aspects of cell membrane macromolecules are needed in order to illuminate regulatory control mechanisms. Study of the genetics of the mammalian cell surface is complex in part because of the large number of genes which can participate in the production of surface structures. In addition, classical genetic operations with their attendant mating procedures are particularly difficult for human situations. For these reasons we have used somatic cell methodologies *in vitro* for genetic analysis of these structures. Our approach involves the use of cultured human and Chinese hamster ovary cells and their somatic cell hybrids. This allows simplification of the study of human cell surface antigens by means of procedures which permit study of membrane macromolecules due to individual or small groups of chromosomes. This chapter describes the experimental developments in the program of cell surface antigens which has taken place in our laboratory only. It does not attempt to cover experimental findings in this field carried out by other investigators.

II. Human–Chinese Hamster Ovary Cell Hybrids

The Chinese hamster ovary cell (CHO-K1) has several characteristics which make it particularly useful for hybridization with human cells (Kao and Puck, 1971): (1) It grows uniformly *in vitro* with a generation time of 11–12 hours and has an absolute plating efficiency of 85% or more; (2) a minimal nutritional medium has been defined for its growth; (3) it has a relatively small number and reasonably constant set of chromosomes (20); (4) when it is hybridized with human cells, human chromosomes are rapidly and preferentially lost; (5) a large variety of stable auxotrophic mutants have been prepared by means of the bromodeoxyuridine (BUDR) near-visible light technique; (6) growth of human–Chinese hamster hybrids prepared with specific auxotrophs can be carried out in appropriate selective media so as to permit isolation of hybrids retaining particular human chromosomes which supply the genetic information which complements the auxotrophic deficiency.

The CHO-K1 auxotrophs are fused with human cells with the use of UV-inactivated Sendai virus or polyethylene glycol and plated in a medium lacking the nutrient required by the auxotroph for growth. The human cells are either nonmultiplying or grow extremely slowly in the culture medium used. Since neither of the parental cells can grow efficiently, only hybrid cells which retain the specific human chromosome which corrects the enzyme deficiency of the auxotroph survive

5. IMMUNOGENETICS OF CELL SURFACE ANTIGENS

and form colonies. After several generations, the hybrids lose most of the unnecessary human chromosomes, thus facilitating determination of the chromosome retained by all of the hybrids. If the loss of this chromosome is accompanied by the loss of the ability to grow in selective medium, it can be concluded that this chromosome contains the genetic information which complements the auxotroph's deficiency. At least one human enzyme is now assigned to each human autosome and to the X chromosome (McKusick and Ruddle, 1977). Therefore, the presence of particular human chromosomes can be demonstrated either by cytogenetic analysis using chromosome banding, or by assaying for human isozymes which can act as specific chromosomal markers. Fig. 1 shows the banded chromosomes of a hybrid prepared with the ade C auxotroph which shows in addition to the Chinese hamster chromosomes a single human chromosome identified as number 21. The presence of human chromosome 21 also resulted in the expression of human superoxide dismutase (soluble) by this hybrid. Table I summarizes the human genes whose chromosomal locations have been ascertained after hybridization of specific CHO-K1 auxotrophs with human cells.

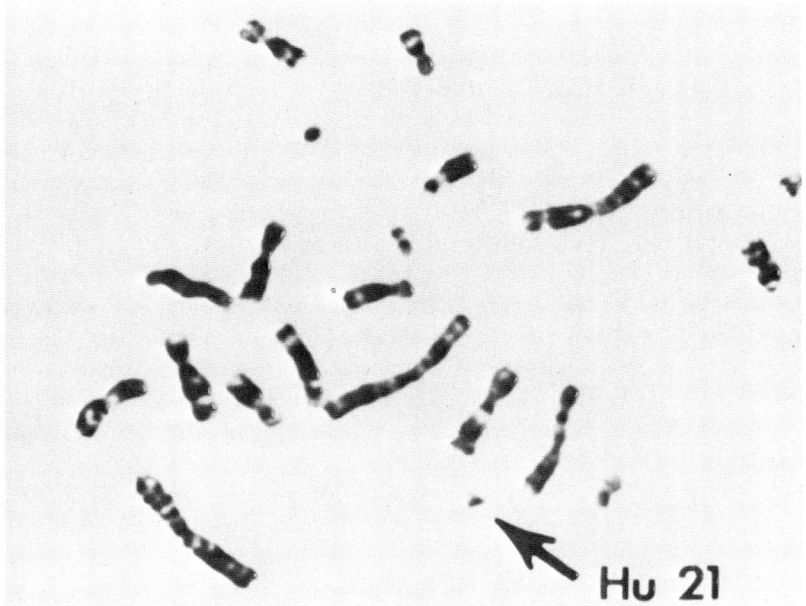

FIG. 1. Banded chromosomes of a human–Chinese hamster cell hybrid containing the CHO chromosomes and human chromosome 21 (Hu 21).

TABLE I

Chromosomal Localizations Carried out with Human–CHO Hybrids

Auxotrophic Mutant	Enzyme deficiency	Human chromosome	References
gly⁻A	Serine hydroxymethyl transferase (EC 2.1.2.1)	12	Jones et al. (1972)
ade⁻B	FGAR[a] amidotransferase (EC 6.3.5.3.)	4 or 5	Kao and Puck (1972)
ade⁻C	GAR[b] synthetase (EC 6.3.4.13)	21	Moore et al. (1977)
pro⁻	Glutamic α-semialdehyde synthetase	10	Jones (1975)

[a] FGAR = formylglycinamide ribonucleotide.
[b] GAR = glycinamide ribonucleotide.

III. Immunologic Specificity of Human and Chinese Hamster Ovary Cell Surface Structures

Antisera raised in rabbits and sheep after immunization with a variety of human cells are lethal to these cells in the presence of complement. Human cells which have been used for immunization include erythrocytes, lymphocytes, cultured fibroblasts, lymphoblasts, S3 HeLa cells, and biopsy samples from various human organs. However, these antisera which are highly toxic to human cells do not affect the viability of the CHO-K1 cell under similar test conditions. Conversely, antisera prepared after injection of CHO-K1 cells into experimental animals are highly lethal to the CHO-K1 cell but do not kill human cells under the defined test conditions (Oda and Puck, 1961). The standard test conditions involve plating a measured number of cells into petri dishes which contain growth medium, antiserum, and complement. We have adopted normal rabbit serum as a standard complement source. Control plates without antiserum or complement, or with neither reagent, receive an identical innoculum of cells. After incubation for 5–7 days, the plates are fixed and stained and the number of colonies scored for each plate. This allows determination of the killing action of different antisera on human and CHO-K1 cells. The results of a typical analysis illustrating the species specificity of the different antisera are shown in Table II. By using a series of petri dishes containing increasing concentrations of antiserum, it is possible to obtain quantitative survival curves.

TABLE II

Demonstration of the Survival by Colony Formation of Various Types of Cells When Treated with 0.2% of a Standard Anti-Human Erythrocyte Serum in the Presence of 2% Normal Rabbit Serum as the Complement Source

Cell tested	Relative plating efficiency (%)[a]
CHO–K1	100
Human S3 HeLa	0
Human sarcoma	0

[a] Compared to a control with no antiserum.

IV. Human Antigens (A_L, B_L, C_L, etc.) Expressed on Human–Chinese Hamster Somatic Cell Hybrids

The immunologic specificity of the antisera described makes possible detection of human cell surface antigens on human–Chinese hamster somatic cell hybrids. The killing of a hybrid cell by antihuman cell serum in the presence of complement is taken as evidence that one or more human antigens are shared by this hybrid and the human cell used to elicit antibody formation. Conversely, if the hybrid is not killed, the cell is not expressing human cell surface antigenic structures which are reacting with the test antibodies and promoting a cytolytic complement–dependent response.

Figure 2 illustrates the results obtained when two different hybrids were tested with the same antiserum produced in rabbits after injection of human erythrocytes. The hybrid containing chromosome 12 as its only human chromosome displays a behavior indistinguishable from that of the parental Chinese hamster ovary cell in showing no sensitivity to killing by antibody plus complement. The hybrid containing chromosome 11 as its only human chromosome is, however, completely killed by low concentrations of this antiserum. It is therefore concluded that chromosome 11 contains information specifying a cell surface structure which is present in immunologically indistinguishable form on the human erythrocyte. Human cell surface markers can be assigned to human chromosomes using methods similar to those described earlier for assigning enzyme markers (Puck *et al.*, 1971). The presence of well-characterized hybrids with single human chromosomes simplify this task as illustrated for the hybrid with chromosome 11. The assignment of human cell surface antigen(s) to chromosome 11

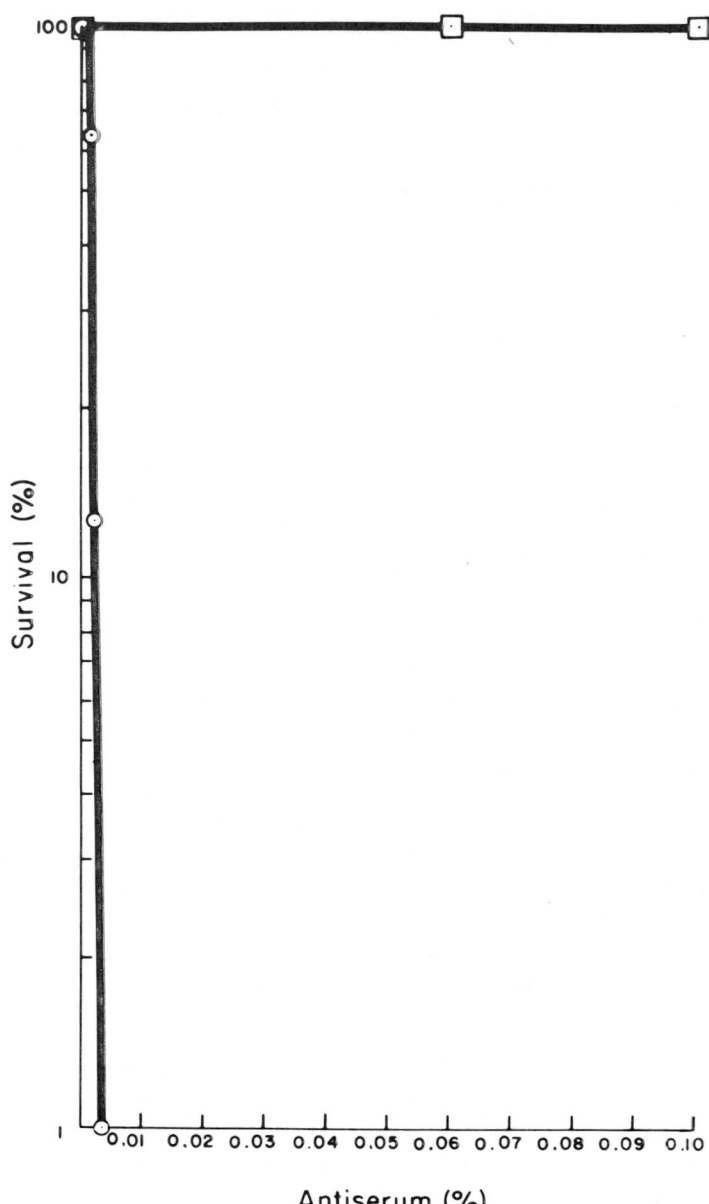

Fig. 2. Demonstration that anti-human erythrocyte serum fails to kill the hybrid containing human chromosome 12 (□) as its only human chromosome but kills the hybrid containing human chromosome 11 (○).

TABLE III

Retention of Killing Antibodies after Exhaustive Adsorption by Various Cells of an Antiserum Prepared by Injection of Human Lymphoblasts into a Horse

Adsorbing cells	Cell killing by specifically adsorbed antiserum			
	S3 HeLa	Hybrid 1 ($A_L{}^+$)	Hybrid 2 ($B_L{}^+C_L{}^+$)	Hybrid 3 ($C_L{}^+$)
None (control)	+[a]	+	+	+
S3 HeLa	0[b]	0	0	0
Hybrid 1	+	0	+	+
Hybrid 2	+	+	0	0
Hybrid 3	+	+	+	0
Hybrid 1 + hybrid 2	+	0	0	0

[a] A plus sign indicates >98% of the cells are killed by <0.1% antiserum.
[b] A zero indicates no killing is obtained under similar conditions.

was demonstrated by standard synteny analysis utilizing the human isozyme marker lactic dehydrogenase A which has previously been assigned to human chromosome 11. Thus, analysis of a large number of primary and secondary hybrid clones showed 161 were concordant for human LDH_A and the cell surface marker designated A_L and no clones were found discordant for the two markers. In addition to A_L other antigen systems shown not to be on chromosome 11 and named B_L, C_L, D_L, etc. have been identified by the following procedure: By exhaustive absorption of a lethal antihuman cell serum with an A_L hybrid, it was possible to remove all lethal antibody activity for A_L hybrids for the serum. The resulting antiserum was then tested with other hybrids. The next one found to be killed by the absorbed antiserum was designated as containing B_L human antigens (Wuthier et al., 1973). This process was repeated and antiserum adsorbed with A_L and B_L hybrids was used to identify hybrids with other human antigens. Table III illustrates typical results for such experiments.

The data show that the S3 HeLa cell contains all three antigens $A_L{}^+$, $B_L{}^+$, and $C_L{}^+$. Antiserum absorbed with hybrids 1 ($A_L{}^+$) and 2 ($B_L{}^+C_L{}^+$) can then be used to look for hybrids containing additional human antigens.

V. Analysis in Depth of the A_L Antigenic System

A. Theoretical Analysis Using Tissues with Some but Not All the Antigenic Activity

The A_L system found in hybrids containing chromosome 11 was selected for special study. If an antiserum can be prepared which selectively reacts with some but not all of the individual cell surface antigens expressed by a given hybrid, it becomes possible to resolve it into component parts. This can be done by preparing an antiserum to a tissue which contains some but not all of the human cell surface antigens of the hybrid. Such a tissue can be recognized by the presence of two characteristics: An antiserum prepared against its cells must be lethal to cells of the given hybrid. Moreover, if the original antiserum active against the hybrid is absorbed exhaustively with cells of the given tissue it is not possible to remove all of the killing power against the given hybrid. This tissue can then be used to adsorb out antibodies to the subset of antigens it does contain from antiserum which reacts with all of the antigens of the system involved. Using such antisera to subsets of individual antigens of a system, it is then possible by mutagenesis and immunoselection to prepare stable clonal stocks derived from the original antigen-positive hybrid which express different specific combinations of the antigens that have been resolved. This process is illustrated in the A_L antigenic system, using human erythrocytes (Jones et al., 1975).

B. The Use of Rabbit Antiserum (Human Red Blood Cell) and the Human Red Blood Cell to Resolve the A_L System into Two Separate Antigens, a_1 and a_2

We demonstrated that it is possible to completely adsorb the killing activity from an antiserum by use of a suspension of cells which contain the appropriate antigen(s) on their cell surface. This was found to be true for the A_L system as illustrated in Table IV.

As can be seen from this table, human cultured HeLa cells, human fibroblasts, and human lymphocytes were all able completely to remove the killing activity from the anti-A_L serum tested. Different numbers of cells were needed to give equivalent absorptions, a fact which reflects in part differences in cell size. These three cell types were all capable of eliciting antibodies lethal to the A_L hybrid. Also as expected, the A_L^+ hybrid was capable of adsorbing all of this killing activity from the test antiserum, but the A_L^- hybrid was not. The human red blood cell, however, showed a more complex set of results. The human red blood cell when injected into a rabbit produced an antiserum which effectively killed A_L^+ hybrids. But, as shown in the table, it was not capable of removing all of the killing activity from the test antiserum. The lethal activity of anti-(human red blood cell)serum could be removed by adsorption with the red blood cell as well as adsorption with HeLa cells, human fibroblasts, human lymphocytes,

TABLE IV

Relative Adsorbing Abilities for A_L Antibodies of Different Human Cell and Cell Hybrids.

Cell type	Ability of cells to adsorb all of the anti-A_L activity from rabbit antihuman cell serum
A_L^+ hybrid	$+^a$
A_L^- hybrid	0^b
Human fibroblast	$+$
HeLa	$+$
Human lymphocyte	$+$
Human RBC	0

^{a} A plus sign indicates that the cells are able to adsorb the anti-A_L activity.
^{b} A zero indicates cells unable to adsorb the anti-A_L activity.

and the A_L^+ hybrid. These results led to the hypothesis that the human red blood cell contains some but not all of the antigenic components of the A_L system. The antigenic sites found on the A_L system and shared by the red blood cell were then defined as a_1 and antiserum prepared in the rabbit after injection of human red blood cells was defined as anti-a_1 serum.

Those antigenic sites belonging to the A_L system not shared by the red blood cell were designated as a_2, and anti-a_2 serum was prepared by injecting HeLa cells into a rabbit and exhaustively adsorbing the resulting antiserum with red blood cells. This means the A_L hybrid cell containing human chromosome 11 as its sole human chromosome is $a_1^+a_2^+$. If this hypothesis is correct, it should be possible by mutagenesis to prepare from A_L cells mutant clones with the additional phenotypes: $a_1^+a_2^-$, $a_1^-a_2^+$, and $a_1^-a_2^-$. This cell was then treated with known mutagenic agents using amounts which reduced the survival to 1–20%. After allowing time for recovery of cellular viability and expression of possible mutation, these cells were distributed in a series of petri dishes containing anti-a_1 serum, anti-a_2 serum, or both anti-a_1 and anti-a_2 serum, in the presence of complement. The majority of cells were killed in each case but occasional colonies were found to form under the selective conditions. These colonies were isolated and grown into new clonal stocks and retested for expression of the a_1 and a_2 antigens. Tests demonstrated that all of the expected clonal forms could be isolated (see Fig. 3). These clones proved to be stable after many generations in culture and possessed the absorptive properties predicted by their phenotype.

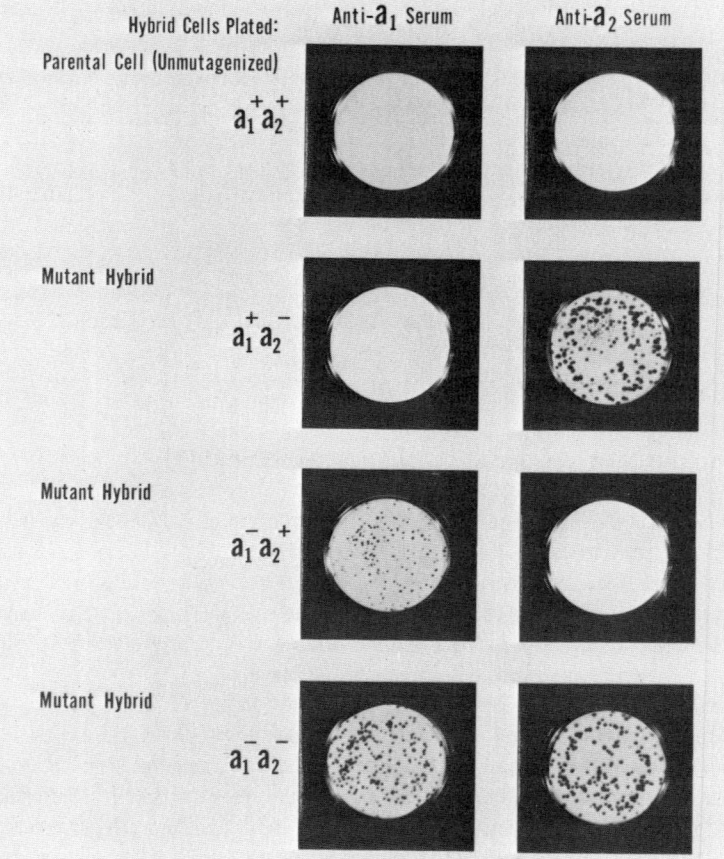

Fig. 3. Demonstration of the new clones obtained from the A_L^+ hybrid (top line) by mutagenesis followed by growth in each antiserum indicated. The resulting clones can be cultivated stably for indefinite periods in the absence of the individual antisera used in their initial selection.

C. Demonstration of a_3

The availability of antisera which have been characterized as anti-a_1 and anti-a_2 makes possible the search for other cell surface markers arising from loci on human chromosome number 11. Clones previously isolated after mutagenesis and immunoselection that have been shown to be $a_1^- a_2^-$ can now be tested with antisera raised in different experimental animals against other human cells. If any of these clones are killed by these antisera, they must possess one or more additional

TABLE V

DEMONSTRATION THAT TWO MUTANT CLONES WHICH SHOW SIMILAR BEHAVIOR WHEN TESTED WITH ANTI $(a_1 + a_2)$ SERUM SHOW DIVERGENT BEHAVIOR WHEN TREATED WITH ANTI-$(a_1 + a_2 + a_3)$ SERUM

Clone[a]	Antiserum		Antigenic designation of clone
	Anti-$(a_1 + a_2)$	Anti-$(a_1 + a_2 + a_3)$	
1	0	+	$a_1^- a_2^- a_3^+$
2	0	0	$a_1^- a_2^- a_3^-$

[a] Clone 1 is killed almost as effectively as the parental hybrid cell (+ reaction); clone 2 is unaffected (0 reaction).

human surface antigens and the new antigenic marker must also be dependent on loci on human chromosome 11. The first new hybrid clone found by this procedure which is resistant to anti-a_1 and anti-a_2 serum but is killed by horse antilymphocyte globulin has been called a_3^+ (Jones and Puck, 1977). The survival behavior of this cell (clone 1) and a cell (clone 2) with an $a_1^- a_2^- a_3^-$ phenotype is shown in Table V. Both clones are resistant to the lethal action of the $a_1^+ a_2^+$ antiserum but the clone defined as a_3^+ is readily killed by the second antiserum, which is then called anti $(a_1^+ a_2^+ a_3^+)$. When the $a_1^- a_2^- a_3^+$ hybrid was tested with antiserum prepared in the sheep after injection of human red blood cells, it was also readily killed. The human red blood cell was also capable of removing anti-a_3 activity by adsorption. Although the rabbit produced only a_1 antibodies when immunized by standard procedures, anti-a_3 activity could be produced after hyperimmunization. Therefore, it is concluded that the human red blood cell is $a_1^+ a_2^- a_3^+$. In neither the sheep anti(human red blood cell)serum nor in the rabbit antiserum after hyperimmunization was any anti-a_2 activity found. Table VI summarizes some representative antisera that have been prepared with specific activities toward a_1, a_2, a_3 antigens. The availability of these antisera allows preparation of further varieties of variant clones with different combinations of a_1, a_2, and a_3 activities. Table VII lists the phenotypes which have been isolated so far.

D. Use of the Variant Clones for Regional Gene Mapping

The clones described in Table VII were also tested by isozyme analysis for their possession of human lactic dehydrogenase A (LDH_A) activity. It seemed reasonable to consider that the phenotypes isolated which had lost all three antigenic markers plus the syntenic human

TABLE VI

Demonstration of Preparation of Representative Antisera with Specific Activities toward a_1, a_2, and a_3 Antigens

Cell used as antigen	Animal host	Cell used for adsorption	Presence in the final preparation of specific antibody against		
			a_1	a_2	a_3
Human RBC	Rabbit	None	$+^a$	0^b	0
	Sheep	None	+	0	+
S3 HeLa	Rabbit	None	+	+	0
	Rabbit	Human RBC	0	+	0
Human lymphoblast	Horse	None	+	+	+
$a_1^- a_2^+ a_3^-$ hybrid	Rabbit	CHO-K1	0	+	0

^{a} A plus sign means the given antiserum at a concentration of 0.1% or less in the presence of standard complement kills 90% or more of a test cell carrying the given antigen.

^{b} A zero means no or virtually no killing is observed under these conditions.

LDH_A isozyme marker may have lost all (or most) of chromosome 11. Clones which lost a_1, a_3, and LDH_A but retained a_2, presumably had suffered chromosomal breaks and deletions. To test this further, detailed cytogenic analysis was performed on selected variants. Five of six clones which had lost all four of the above markers were found to have lost the entire chromosome while the sixth contained only a small unstable centromeric fragment (Kao *et al.*, 1976). Cytogenetic analysis

TABLE VII

Examples of the Varieties of Stable Variant Clones Obtained by Mutagenesis of $a_1^+ a_2^+ a_3^+$ Hybrids in Single or Multiple Steps^a

Clone number	Antibody activity used in original clone isolation	Presence of specific antigen		
		a_1	a_2	a_3
Parent	None	+	+	+
1	a_1	−	+	+
2	a_2	+	−	+
3	$a_1 + a_2$	−	−	+
4	$a_1 + a_3$	−	+	−
5	$a_1 + a_2 + a_3$	−	−	−

^{a} The $a_1^+ a_2^+ a_3^+$ hybrid cell population was treated with a standard mutagen, and then plated in five different antisera each of which was lethal for the antigens listed in the second column.

of a series of other clones with different combinations of marker loss revealed deletions which when correlated with the phenotypes could be used to provide regional mapping information concerning the three antigen markers, human LDH_A, and a fifth marker human acid phosphatase 2 (ACP_2) which has also been assigned to chromosome 11 (Jones and Kao, 1978). Table VIII summarizes the breakpoints identified and the phenotypes of the clones which have led to the following regional mapping assignments:

a_1 and a_3 11p13 → 11pter
LDH_A 11p12 → 11p13
ACP_2 11p11 → 11p12
a_2 11q13 → 11qter

These map assignments are illustrated in Fig. 4. These clones should provide a useful regional clone panel for rapid regional mapping of other genes assigned to chromosome 11.

As shown in Fig. 4, LDH_A is placed proximal and ACP_2 distal to a_1 and a_3. Since the frequency with which syntenic loci are lost singly and in combination should provide a measure of their linear order on the chromosome, we should expect a higher frequency of $a_1^-a_3^-$ clones to have lost LDH_A than ACP_2. Similarly, we should expect $LDH_A^+ACP_2^-$ clones to occur at a lower frequency than $LDH_A^-ACP_2^+$ clones because multiple breaks resulting in interstitial deletions would presumably be required to yield the former phenotype. To test this $a_1^-a_3^-$ clones were selected from X-irradiated cells and their phenotype determined. Sixty-four percent of 34 $a_1^-a_3^-a_2^+$ clones had lost LDH_A and only 26% had lost ACP_2. Thirty-eight percent were $LDH_A^-ACP_2^+$ and only 9%

TABLE VIII

Characteristics of the Regional Clone Panel of Human Chromosome 11

Clone	Deletion in chromosome 11	Phenotype[a]				
		a_1	a_2	a_3	LDH_A	ACP_2
Parent J1	None	+	+	+	+	+
J1-23	pter → p13	−	+	−	+	+
J1-10	pter → p12	−	+	−	−	+
J1-7	pter → p11	−	+	−	−	−
J1-11	qter → q13	+	−	+	+	+

[a] The phenotypes of 5 clones containing specific marker losses, produced by mutagenesis and selection, are compared to that of the unmutagenized parental hybrid cell. (+) Clone expresses phenotypic marker; (−) clone does not express marker.

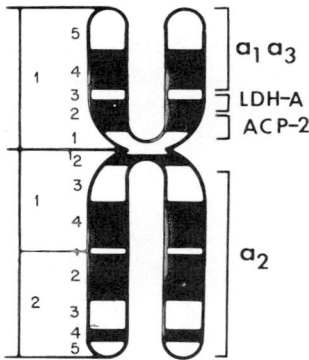

FIG. 4. Diagram summarizing the regional map of five genes on human chromosome 11.

were $LDH_A{}^+ACP_2{}^-$. These results are consistent with the map order determined by cytogenetic analysis. Quantitative determination of co-loss and co-retention frequencies of syntenic markers in a larger series of mutant clones should provide definitive data which could be used to calculate the relative map distances for these genes on chromosome 11.

E. COMPLEMENTATION ANALYSIS

Karyotype analysis of a set of clones with the phenotype $a_1{}^-a_2{}^+a_3{}^+$ revealed the presence of a normal chromosome 11 which is consistent with this class of variants arising by single gene mutation. A series of independently isolated variants with this phenotype were subjected to further genetic analysis in order to determine for the a_1 locus: (1) the dominance or recessiveness of mutant phenotypes; (2) the number of complementation groups affecting this genetic locus; and (3) the existence of essential participation by Chinese hamster genes in the expression of the a_1 antigen.

Three types of crosses were performed to obtain this information: (1) The $a_1{}^-a_2{}^+a_3{}^+$ variants were hybridized with the parental $a_1{}^+a_2{}^+a_3{}^+$ hybrid; (2) The $a_1{}^-a_2{}^+a_3{}^+$ variants were hybridized with each other in pairwise combinations; (3) The $a_1{}^-a_2{}^+a_3{}^+$ variants were hybridized with the parental CHO-K1 cell.

It was, of course, necessary to devise a selective system and place auxotrophic markers on the cells involved in these crosses so that only the new hybrids would survive after fusion (Moore, 1976). These could then be isolated and subsequently tested for antigen expression. In a sense these new hybrids are double hybrids as they are prepared from hybridization of cells that are already human–Chinese hamster hybrids. The new hybrids were also examined karyotypically and found to contain the expected 35–40 Chinese hamster chromosomes. Two chromosomes 11 were identified in crosses of types 1 and 2 and one in type 3, as was expected. Figure 5 shows the survival curves obtained when two of the a_1^- parental variants and a hybrid formed after their fusion were tested with anti-a_1 serum, demonstrating the existence of complementation. In Table IX the results of the antigen analysis of the hybrids resulting from selected crosses are given. These data show that both mutants are recessive because hybrids made with the antigen-positive parent still express the a_1^+ phenotype.

In the cross-designated "2" mentioned previously, it was found that hybrids formed from two different a_1^- forms could regain the a_1^+ phenotype. Therefore, the two a_1^- variants employed must belong to different complementation groups.

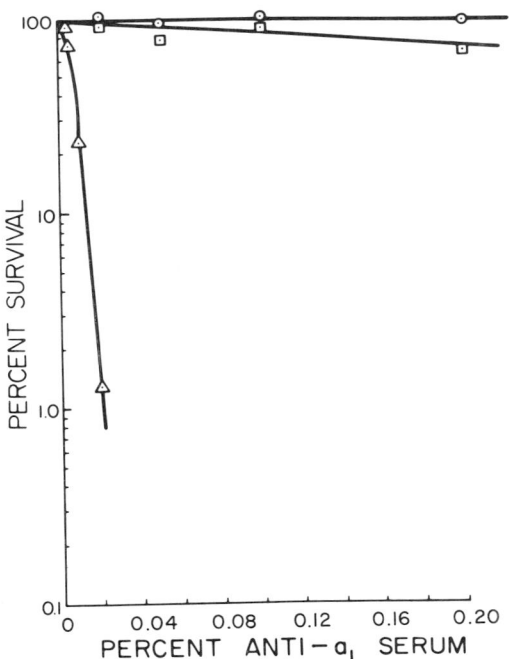

FIG. 5. Survival curves showing the behavior of two a_1^- variants (○, □) and a hybrid formed after their fusion (△) to anti-a_1 serum in the presence of standard complement. The a_1^- parental cells exhibit no killing but the hybrid exhibits extensive killing which is taken as evidence that complementation has occurred and that the two a_1^- variants resulted from mutation in different genetic loci.

TABLE IX

Complementation Analysis of Two Independently Isolated $a_1^-a_2^+a_3^+$ Variants with the Parental A_L^+ Hybrid, with CHO-K1, and with Each Other

Variant[a]	Parent $(a_1^+a_2^+a_3^+)$	Variant 1 $(a_1^-a_2^+a_3^+)$	Variant 2 $(a_1^-a_2^+a_3^+)$	CHO–K1 $(a_1^-a_2^-a_3^-)$
No. 1 $(a_1^-a_2^+a_3^+)$	+[b]	−[c]	+	−
No. 2 $(a_1^-a_2^+a_3^+)$	+	+	−	+

[a] When variant 1 was hybridized with itself and variant 2 hybridized with itself, no change in marker expression resulted, as is expected if the system behaves in straightforward genetic fashion.

[b] A plus sign indicates the phenotype of the hybrid resulting from the cross of the two indicated cells was a_1^+ (complementation occurred).

[c] A minus sign indicates the phenotype of the resulting hybrid was a_1^- (no complementation occurred).

As shown in Table IX, complementation was obtained not only when an a_1^- hybrid was hybridized with another a_1^- hybrid, but also when particular a_1^- hybrids were further hybridized with the parental Chinese hamster cell, CHO-K1. The genetic defect present in variant 2 must therefore reside in the Chinese hamster contribution to the hybrid genome. These experiments are being extended. The results obtained to date show that all the a_1^- mutants analyzed so far are recessive; at least four complementation groups are involved in expression of a_1^+ competence; and certain Chinese hamster genes are required for the expression of the a_1 antigen in hybrids containing the single human chromosome number 11. The need for multiple genes for a_1 expression might reflect the involvement of regulatory processes, or may be due to the need for multiple genes to achieve the final chemical structure of the a_1 antigen.

F. Biochemical Characterization of the a_1 Cell Surface Antigen

The cell surface antigens require chemical as well as immunological characterization. Because a_1 is carried on the human red blood cell, it is possible to determine whether a_1 is immunologically related to any of the red cell membrane components which have already been characterized. It was found that in the presence of complement, antiserum directed against the glycophorin fraction of the human red blood cell kills only those hybrids with chromosome 11 which express the a_1 antigen (Moore et al., 1976). Furthermore, addition of the glycophorin fraction can completely inhibit the killing of the a_1^+ hybrids by antiserum against a_1. It has been shown that the glycophorin preparation used in these experiments is not a single homogeneous protein but contains a related set of glycoprotein molecules (Furthmayr, 1978). Recent experiments have shown that the original preparation also contained a macroglycolipid and that antibodies specific for the a_1 antigen bind to the macroglycolipid but not to purified glycoprotein molecules. It may therefore be concluded that the macroglycolipid fraction contains the a_1 activity.

Another current approach to elucidate the chemical nature of the antigens is to treat the hybrid cells with agents designed to bind to or alter specific chemical groups. In one such series of experiments purified plant lectins which specifically interact with carbohydrate moieties of glycoproteins and glycolipids were tested to see if they would inhibit the killing action of anti-a_1, a_2, and a_3 sera for cells containing the respective antigens (Jones et al., 1979). Concentrations of each lectin were used which had previously been shown to have no effect by themselves on the survival of the cells in the absence of antiserum. As shown in Table X, certain lectins were capable of specifically inhibiting the antibody killing for each of the antigens. These results suggest that certain lectins and antibodies are binding to the same or sterically related sites.

TABLE X

Effect of Different Lectins on Inhibiting Antibody Killing[a]

| | | Survival | | |
| | | Antigen | | |
Lectin	Sugar specificity	a_1	a_2	a_3
Lentil (LCA)	α-D-Mannose	+[b]	−	+
Pea	D-Mannose, D-Glucose	−[c]	−	−
Pokeweed	—	−	−	−
Kidney bean (PHA)	N-Acetyl galactosamine	−	+	+
Concanavalin A (Con A)	α-D-Mannose	+	+	+
Wheat germ agglutinin (WGA)	N-Acetyl glucose amine)	−	−	+

[a] Concentrations of specific antibody and complement sufficient to give 95–99% killing for each indicated antigen were used.
[b] A plus sign indicates the lectin was capable of inhibiting this killing.
[c] A minus sign indicates no change was observed in survival in the presence of lectin.

G. DIRECT DETECTION OF ANTIGENS ON TISSUE CELLS BY ANTIBODY TITRATION AND BY USE OF IMMUNOFLUORESCENCE AND THE HORSERADISH PEROXIDASE REACTION

Once an effective antibody preparation has been achieved for a given cell surface antigen, it can be used to explore tissue cells directly without the need for cultivation *in vitro*. Two different approaches can be utilized. A given tissue can be dispersed into a single cell suspension by the use of very gentle trypsinization, treatment with cationic sequestering agents, or high concentrations of sugars. The cells so produced then can be used to absorb specific antibodies from a standardized antiserum. The amount of specific cell-killing activity remaining in the preparation is then titrated by a single cell survival curve, using the standard cell containing the antigen in question. Alternatively, the cells can be used to immunize experimental animals and the resulting antiserum titrated against standard hybrid cells. By this means it has been shown that an antiserum prepared by injection of adult human brain into rabbits is capable of killing cells which have either the a_1^+ or the a_2^+ antigens. Moreover, when this experiment was repeated with a 3-month fetal human brain biopsy, the antiserum resulting still killed cells with the a_2^+ antigen but failed to kill cells containing the a_1^+ but not the a_2^+ antigen. This latter experiment must be regarded as preliminary only, because while the antiserum behavior has been carefully confirmed, we have not yet received additional fetal specimens which would make possible repeated tests of the immunization procedure.

This procedure has several disadvantages: (1) It must be ascertained that the method used to prepare a monodisperse cell suspension will not itself decrease the concentration of antigen available at the cell surface; (2) the cell suspension obtained in this way may not be homogeneous but may have a variety of different cell types present so that one may secure a value which represents an average for the antigen present per cell for a variety of cells of unknown distribution; (3) complications may be introduced by the presence of extraneous antibody activities due to particular tissue cells; and (4) finally, one derives no information about any relationship between the presence of the given antigen on cell surface and the geometrical distribution of the cells in the particular tissue under study.

An alternative procedure is to secure direct visualization of the antigen distribution in the cells of a given tissue by use of immunofluorescence or the horse radish peroxidase reaction. This method is less readily susceptible to quantitative determination of the amount of antigen present in the cell population under study, but it can pinpoint presence or absence of antibody in particular cells of a given tissue. Representative experimental results are shown in Fig. 6. It is important to note that the immunofluorescence and horseradish peroxidase methodologies can reveal the presence of antigens which bind antibodies that do not fix complement and therefore which could be missed by methodologies which depend on cell killing alone.

These methods appear to offer the interesting possibility of detection whether changes in particular cell surface antigens occur as characterizing events in different diseases. They may also demonstrate particular stages in mammalian development at which specific cell antigens arise or disappear, and so appear to offer considerable promise in study of the developmental process.

It is obvious that this approach will be more powerful the more specific are the antibodies used in these experiments. If an antiserum contains a mixture of antibodies it may yield ambiguous results, or results with relatively poor resolving power because there may always be some cell surface antigen which is reactive to some component of the mixture. The use of monoclonal antibody preparations promises to be particularly helpful for such applications.

VI. Use of A_L System to Detect Environmental Mutagens

These antigens and the hybrids containing them offer a system which might be particularly useful for the screening of environmental agents capable of causing mutagenesis and cancer. A convincing body of evidence has been accumulated, demonstrating that a considerable amount of cancer in our society is caused by environmental agents [World Health Organization (WHO), 1964; Cairns, 1975]. It also has become clear that many if not most of such agents which cause malignancy do so by bringing about changes in the genome of the affected somatic cells. As a result of these advances, a variety of systems have been proposed for use in screening environmental agents. They offer promise for monitoring of the environment so as to make possible significant prevention of malignant exposure in human populations. Ames called attention to the need for such monitoring systems and provided one which is simple and effective (Ames et al., 1975). Subsequently, a variety of other tests have also been proposed.

However, all of the *in vitro* tests for mutagenesis known to us emphasize measurement of single gene mutations in particular biological systems. But human cells are prone to other kinds of genetic damage involving chromosomal defects like aneuploidy, chromosomal deletions, and structural malformation such as translocations which often

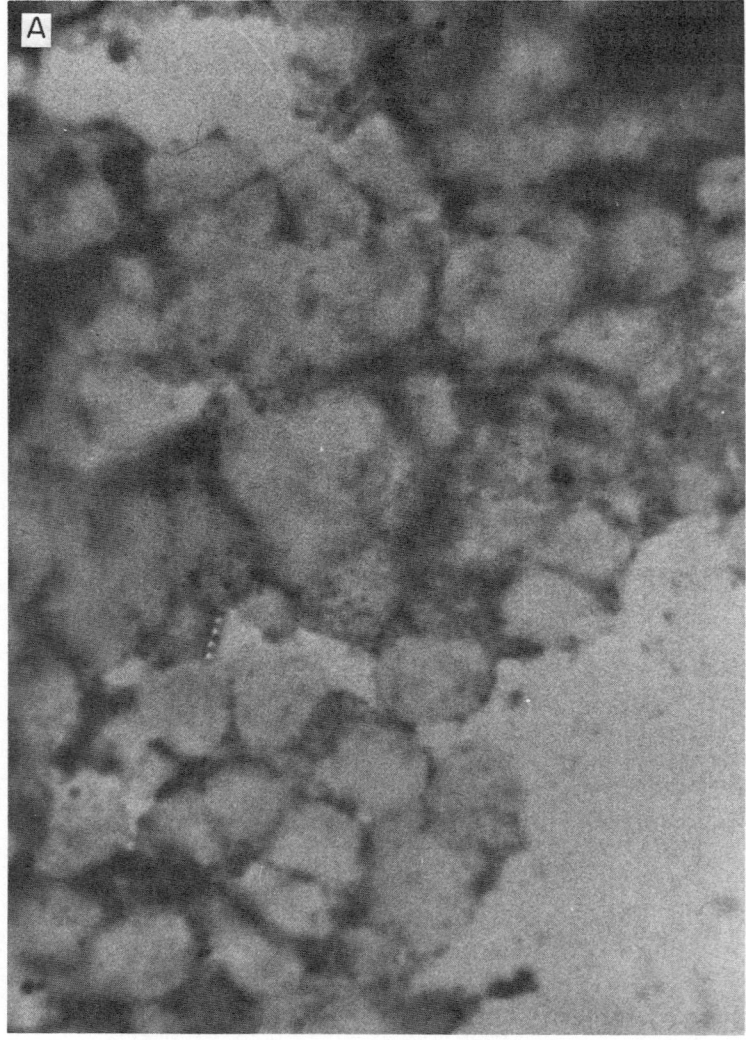

Fig. 6. Typical results obtained by the immuno-horseradish peroxidase reaction. (A) Treatment of human brain cells with an anti-a_2 serum (rabbit, antiserum against $a_1^- a_2^+ a_3^-$ cell adsorbed with CHO-K1.

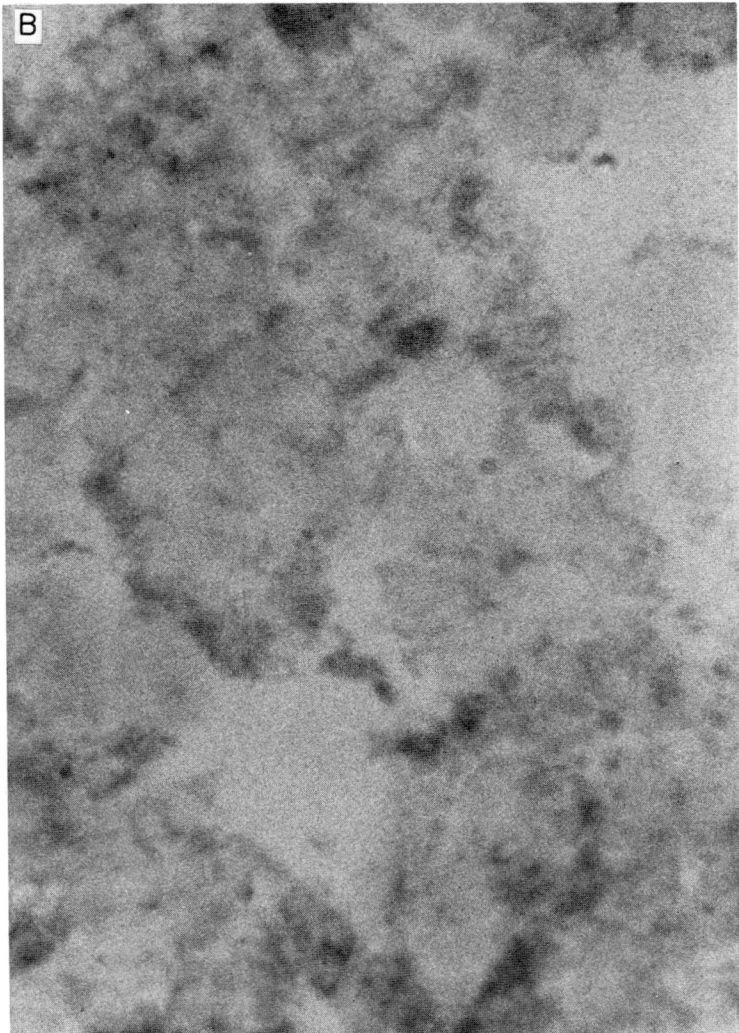

FIG. 6 (continued). (B) Control situation in which the same cell has been treated with normal rabbit serum.

occur subsequent to the deletion event. Such lesions would fail to be recorded in any screening test which utilizes marker genes carried on chromosomes which also contain large numbers of other genes needed for cell reproduction. Thus, multigene deletions or complete loss of such essential chromosomes would cause failure of the affected cell to form a colony and the change in a scored marker would fail to be noted. In most if not all of the *in vitro* tests used to screen for mutagenesis the marker genes employed do indeed occur on chromosomes containing many other loci needed for cell reproduction. Thus, agents which cause chromosomal abnormalities stand a high probability of being missed in such screening tests. Chromosomal errors appear to be highly significant for human cancer and chromosomal anomalies constitute the most frequent cause of *de novo* (not inherited) genetic disease in man. Such chromosomal abnormalities produce serious disease in approximately 0.5–1% of all live human births and a very significant proportion of fetal wastage through spontaneous abortion. These considerations demonstrate the need for a screening test capable of detecting chromosomal loss and deletions as well as single gene mutations. Such a test could presumably help to prevent exposure of human populations to agents important in the production of both cancer and genetic disease.

The use of human cell surface antigens on hybrid cells containing particular human chromosomes appears to offer a means of securing just this kind of screening system. Consider the hybrid of the CHO-K1 cell which contains human chromosome 11. Populations of such cells exposed to suspected mutagens then can be plated in antiserum against the a_1 marker. Only those cells which have been mutated to loss of this marker will produce colonies. Any such colonies developing can then be tested for retention of the a_3 and LDH_A markers. Loss of additional markers on the same arm of this chromosome would indicate with high probability that deletions had occurred. Finally, if all of the markers are lost including the a_2 marker which is on the other arm of chromosome 11, it could be concluded that all or most of the chromosome has been lost. Thus, the use of these markers makes possible detection of the action of agents like Colcemid which had been shown earlier to produce nondisjunction in mammalian cells. Such agents are not mutagenic in the limited sense of the word as it is usually used in microbial genetics. However, since they can cause aneuploidy in human cells they appear to be of great importance in human disease. This approach embodies the following features for use in monitoring environmental agents: (1) Mutagenesis is scored by marker genes on a chromosome unnecessary for cell division, so that genetic damage like the production of large deletions is counted whereas it would be missed if the chromosome carrying the scored genes contains other loci essential for reproduction. (2) Forward mutation to loss of gene function in the cell surface antigens is measured as the primary screening action so that almost any defect anywhere within the genes under study can be detected. (3) The hemizygous state of the human chromosome means recessive mutations are detectable and that problems associated with ploidy are circumvented. (4) The system permits simultaneous detection and quantitation of three kinds of events which cause loss of antigen expression: (a) those arising from single gene mutation; (b) changes resulting from chromosome breaks including small and large deletions; (c) loss of an entire chromosome. We have described earlier the evidence for these three types of events when clones were selected for the loss of a_1 and then subjected to isozyme, karyotype, and complementation analysis. (5) Several genes, both human and Chinese hamster, are involved in expression of the a_1 antigen. Thus, a large target is available for mutagenesis. (6) The system provides a measure of the action of mutagens on human genes, carried on a human chromosome so that a high degree of relevance to the human situation is secured.

This system has been employed to quantitate the action of X-irradiation and N-methyl-N'-nitro-N-nitrosoguanidine (MNNG) in producing the spectrum of mutagenic events described (Waldren et al., 1979). The $A_L{}^+$ hybrid was treated with increasing doses of each agent and immunoselection was carried out in anti-a_1 serum. The number of surviving colonies was scored. Both agents produced an increase in the number of mutant colonies. Figure 7 shows the dose response curve for production of a_1 mutants by X-irradiation.

Presumably, genetic markers on other human chromosomes incorporated into hybrids could also be utilized for this purpose. Hybrids

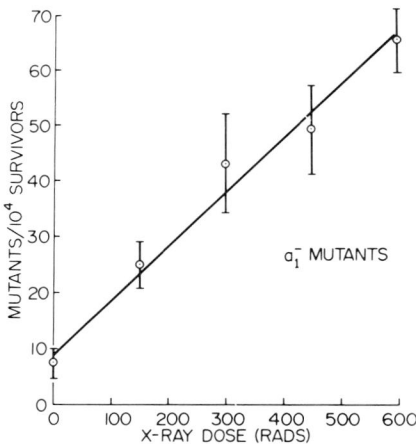

FIG. 7. Dose response curve for the production of $a_1{}^-$ mutants by X-irradiation. Vertical bars represent standard error of the mean.

containing several other human chromosomes have been prepared. At least some of these display human cell surface antigens which cause the hybrids to be killed by appropriate antiserum in the presence of complement.

VII. Cell Surface Antigens of Wide and of Limited Distribution among Different Tissues

The methods described here can operate effectively for cell surface antigens which have a reasonably wide tissue distribution. Thus a_1 is present on erythrocytes and on cells of many other tissues. However, if a particular antigen is found only on a very limited number of tissue cells its detection on such cells may be difficult. More important is the fact that it may be difficult or impossible to secure a long-term culture expressing the given antigen, which could be used for genetic–biochemical studies of the kind described. In order to carry out such genetic analysis of tissue antigens that may be localized to particular differentiated cells more complex procedure may be required. The principal difficulty is due to the fact that many cells which carry specific differentiation properties lose these when they are cultivated *in vitro* for periods greater than several weeks or months. Such short periods of cell cultivation are usually insufficient to carry out the necessary genetic biochemical analysis.

A possible way to get around these limitations is suggested by the following experiment. Cells of a Chinese hamster brain cultivated *in vitro* present the morphological characteristics of a typical undifferentiated fibroblast. However, when such cells are treated with cyclic AMP derivatives within a short space of time, they extend dendritic processes which appear to hook up and form a network that resembles that of a typical neuronal complex.

Cells displaying this property cannot be cultivated for more than several weeks *in vitro*. However, when such cells are hybridized with the CHO-K1, hybrid clones can be isolated which exhibit the dendritic response when treated with agents which increase the intracellular concentration of cyclic AMP. These cultures can be maintained for long periods *in vitro*.

Finally, we have repeated the hybridization process using cells from a human brain obtained in the course of necessary surgical procedures. Again clones were obtained exhibiting the characteristic dendrite-like response to cyclic AMP derivatives. However, because of the propensity of human–Chinese hamster cell hybrids to eject human chromosomes, it was possible to find a reactive hybrid which had retained only eight human chromosomes. Repeating this experiment on a larger scale might make it possible to identify the human chromosomes which are necessary for the retention of the properties which resemble those manifested by brain cells. Experiments are now underway to determine whether such cyclic AMP-stimulated hybrids also exhibit specific brain cell surface antigens. If such macromolecules can indeed be demonstrated, the general principles outlined in this chapter could be applied to genetic biochemical study of the determinants of such brain-specific reactions. It is conceivable that if this approach is successful it may be applicable to a variety of differentiation-specific macromolecules.

VIII. Discussion

Cell surface macromolecules take part in a host of specific reactions such as specific active transport processes, specific binding of hormones, and drugs in the bringing about of specific attachment between cells of particular kinds and in permitting attachment of specific viruses to the cell membrane. Presumably these membrane structures act to acquire information from the molecular constituents of the cell-bathing fluid and from those on the surfaces of neighboring cells. Presumably, such information must be transmitted directly or indirectly to the nucleus if change in gene expression is to be effected. One such intracellular communication system has already been identified. It consists of the specific steroid-binding proteins of the cell cytosol which, on attachment of a particular steroid, become attached to specific regions in the nuclear chromatin. New patterns of gene expression and protein synthesis then result (Jensen, 1975; O'Malley et al., 1977). It has been postulated that another such communication system consists of the microfibrillar apparatus of the cell. Several investigators have independently proposed that microtubules and microfilaments may be directly or indirectly connected to specific cell surface macromolecules. When these surface elements are removed through the action of trypsin or covered up by the action of compounds like lectins or by specific antibodies, the microfibrillar attachments are dissolved or altered. The shape of the cell then changes dramatically and presumably other functions are also altered. It is possible to envisage that any state of cell differentiation is normally associated with three different related parameters: a particular combination of cell surface macromolecules, a particular form of a communication system by which information gathered at the membrane is transmitted to the nucleus, and a particular combination of sequestered and exposed genes.

We have shown that some cell surface antigens with genetic loci present on human chromosome 11 are present on cells of many tissues while others have a more limited distribution. Thus a_1 is present on human erythrocytes, fibroblasts, and lymphocytes, while a_2 is absent from human erythrocytes but present on fibroblasts and lymphocytes. This mode of distribution suggests that normal differentiation may involve a particular combination of cell surface antigens characterizing each cellular differentiation state. This scheme would utilize the combinatorial flexibility of a relatively small number of cell surface antigen combinations in cells of one or more tissues. Study of the genetics and biochemistry of cell surface antigen distribution on cells of different tissues appears to offer the possibility of important insights into normal and pathologic cellular differentiation phenomena.

IX. Concluding Remarks

A method for analysis of human cell surface antigens is described which makes possible identification of specific cell membrane macromolecules and simultaneous identification of the human genes controlling their biosynthesis. The method involves construction of stable hybrids between human and Chinese hamster ovary cells which retain all of the standard CHO chromosomes plus one or a few human chromosomes. The lack of appreciable immunologic cross-reaction between the human and Chinese hamster cell surface antigens makes possible rapid and precise identification of human antigens on the hybrid membranes whose genetic loci must then be carried on the contained human chromosomes. A variety of different antigenic interactions have been demonstrated with several hybrids. The hybrid containing chromosome number 11 as its only human chromosome was selected for intensive study. Three different human cell surface antigens were found to be expressed by this hybrid and variants were prepared by mutagenesis expressing different combinations of these markers. Genetic loci controlling expression of these different antigens have been mapped regionally on chromosome number 11 and their presence on human tissues has been explored by means of immunofluorescence and the horseradish peroxidase reaction. Complementation analysis has been carried out for one of these antigens and it has been demonstrated that several genes are required for its expression. Analysis of the biochemical structure of these antigens is in progress. The fact that the genetic loci controlling the expression of these antigens are contained on both arms of chromosome 11 has made possible use of this hybrid for a particularly effective system of detection of environmental mutagens and carcinogens. The technique so developed appears to afford advantages over any other *in vitro* monitoring system known to us in that it also permits detection of genetic insults which result in multigene deletions or even loss of the entire chromosome. Thus, even nondisjunctional events which are responsible for serious genetic disease in almost 1% of all live human births can be detected by means of this system.

Further developments with this system are designed to detect a variety of differentiation-specific cell surface antigens and to analyze genetic biochemical control of the expression of these cell membrane moieties.

ACKNOWLEDGEMENTS

This work was supported by Grant No. CA-18734 from the National Institutes of Health; by Grant No. PCM-76-80299 from the National Science Foundation; by Grant No. CA-20810 from the National Institues of Health, Cancer Institute; by Grant HD-02080 from the National Institutes of Health, Child Health and Human Development Institute; and by a grant from the Louis B. Mayer Foundation. This is contribution No. 287 from the Eleanor Roosevelt Institute for Cancer Research and Florence R. Sabin Laboratories for Developmental Medicine.

REFERENCES

Ames, B. N., McCann, J., and Yamaski, E. (1975). *Mutat. Res.* **31**, 347–364.
Cairns, J. (1975). *Nature (London)* **225**, 197–200.
Furthmayr, H. (1978). *Nature (London)* **271**, 519–524.
Jensen, E. V. (1975). *In* "Advances in Pathobiology" (D. W. King, ed.), Vol. 1: Cell Membranes: Structure, Receptors, Transport (C. M. Fenoglio and C. Borek, eds.), pp. 48–55. Stratton Intercon., New York.
Jones, C. (1975). *Somatic Cell Genet.* **1**, 345–354.
Jones, C., and Kao, F. T. (1978). *Hum. Genet.* **143**, 1–10.
Jones, C., Puck, T. T. (1977). *Somatic Cell Genet.* **3**, 407–420.
Jones, C., Wuthier, P., Kao, F. T., and Puck, T. T. (1972). *J. Cell. Physiol.* **80**, 291–298.
Jones, C., Wuthier, P., and Puck, T. T. (1975). *Somatic Cell Genet.* **1**, 235–246.
Jones, C., Moore, E. E., and Lehman, D. W. (1979). *In* "Advances in Pathobiology" (D. W. King, ed.), Cell Membranes (C. M. Fenoglio and D. W. King, eds.). Stratton Intercon., New York. (in press)
Kao, F. T., and Puck, T. T. (1971). *Nature (London)* **228**, 329–332.
Kao, F. T., and Puck, T. T. (1972). *Proc. Natl. Acad. Sci. U.S.A.* **69**, 3273–3277.
Kao, F. T., Jones, C., and Puck, T. T. (1976). *Proc. Natl. Acad. Sci. U.S.A.* **73**, 193–197.
McKusick, V. A., and Ruddle, F. H. (1977). *Science* **196**, 390–405.
Moore, E. E. (1976). Ph.D. Thesis, University of Colorado.
Moore, E. E., Jones, C., and Puck, T. T. (1976). *Cytogenet. Cell Genet.* **17**, 89–97.
Moore, E. E., Jones, C., Kao, F. T., and Oates, D. C. (1977). *Am. J. Hum. Genet.* **29**, 389–396.
Oda, M., and Puck, T. T. (1961). *J. Exp. Med.* **113**, 599–610.
O'Malley, B. W., Coty, W. A., Schwartz, R. J., and Schrader, W. T. (1977). *In* "Advances in Pathobiology" (D. W. King, ed.), Vol. 6: Cancer Biology 4: Differentiation and Carcinogenesis (C. Borek, C. M. Fenoglio, and D. W. King, eds.), pp. 79–96. Stratton Intercon., New York.
Puck, T. T., Wuthier, P., Jones, C., and Kao, F. T. (1971). *Proc. Natl. Acad. Sci. U.S.A.* **68**, 3102–3106.
Waldren, C., Jones, C., and Puck, T. T. (1979). *Proc. Natl. Acad. Sci. U.S.A.* **76**, 1358–1362.
WHO (1964). Technical Report, Ser. No. 276.
Wuthier, P., Jones, C., and Puck, T. T. (1973). *J. Exp. Med.* **138**, 229–244.

CHAPTER 6

CELL SURFACE AND EARLY STAGES OF MOUSE EMBRYOGENESIS

François Jacob

SERVICE DE GÉNÉTIQUE CELLULAIRE DU COLLÈGE DE FRANCE
ET DE L'INSTITUT PASTEUR
PARIS, FRANCE

I. Introduction	117
II. The Teratocarcinoma System	118
III. Glycopeptides	119
IV. Lectin Receptors	123
V. Surface Antigens	125
A. F9 Antigens	125
B. At Least Two Distinct Specificities in F9 Antigen	127
C. Monoclonal Antibodies Produced by Hybrid Cells	130
VI. Surface Structures and Cellular Interactions in Early Development	132
VII. Concluding Remarks	134
References	135

I. Introduction

Little is known as yet about the principles underlying the development of the embryo. It seems likely, however, that the unfolding of the genetic program of development calls for some kind of a dialogue between the genome and the cell surface. The latter structure is indeed involved in various processes, such as interactions with other cells and reactions with extracellular factors, which, in all likelihood, act as signals to elicit, through unknown mechanisms, changes in gene expression. These changes, in turn, may modify the structure of the cell surface and its properties. Sequences of such reciprocal influences between genome and cell surface must, in some way, drive cellular lineages of the embryo along paths of development and differentiation.

Since embryonic development proceeds by a series of cellular events, it must be studied at the level of both the whole embryo and the cell. In the early stages of mammalian embryonic development, investigations at the cellular level have been hampered by the scarcity of the material and its heterogeneity. It is possible to circumvent some of these difficulties by the use of mouse teratocarcinoma as a model system for studying certain aspects of embryonic development and cell differentiation (see Graham, 1977; Jacob, 1975, 1978; Martin, 1975; Sherman and Solter, 1975). In particular, it is the use of the teratocarcinoma system which has made possible the study of cell surface properties in the early stages of embryonic development (Edidin *et al.*, 1971; Artzt *et al.*, 1973; Stern *et al.*, 1975). This chapter summarizes the work on cell surface of early embryonic cells done at the Institut Pasteur during recent years.

II. The Teratocarcinoma System

The teratoma system of the mouse was first described by Stevens (Stevens and Little, 1954) and analyzed in detail by Stevens (1967a) and by Pierce (1967). Testicular teratomas occur spontaneously in the mouse strain 129 and appear to derive from primitive germ cells (Stevens, 1967b). Ovarian teratomas are frequent in another mouse strain LT where they result from a high incidence of spontaneous parthenogenesis (Stevens and Varnum, 1974). In addition, teratomas can be induced with a high frequency in a variety of inbred strains by grafting 4- to 7-day embryos into the testes of syngeneic males (Stevens, 1970). Whether spontaneous or induced, a fraction of these teratomas can be serially transplanted into syngeneic adult mice.

In contrast to most other tumors, teratomas contain a wide variety of cell types corresponding to derivatives of the three embryonic germ layers. When transplantable, they are called teratocarcinomas and contain, in addition, embryonic-like cells called embryonal carcinoma (EC). These EC cells possess two important properties: (1) They have many characteristics in common with early embryonic cells and can differentiate into derivatives of the three germ layers (Kleinsmith and Pierce, 1964; Kahan and Ephrussi, 1970; Rosenthal *et al.*, 1970); and (2) they are malignant while their differentiated derivatives are, with a few exceptions, nonmalignant (Pierce, 1967).

Teratocarcinoma-derived cells can be obtained in culture either from solid tumors or from the so-called embryoid bodies present in the ascites formed after intraperitoneal injection of tumor fragments. By their shape, these bodies recall preimplantation stages of the mouse embryo. Some are cystic; others are solid, composed of a layer of primitive endoderm around a core of EC cells. When put in culture, the latter grow and differentiate into a variety of cell types. From such cultures, it is possible to obtain stable lines, either of EC cells or of cells at a late state of differentiation, such as fibroblasts, myoblasts, and heart muscle cells.

A number of EC cell lines have been established in culture in various laboratories. Upon injection into syngeneic mice, these cells produce tumors which, in most instances, contain a variety of cellular types formed by the differentiation of EC cells. Some of these EC cell lines are even able to differentiate in culture. It has been unambiguously demonstrated that a single cell can produce derivatives of the three germ layers (Lehmann et al., 1974; Martin and Evans, 1975; Nicolas et al., 1975). There is now considerable morphological (Pierce and Beals, 1964), biochemical (Bernstine et al., 1973), biological (Kleinsmith and Pierce, 1964; Brinster, 1974; Mintz and Illmensee, 1975; Papaioannou et al., 1975), and serological (Artzt et al., 1973) evidence documenting similarities between EC cells derived from teratocarcinoma and "pluripotent" cells of early embryos. Conversely, no significant dissimilarity has yet been demonstrated. All this warrants the use of EC cells for studying certain aspects of the early embryo, in particular the cell surface. When specific reagents have been developed for EC cells, it becomes possible to detect on the early embryo itself surface structures which otherwise could hardly have been found. We shall thus consider in turn different approaches to cell surface structures of early embryo.

III. Glycopeptides

Glycopeptides isolated from EC cells differ from those found on adult somatic cells and exhibit unusual properties (Muramatsu et al., 1978, 1979). Such glycopeptides were prepared by labeling the cells with radioactive sugars, followed by extensive digestion with Pronase of either the whole cells or of purified membrane fractions. In both cases the results were very similar. Upon Sephadex G-50 column chromatography, two main classes of Pronase-digested glycopeptides could be distinguished (Fig. 1).

1. A high-molecular-weight fraction which is eluted near the excluded volume of the column. It contains fucose, galactose, and glucosamine, but little or no mannose. Various experiments have excluded the possibility that such components could correspond to glycosylaminoglycans, glycolipids, conventional mucin-type glycopeptides, or products of incomplete Pronase digestion.

2. A low-molecular-weight fraction which is eluted in a well-retarded position. In this fraction at least two components can be distinguished according to the nature of the label. Mannose-labeled material has a molecular weight around 1500 and presents, in its majority, chemical properties similar to those of the so-called "high-mannose" glycopeptides described in adult cells (see Kornfeld and Kornfeld, 1976). Fucose- or galactose-labeled material has a molecular weight

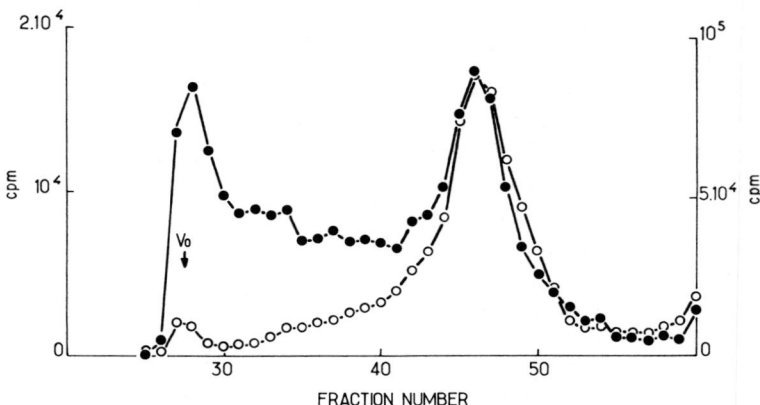

FIG. 1. Sephadex G-50 column chromatography of glycopeptides synthesized by EC cells and their differentiated derivatives. PCC3 cells, either in exponential growth (—●—●—) or after complete *in vitro* differentiation (—○—○—) were labeled with [^{3}H]-fucose. After extensive digestion with Pronase, the preparations were chromatographed on a Sephadex G-50 column (for experimental details, see Muramatsu *et al.*, 1979). Arrow indicates position of Blue Dextran. Material eluted between tubes 45 and 50 corresponds to an apparent MW of about 2000.

around 2500–3000. However, by its chemical properties—neutral, low affinity to concanavalin A, and resistance to endoglycosidases—this material differs from the so-called "complex" glycopeptides of adult cells (see Kornfeld and Kornfeld, 1976). Small glycopeptides thus appear also to present some unusual properties in EC cells.

With fucose or galactose as a label, the large components found near the excluded volume upon G-50 Sephadex column chromatography represent an important fraction of glycopeptides from EC cells. The experimental evidence indicates that at least some of these large glycopeptides are located on the cell surface. (1) From work on somatic cells, fucosyl glycopeptides are known to be mostly located in plasma membranes (Atkinson and Summers, 1971; Gahmberg, 1971); (2) a significant fraction of the glycopeptides prepared from cell surface material released by mild trypsin digestion has a large molecular weight; and (3) as will be discussed, such surface structures as "F9 antigen" or receptor sites for fucose-binding proteins and peanut agglutinin are glycoproteins which, upon Pronase digestion, release large glycopeptides.

The presence of these large glycopeptides was investigated in different cell types. They were found in large amounts in a series of EC cell lines, derived either from 129 or from C3H mice. They disappeared rapidly and almost completely during *in vitro* differentiation of EC cells. They were scarcely detectable in differentiated derivatives of EC cells, such as parietal yolk sac cells, fibroblast-like cells, or myoblasts. Fucosyl or galactosyl glycopeptides with such a high molecular weight have not been reported in significant amounts in whole cells, plasma membranes, or cell-surface material from a number of normal and malignant cells which have been investigated (Ogata *et al.*, 1976; Buck *et al.*, 1970, 1974; Muramatsu *et al.*, 1973). However, intracellular membranes of fibroblasts have been reported to release significant amounts of large fucosyl glycopeptide upon Pronase digestion (Buck *et al.*, 1974), as well as the heart of early chicken embryo (Manasek, 1976).

The presence of large glycopeptides was also investigated on preimplantation mouse embryos (Fig. 2). Two-cell stage embryos were collected, pooled, and cultured with radioactive fucose for 48 hours. They had then reached the late morula or early blastocyst stage and were harvested. Glycopeptides were prepared and analyzed as previously. Large glycopeptides were indeed found to represent a major component in such preimplantation embryos. After implantation, the embryos can be dissected and grown *in vitro* for a few hours in the presence of radioactive fucose; glycopeptides can then be prepared and analyzed as previously. In 6-day-old embryos, the glycopeptide profile is very similar to that found in EC cells and in preimplantation embryo, the large glycopeptides eluted near the excluded volume of the column representing an important fraction (about $\frac{1}{3}$). In the glycopeptides synthesized by 7-, 8-, and 9-day-old embryos, the relative contribution of high-molecular-weight glycopeptides decreases progressively. By day 10, the large majority of the radioactivity is located in the low-molecular-weight fraction. Neither the yolk sac of 8-day embryos, nor various organs dissected from day 10 or day 12 embryos retain the capacity to synthesize significant amounts of large fucosyl glycopeptides. After day 10, the profile of the glycopeptides synthesized by the embryo resembles that found in adult somatic cells or in differentiated derivatives of EC cells (T. Muramatsu, H. Condamine, and G. Gachelin, unpublished).

Large glycopeptides have also been detected on several human teratocarcinomas in culture, but not on a series of other human tumors (T. Muramatsu, G. Gachelin, and M. Fellous, unpublished). Such similarities between human and murine teratocarcinomas can be expected since early stages of embryonic development appear to be similar in various mammals. The presence of high-molecular-weight glycopeptides—the exact structure and function of which still remain un-

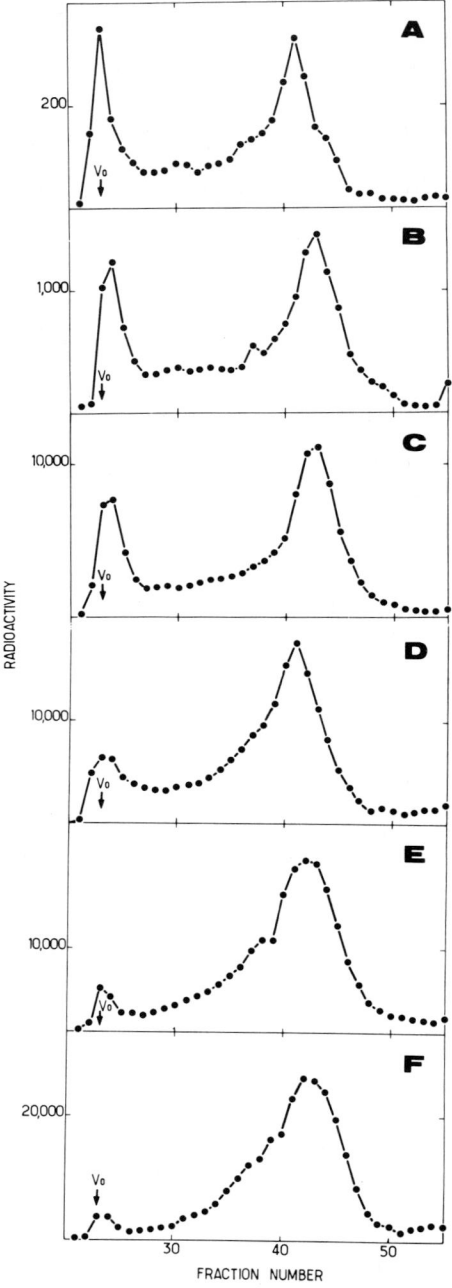

FIG. 2. Sephadex G-50 column chromatography of glycopeptides synthesized by mouse embryos at various stages. At various stages of development, embryos were dis-

known—might well represent a common feature of early mammalian embryos.

IV. Lectin Receptors

Four lectins, labeled with ^{125}I or fluorescein, were assayed for binding to EC cells: concanavalin A (Con A; binding inhibited by D-mannosyl residues); wheat germ agglutinin (WGA; N-acetylglucosamine); peanut agglutinin (PNA; D-galactose); and isofucose-binding proteins from lotus (FBP; L-fucose).

Receptors to Con A and to WGA are of common occurrence in most glycoproteins. They are indeed found on all adult cell types. They are also found on EC cells (Gachelin et al., 1976a) and on preimplantation embryos (Brownell, 1977). These receptors remain present at the surface of the various cell types which appear during in vitro differentiation of EC cells (G. Gachelin, unpublished).

More interesting are the receptors to PNA and FBP which are rarely found in significant amount on adult cell types (Reisner et al., 1977). Both lectins bind to EC cells (about 5×10^6 receptors per cell) in a specific way. In immunofluorescence tests, the fraction of the EC population exhibiting receptors depends on the cell line. With some EC lines (such as F9), 100% of the cells are labeled; with some others (such as the in vitro differentiating line PCC3), only 50% of the populations exhibit PNA or FBP receptors. This result is very similar to that found with F9 antigen. Actually, a series of double-labeling experiments indicate that, in populations of PCC3 cells, the same cells possess F9 antigen and receptors to both lectins. It thus appears that the cultures of such cells consist of two subpopulations in equal amount, in one of which only the three surface markers can be detected. These two populations can be separated and upon subculture each one appears to produce again mixed cultures. The reason for this variation is unknown; it might be linked to the cell cycle (J. F. Nicolas and G. Gachelin, unpublished).

When PCC3 cells are put in culture under conditions allowing in vitro differentiation, the fraction of cells expressing receptors to PNA or FBP drops rapidly as differentiation proceeds. After a few days,

---

sected, pooled, and labeled in vitro with [^{3}H]fucose for 6 hours. After extensive digestion with Pronase, the preparations were chromatographed on a Sephadex G-50 column. Arrows indicate position of Blue Dextran. Material eluted around tube 45 corresponds to an apparent MW of about 2000. (A) 6 embryos harvested at day 6; (B) 21 embryos at day 7; (C) 16 embryos at day 8; (D) 5 embryos at day 9; (E) 7 embryos at day 9 cultured for 24 hours instead of 6; (F) 3 embryos at day 10. From T. Muramatsu, G. Gachelin, and H. Condamine, unpublished.

when the majority of the cells are no longer in the EC state, the fraction of cells positive for PNA or FBP test is only 1–2%. This result is similar to that observed with F9 antigen which also disappears rapidly in the course of *in vitro* differentiation.

Like EC cells, the cells of preimplantation embryos bind PNA and FBP (Fig. 3) as well as Con A and WGA. [Binding of FBP was carefully investigated because of a report by Brownell (1977) who found no binding of FBP to preimplantation embryos. The reason for the discrepancy between these results and ours is unknown.] Preliminary experiments show that, after implantation at day 10, the large majority of the cells still bind Con A and WGA, but no longer PNA or FBP. These results indicate that rather drastic changes occur in the cell surface during early stages of cellular differentiation. Furthermore, the changes observed during *in vitro* differentiation of EC cells and during *in vivo* development of the embryo appear to be similar (M. Damonneville and G. Gachelin, unpublished). While the capacity to bind both PNA and FBP disappears during *in vitro* differentiation of EC cells, the ability to incorporate galactose and fucose per cell remains essentially unaltered. This indicates a change in the structure of surface polysac-

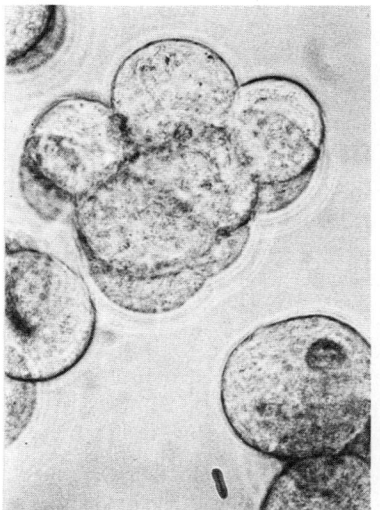
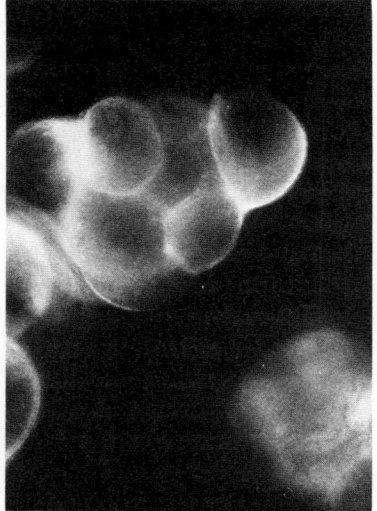

FIG. 3. Fixation of fucose-binding proteins onto preimplantation embryo. After removal of zona pellucida with Pronase, C57BL/6 morulae were treated with fluorescein-isothiocyanate (FITC) FITC-fucose-binding proteins (400 µg/ml). All blastomeres appear to be stained; staining disappears in presence of L-fucose ($10^{-2}\,M$). Left: phase; right: fluorescence. Magnification: ×850. From G. Gachelin, unpublished.

charides, probably as a result of a change in enzymatic activities during early differentiation.

Preliminary attempts to characterize PNA and FBP receptors were performed by labeling EC cells with radioactive fucose, galactose, or isoleucine and lysing the cells with a nonionic detergent. One of the lectins was then added, followed by a specific anti-lectin serum; the immunoprecipitate was characterized by sodium dodecyl sulfate (SDS)-polyacrylamide gel electrophoresis. Controls were treated in a similar way, but in the presence of the competing sugar. With either PNA or FBP, the precipitate was found to contain some glycolipids and a majority of glycoproteins. By their molecular weight (MW), these glycoproteins are distributed in two main classes: some have a MW around 35,000–42,000 and bind both PNA and FBP; the others of MW around 90,000 bind only PNA, although they contain both galactose and fucose.

These immunoprecipitated lectin receptors were extensively digested with Pronase, and the glycopeptides analyzed on Sephadex G-50. Under these conditions, receptors to FBP and receptors to PNA yielded almost exclusively high-molecular-weight glycopeptides of the type previously described. Large glycopeptides and receptors to PNA or FBP thus appear to be correlated and characteristic of cellular surface in early embryonic stages (G. Gachelin and T. Muramatsu, unpublished).

V. Surface Antigens

Embryonal carcinoma cells provide a rather homogeneous material in amount sufficient for immunization. With the antisera thus obtained, it has been possible to detect and study on early embryos some specific surface antigens which otherwise would have been extremely difficult to recognize. Recently, the study of cell surface antigens has been renewed by the possibility of producing monoclonal antibodies (Köhler and Milstein, 1975). Before discussing the use of this technique for the analysis of embryonic development, we shall first consider the results obtained with conventional serological techniques. A part of the work at the Institut Pasteur has already been reviewed (Jacob, 1977; Gachelin, 1978) and will only be summarized here, this discussion being mainly devoted to more recent results.

A. F9 ANTIGENS

Embryonal carcinoma cells appear to be completely devoid of the products of the major histocompatibility (MHC) H-2 complex of the mouse. They react neither with specific allogeneic anti-H-2 sera (Artzt and Jacob, 1974), nor with specific anti-H-2 killer T lymphocytes (Formann and Vitetta, 1975). They do not react either with a xenogeneic serum from a rabbit immunized with purified mouse H-2 proteins, a serum which detects not only the products of many mouse *H-2* haplotypes, but also of MHC from various mammals (Morello *et al.*, 1978). Similarly, EC cells do not react with a rabbit anti-mouse β_2-microglobulin serum (Dubois *et al.*, 1976). It thus appears unlikely that the surface of EC cells contains either a precursor or a masked form of H-2 products.

Hyperimmunization of syngeneic 129 male mice with irradiated cells from EC line F9 produces antisera very active on F9 cells either in direct cytotoxicity tests in the presence of complement or in indirect immunofluorescence or peroxidase tests (Artzt *et al.*, 1973). For several years, whether such syngeneic antisera could detect one or several antigenic molecules on the surface of F9 cells remained unclear and the structure(s) recognized by these antisera was referred to as the "F9 antigen" (see Jacob, 1977, 1978; Gachelin, 1978).

The F9 antigen was thus found to be present on all EC cell lines tested, whether derived from 129 or other mouse strains. In cultures of F9 cells, the F9 antigen is present on 100% of the cell populations. However, in other lines, such as PCC3 which differentiates *in vitro*, it is detected only on 50–60% of the population as previously mentioned. The F9 antigen is absent from differentiated derivatives of EC cells such as yolk sac cells, fibroblasts, myoblasts, or myocardial cells, most of which display H-2 antigens at their surface. During *in vitro* differentiation of EC cells PCC3, the fraction of cells expressing the F9 antigen drops rapidly while H-2 antigens appear on an increasing fraction of cells.

In the adult mouse, the F9 antigen is not found on any of the somatic cells tested. It is expressed, however, at every stage of spermatogenesis from gonocytes to spermatozoa (Artzt *et al.*, 1973; Gachelin *et al.*, 1976b).

In the embryo, the F9 antigen can be detected a few hours after fertilization and increases in amount up to the morula stage. In the blastocyst, it is present on both types of cells: trophectoderm and inner cell mass (ICM). After implantation, the F9 antigen remains expressed on embryonic ectoderm, the cells from which most tissues of the embryo proper derive. It disappears from the embryo at about day 9 after fertilization (Buc-Caron *et al.*, 1978).

The F9 antigen is expressed on morulae and sperm of all mice tested, whether inbred or not. Furthermore, a material reacting with anti-F9 antisera is found on morulae and sperm of all mammals tested so far, from kangaroo to man (Buc-Caron et al., 1974). In man, for instance, an antigen very close to F9 can be detected on the germ line at various stages of spermatogenesis (Gachelin et al., 1976b) as well as on teratocarcinoma (Hogan et al., 1977; Holden et al., 1977). In contrast, anti-F9 sera do not react with material from the few nonmammals investigated so far, such as chicken, axolotl, and frog.

By its restricted distribution on male germ line and early embryo, the F9 antigen is reminiscent of those antigens attributed to the T/t locus of the mouse, i.e., a segment of chromosome 17 known to control several steps of embryonic development as well as sperm production and function (for details about T/t complex, see Gluecksohn-Waelsch and Erickson, 1970; Bennett, 1975; Klein and Hammerburg, 1977). Two lines of evidence support the hypothesis of a relationship between F9 antigen and T/t complex. (1) Absorption experiments with sperm followed by measurement of residual cytotoxic activity indicate that sperm cells from heterozygotes for one particular t haplotype (t^{w32} or t^{12}), but not for others, contain half the amount of F9 antigen as sperm cells from wild type. This suggests a direct relationship between F9 antigen and a wild allele at the t^{w32} locus (Artzt et al., 1974; K. Artzt, unpublished); (2) in crosses producing homozygotes for certain haplotypes (t^{w32} and t^{w5}), but not for others (t^{w18} and T), a fraction of the morulae does not express F9 antigen in indirect immunofluorescence tests. Since the F9 antigen is not expressed by two distinct t haplotypes, it can hardly be considered as the product of a particular wild allele of the T/t complex. However, the segregation of F9 antigen in these crosses suggests that its expression might be controlled by a gene belonging, or closely linked, to the T/t complex (Kemler et al., 1976).

B. At Least Two Distinct Specificities in F9 Antigen

Some discrepancy had thus appeared in the results of experiments involving t haplotypes. On the one hand, when *cytotoxic activity* of anti-F9 serum was absorbed with sperm, the t^{w5} haplotype appeared to express F9 antigen normally. On the other hand, when *indirect immunofluorescence* activity of anti-F9 serum on morulae was investigated, the t^{w5} haplotype was found not to express F9 antigen. A similar discrepancy appeared with compound heterozygotes t^{w32}/t^{w5}. Morulae t^{w32}/t^{w5} did not react with anti-F9 serum in indirect immunofluorescence tests, while sperm from males of the same genotype did absorb completely cytotoxic anti-F9 activity (M. Fellous and R. Kemler, unpublished).

Because of difficulties met in attempts to characterize immunochemically the F9 antigen and of the report that teratocarcinoma-bearing mice can produce antitumor antibodies of various classes (Isa and Sanders, 1975), the types of immunoglobulins associated with anti-F9 activity were studied. A series of anti-F9 antisera, prepared by immunizing either syngeneic 129, or C57BL/6 mice, or 129 × C57BL/6 hybrids were analyzed by ultracentrifugation to separate 7S and 19S Ig, and characterized with specific rabbit antisera directed against particular classes or subclasses of mouse immunoglobulins (M. Damonneville and G. Gachelin, unpublished). In all cases, specific binding activity detected by immunofluorescence was found to be associated with both IgM and IgG1 (also with a trace of IgG2a, but not with IgG2b, IgG3, or IgA). Nearly all cytotoxic activity, however, turned out to be associated with IgM. The IgM and IgG1 components of anti-F9 serum do not detect the same antigenic specificities, since massive absorption of anti-F9 serum with sperm removes all specific IgM and cytotoxic activity, but not IgG1 and the associated immunofluorescence.

Because of these two F9 antigenic specificities, the cells known to react with syngeneic anti-F9 serum were reinvestigated for their capacity to bind the IgM and IgG components of the serum. Most of the EC cells used in this laboratory (F9, PCC3, PCC4) as well as preimplantation embryos react with both anti-F9 IgM and IgG1. Two other EC lines, LT1 and PCS-7, appear to react with IgM, but not or very weakly with the IgG1 component (J. F. Nicolas, unpublished experiments).

The existence of two F9 antigenic specificities—which might be called F9M and F9G—raises the problem of their chemical nature and of their presence on the same or on different molecules. Preliminary attempts to characterize F9 antigen by immunoprecipitation followed by electrophoresis had indicated a structure similar to that of the K and D products of the H-2 complex, i.e., a protein composed of two chains: a heavy one of MW 44,000 and a light one of MW 12,000 (Vitetta et al., 1975), which in H-2 antigens corresponds to β_2-microglobulin. However, recent serological analysis has shown that the light chain of F9 is not β_2-microglobulin (Dubois et al., 1976) and that F9 cells do not react with the serum of a rabbit immunized with purified mouse H-2 proteins, a serum which detects the MHC products of various mammals (Morello et al., 1978). If there exists some relation between H-2 and some F9 component, it must, therefore, be a loose one.

Recent experiments were performed with the aim of characterizing the structures recognized by the IgM or IgG components of anti-F9 serum. Immunoprecipitation with IgG followed by SDS electrophoresis gave, as major component, a peak of MW around 42,000–43,000, which can be labeled with radioactive fucose, galactose, galactosamine, or amino acids, but not with mannose (Fig. 4). Pronase digestion of such a component labeled with fucose released a single peak of large glycopeptide which, upon Sephadex G-50 column chromatography, migrated near the excluded volume of the column (Fig. 5) (G. Gachelin and T. Muramatsu, unpublished). These results, however, must be interpreted with caution for two reasons. On the one hand, EC cells lysed with nonionic detergents contain active proteases which may modify the profile of proteins during precipitation experiments. On the other hand, it is not excluded yet that anti-F9 IgG would detect a glycolipid

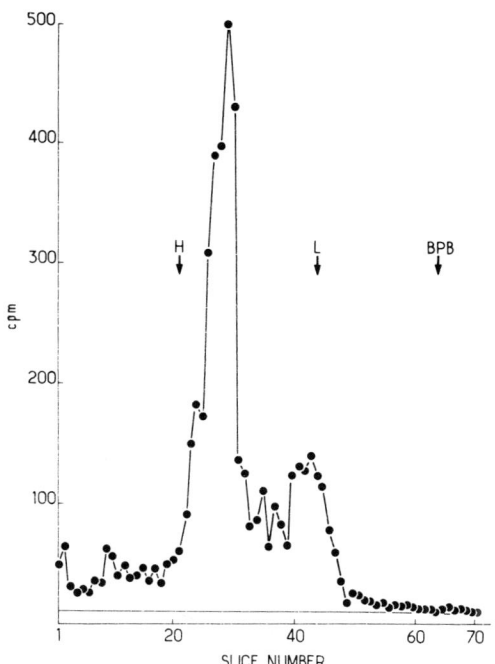

Fig. 4. SDS-polyacrylamide gel electrophoresis of F9G antigen. F9 cells labeled with [^{3}H]fucose were lysed with NP-40. To this preparation were successively added mouse anti-F9 serum and rabbit anti-mouse IgG1. Immune complexes were collected by centrifugation, washed, dissolved in SDS + 1% mercaptoethanol, and analyzed by SDS-polyacrylamide gel electrophoresis. Control was done with anti-F9 serum absorbed with F9 cells. After electrophoresis, no significant counts above background were obtained with the control (which is not shown). Internal markers were: H = heavy chain of mouse IgG; L = light chain of mouse IgG; BPB = bromphenol blue. From G. Gachelin and T. Muramatsu, unpublished.

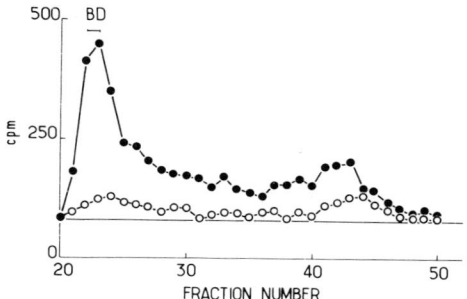

FIG. 5. Sephadex G-50 column chromatography of glycopeptides recovered from F9G antigen. Immune precipitate as described in Fig. 4 was extensively digested with Pronase and the preparation chromatographed on a G-50 Sephadex column. (—●—●—), precipitate with anti-F9 serum; (—O—O—), control precipitate with anti-F9 serum absorbed with F9 cells. BD = Blue Dextran. Counts around tube 43 correspond to an apparent MW of 2000. From G. Gachelin and T. Muramatsu, unpublished.

able to bind some proteins. In any case, the structure of the antigen detected by the IgG component of anti-F9 serum appears to be very different from that of H-2 antigens. Attempts to characterize the antigen reacting with the IgM component of anti-F9 serum by immunoprecipitation have not yet succeeded.

C. MONOCLONAL ANTIBODIES PRODUCED BY HYBRID CELLS

Immunization with such complex antigens as whole cells results in the production of complex sera, even in syngeneic animals as shown in the previous section. The heterogeneity of antibodies present in such conventional sera makes especially difficult and ambiguous the analysis of cell surface antigens. Some of these difficulties disappear with the use of monoclonal antibodies prepared according to Köhler and Milstein (1975). Each of these monoclonal antibodies is directed against a particular antigenic determinant and can therefore be used directly as a specific reagent without complicated absorptions. By carefully selecting the cell types used as immunogens and the animals to be immunized, it should be possible to obtain batteries of such reagents for an analysis of cell surface in the developing embryo.

At the Institut Pasteur, a first set of monoclonal antibodies against EC cells was prepared by fusing the cells of a mouse myeloma line with the splenocytes of rats previously immunized with F9 cells (R. Kemler and D. Morello, unpublished). Individual hybrid cell lines were selected and assayed in a radioactive binding assay and a cytotoxicity test for production of antibodies active on F9 cells. Out of 93 hybrid lines obtained, 28 turned out to produce antibodies active on F9 cells. Five of these were chosen for subcloning and further analysis, the

TABLE I

MONOCLONAL ANTIBODIES AGAINST EC CELLS F9

		ECMA No.[a]		
	Target	1	2	3
EC cells[b]	F9, from 129 mice	+	+	+
	PCC4, from 129 mice	+	+	+
	C17-S1, from C34 mice	−	+	+
PCC3 differ-entiating in vitro[c]	Day 1	2.5	70	70
	Day 2	NT	70	50
	Day 8	4	6	3
	Day 14	1	1	2
Embryo 129[d]	2-Cell	−	+	+
	Morula	−	+	+
	Blastocyst { Trophectoderm	+	+	+
	{ ICM	+	+	+
Sperm	From 129 mouse[b]	+	+	+
Lymphocytes	From 129 mouse	+	+	−

[a] A plus sign indicates a positive reaction; a minus sign, no reaction; NT = nontested. (From R. Kemler and D. Morello, unpublished.)

[b] Direct cytotoxicity in presence of complement or absorption followed by cytotoxicity on F9 cells.

[c] Percent of cells labeled in indirect immunofluorescence.

[d] Indirect immunofluorescence.

antibodies they produce being called ECMA 1 to 5 (for Embryonal Carcinoma Monoclonal Antibody). Four of these ECMAs were IgM and had cytotoxic activity on F9; one was IgG and not cytotoxic. Concentrated culture supernatants were then tested by absorption, radioactive binding assay, or indirect immunofluorescence on various EC cells, on some differentiated cell types, as well as on preimplantation embryos.

Each ECMA was thus found to detect a different antigen, with a characteristic distribution. Three ECMAs, all IgM, reacted with preimplantation embryos (Table I): one only at blastocyst stage, with both trophectoderm and ICM cells; the two others at all stages from 2-cell on. The two latter, which both react with some 60–70% of growing EC cells PCC3, disappear rapidly as these cells differentiate *in vitro*. Yet the antigens detected by these two ECMAs are different since one is expressed on a rather wide spectrum of differentiated types while the other has a much narrower distribution, similar to that found for the F9M antigen (Kemler *et al.*, 1979).

VI. Surface Structures and Cellular Interactions in Early Development

The possible role of surface antigens can be investigated through the effect of specific antisera on the *in vitro* development of preimplantation embryos. A series of antisera was used for such experiments: syngeneic mouse anti-F9 serum; sera from rabbits immunized either with embryonic liver cells, or brain homogenate, or EC cells F9. The anti-liver and anti-brain sera react with some unknown surface structures present in both F9 cells and blastomeres. From the rabbit anti-F9 serum, two types of antibodies could be distinguished and separated: one, referred to as "antilymphocytes," detects an antigen present on the surface of F9 cells, blastomeres, and lymphocytes; the other, referred to as "anti-F9," detects a structure present only on blastomeres and F9 cells.

Two-cell-stage mouse embryos were grown *in vitro* in the presence of one of these sera at various concentrations. No effect whatsoever was observed, the embryos reaching the blastocyst stage. Since multivalent antibodies can induce side effects, such as antigen redistribution or cross-linking of cells, monovalent Fab fragments were prepared from IgG of each rabbit preparation and assayed on early development. These preparations had no effect, except for Fab fragments from the rabbit "anti-F9" preparation. When anti-F9 Fab are added to embryos in culture, cleavage proceeds normally, but the blastomeres do not undergo compaction to form a morula. Instead the embryos, containing from 20 to 40 cells, form some kind of grapelike structures in which individual cells can easily be distinguished in contrast to cells of normal, compact morulae. In the presence of Fab, these grapelike structures do not form blastocysts and the cells die in culture after a few days. The effect, however, remains reversible for some time: if the structures formed after 2 days of culture in presence of anti-F9 Fab fragments are washed and put in culture without Fab, they form compact morulae within 3–5 hours. In 5–7 hours, these morulae produce blastocysts which, upon reimplantation into pseudopregnant mothers, give rise to viable newborn with the same efficiency as normal, control blastocysts (Babinet *et al.*, 1977; Kemler *et al.*, 1977).

It is thus clear that rabbit anti-F9 Fab fragments have no toxic effect on the embryo. They appear to act by loosening cellular interactions, thereby preventing morula compaction and consequently blastocyst formation. Although compaction appears to be programmed to begin around the 8-cell stage, it can be reversibly altered by addition of anti-F9 Fab up to the beginning of blastocoel formation. Only then does some irreversible event occur which prevents Fab to act any longer. This event might be related to the appearance of desmosomes, which are first observed between the outer cells of the embryo at the onset of blastocoel formation (Ducibella, 1977), and the concomitant establishment of a barrier to the penetration of antibody molecules (McLaren and Smith, 1977; Handyside, 1978).

Morula compaction is known to be reversibly inhibited *in vitro* by a decrease in Ca^{2+} concentration of the medium or by addition of cytochalasin B (Ducibella and Anderson, 1975). Anti-F9 Fab thus appears to affect development in the same way as both these treatments. The effect of cytochalasin B is difficult to interpret because its mode of action is not yet completely understood. Calcium ion is generally considered to be involved in intercellular adhesion by complexing with glycoproteins of apposing cell surfaces or with a divalent intermediate glycoprotein (Kuhns *et al.*, 1974; Edelman, 1976). It seems likely that an interaction between Ca^{2+} and the surface structure detected by rabbit anti-F9 antibodies is required for close blastomere adhesion and junction formation accompanying morula compaction.

The decompacting effect of rabbit anti-F9 Fab is not restricted to morulae. It is also observed with early ICM as shown by experiments in which ICM isolated from blastocysts by immunosurgery (Solter and Knowles, 1975) are put in culture in the presence of anti-F9 Fab. Depending on the time blastocysts are harvested and the way they are handled after immunosurgery, two kinds of ICM can be prepared. Early ones consist of a mass of compacted cells without detectable endodermal cells. Late ones contain a core of compacted cells surrounded by a layer of endodermal cells. Addition of anti-F9 Fab to early ICM results in a rapid decompaction and the production of grapelike structures similar to those formed by morulae. Late ICM, in contrast, appear to be insensitive to anti-F9 Fab: they remain intact and proceed in their growth and differentiation (R. Kemler and C. Babinet, unpublished).

Anti-F9 Fab fragments are also active on EC cells. When PCC3 cells, which can differentiate in culture, are grown in the presence of rabbit anti-F9 Fab, the cells round up and appose less closely to each other. They multiply at about the same rate as control cells, but no cytodifferentiation is observed. The process is reversible: upon removal of anti-F9 Fab fragments, the cells resume close adhesion and differentiated cells appear rapidly (J. F. Nicolas, unpublished). Embryonal carcinoma cells are able to cooperate metabolically with other EC cells, probably through the formation of gap junctions, but not with differentiated cells (Nicolas *et al.*, 1978). Metabolic cooperation between EC cells is either suppressed by anti-F9 Fab fragments, or largely decreased depending on the cell line, while remaining unaffected by Fab from other antisera. The decrease in cellular adhesion resulting from the effect of anti-F9 Fab probably prevents or decreases strongly the formation of intercellular junctions.

Anti-F9 Fab thus appears to interfere with differentiation of morula and of ICM as well as of EC cells. It seems likely that the surface structure detected by rabbit anti-F9 antibodies is involved in close cell–cell adhesion at various stages of early development, thereby allowing other, more specific events to occur. The chemical nature of this surface structure is not yet known.

VII. Concluding Remarks

In recent decades, the study of embryonic development and cellular differentiation has shifted in emphasis: from description of changing morphology at the level of tissue and cells, it has turned to the study of the molecular mechanisms which underlie these complex processes. One of the main targets of this new outlook has been the cell surface. By its importance in such events as cellular division, migration, and recognition, the surface of cells must necessarily play a central role in regulating the cellular interaction involved in the orderly development of the genetic program during embryogenesis.

In the early stages of mouse embryonic development, a molecular approach has been made possible by the use of mouse teratocarcinoma-derived cell lines. The results obtained by chemical as well as immunological analysis have shown that surface structures detected on EC cells are also often present on the cells of the early embryo. Furthermore, human and mouse teratocarcinomas appear to have many features in common. These findings, in a way, are a mere confirmation, at the molecular level, of Karl von Baer's observation that, among related organisms, early stages of development remain similar, differences showing up at later stages.

Different lines of investigation have unravelled the presence of several markers on the surface of EC cells: a set of unusually large glycopeptides apparently released from various glycoproteins, the receptor sites of the lectins PNA and FBP of lotus seeds, as well as F9 antigens. All these markers disappear rapidly and almost completely during *in vitro* differentiation of EC cells. Their presence on early embryonic stages followed by rapid disappearance during development can also be demonstrated directly for the antigens and the lectin receptor sites. The presence of the large glycopeptides in preimplantation embryos as well as their relative decrease during development have also been established. Yet their location on the cell surface is technically more difficult to prove. It seems a likely location, however, since in EC cells the F9G antigen as well as the receptor sites for FBP and

PNA are glycoproteins releasing large glycopeptides upon Pronase digestion. Consistent results are thus obtained with the different technical approaches.

Since they disappear rapidly from the embryo, these markers can be used to characterize early embryonic cells. For obvious reasons, few biochemical markers are available for the early states of differentiation. The distribution of some surface markers might allow one to specify such early states in a way similar to what is done with lymphocytes subsets (Boyse and Old, 1969). For example, the surface antigen PCC4, detected by syngeneic sera directed against multipotential EC cell PCC4, is expressed on ICM cells in the blastocyst, but not on morulae nor on trophectoderm cells (Gachelin et al., 1977). It is likely that monoclonal antibodies will provide a unique repertoire of specific reagents for characterization of cell state.

The effect of rabbit anti-F9 Fab fragments shows conclusively the importance of surface components in the first steps of embryonic development. It raises a number of questions: What is the structure of the molecule combining with these rabbit antibodies? What is its role in cellular interaction? How do such surface molecules relate to and communicate with the epigenetic system controlling gene expression? Before understanding how two like cells of an early embryo become different, it will be necessary to have answers to these questions.

ACKNOWLEDGMENTS

The work done at the Institut Pasteur was supported by grants from the Centre National de la Recherche Scientifique, the Délégation Générale à la Recherche Scientifique et Technique (No. 77.7.0966), the Institut National de la Santé et de la Recherche Médicale (No. 76.4.311.AU), the National Institute of Health (CA 16355), the Fondation pour la Recherche Médicale Française, and the Fondation André Meyer.

REFERENCES

Artzt, K., and Jacob, F. (1974). *Transplantation* **17**, 633–634.
Artzt, K., Dubois, P., Bennett, D., Condamine, H., Babinet, C., and Jacob, F. (1973). *Proc. Natl. Acad. Sci. U.S.A.* **70**, 2988–2992.
Artzt, K., Bennett, D., and Jacob, F. (1974). *Proc. Natl. Acad. Sci. U.S.A.* **71**, 811–814.
Atkinson, P. H., and Summers, D. F. (1971). *J. Biol. Chem.* **246**, 5162–5175.
Babinet, C., Kemler, R., Dubois, P., and Jacob, F. (1977). *C. R. Acad. Sci. Paris* **284**, 1919–1922.
Bennett, D. (1975). *Cell* **6**, 441–454.
Bernstine, E. G., Hopper, M. L., Grandchamp, S., and Ephrussi, B. (1973). *Proc. Natl. Acad. Sci. U.S.A.* **70**, 3899–3903.
Boyse, E. A., and Old, L. J. (1969). *Annu. Rev. Genet.* **3**, 269–289.
Brinster, R. L. (1974). *J. Exp. Med.* **140**, 1049–1056.
Brownell, A. G. (1977). *J. Supramol. Struct.* **7**, 223–234.
Buc-Caron, M. H., Gachelin, G., Hofnung, M., and Jacob, F. (1974). *Proc. Natl. Acad. Sci. U.S.A.* **71**, 1730–1733.

Buc-Caron, M. H., Condamine, H., and Jacob, F. (1978). *J. Embryol. Exp. Morphol.* **47**, 149–160.
Buck, C. A., Glick, M. C., and Warren, L. (1970). *Biochemistry* **9**, 4567–4576.
Buck, C. A., Fuhrer, J. P., Soslau, G., and Warren, L. (1974). *J. Biol. Chem.* **249**, 1541–1550.
Dubois, P., Fellous, M., Gachelin, G., Jacob, F., Kemler, R., Pressman, D., and Tanigaki, N. (1976). *Transplantation* **22**, 467–473.
Ducibella, T. (1977). *In* "Development in Mammals" (M. H. Johnson, ed.), Vol. 1, pp. 5–30. North-Holland Publ., Amsterdam.
Ducibella, T., and Anderson, E. (1975). *Dev. Biol.* **47**, 45–58.
Edelman, G. M. (1976). *Science* **192**, 218–226.
Edidin, M., Patthey, H. L., McGuire, E. J., and Sheffield, W. D. (1971). *In* "Embryonic and Fetal Antigens in Cancer" (N. G. Anderson and J. H. Coggin Jr., eds), pp. 239–248. Oak Ridge National Laboratory, Oak Ridge, Tenn.
Forman, J., and Vitetta, E. S. (1975). *Proc. Natl. Acad. Sci. U.S.A.* **72**, 3661–3665.
Gachelin, G. (1978). *Biochim. Biophys. Acta* **516**, 27–60.
Gachelin, G., Buc-Caron, M. H., Lis, H., and Sharon, N. (1976a). *Biochim. Biophys. Acta* **436**, 825–832.
Gachelin, G., Fellous, M., Guénet, J. L., and Jacob, F. (1976b). *Dev. Biol.* **50**, 310–320.
Gachelin, G., Kemler, R., Kelly, F., and Jacob, F. (1977). *Dev. Biol.* **57**, 199–209.
Gahmberg, C. G. (1971). *Biochim. Biophys. Acta* **249**, 81–95.
Gluecksohn-Waelsch, S., and Erickson, R. P. (1970). *In* "Current Topics in Developmental Biology" (A. A. Moscona and A. Monroy, eds.), Vol. 4, pp. 281–316. Academic Press, New York.
Graham, C. F. (1977). *In* "Concepts in Mammalian Embryogenesis" (M. I. Sherman, ed.), pp. 315–394. MIT Press, Cambridge, Massachusetts.
Handyside, A. H. (1978). *J. Embryol. Exp. Morphol.* **45**, 37–53.
Hogan, B., Fellous, M., Avner, P., and Jacob, F. (1977). *Nature (London)* **270**, 515–518.
Holden, S., Bernard, O., Artzt, K., Whitmore, W. F. Jr., and Bennett, D. (1977). *Nature (London)* **270**, 518–520.
Isa, A. M., and Sanders, B. R. (1975). *Transplantation* **20**, 296–302.
Jacob, F. (1975). *In* "The Early Development of Mammals." Cambridge Univ. Press, London and New York, pp. 233–241.
Jacob, F. (1977). *Immunol. Rev.* **33**, 3–32.
Jacob, F. (1978). *Proc. R. Soc. London B* **201**, 249–270.
Kahan, B. W., and Ephrussi, B. (1970). *J. Natl. Cancer Inst.* **44**, 1015–1029.
Kemler, R., Babinet, C., Condamine, H., Gachelin, G., Guénet, J. L., and Jacob, F. (1976). *Proc. Natl. Acad. Sci. U.S.A.* **73**, 4080–4084.
Kemler, R., Babinet, C., Eisen, H., and Jacob, F. (1977). *Proc. Natl. Acad. Sci. U.S.A.* **74**, 4449–4452.
Kemler, R., Morello, D., and Jacob, F. (1979). *In* "Cell Lineage, Stem Cells and Cell Determination" (N. Le Douarin ed.). North-Holland Elsevier, Amsterdam (in press).
Klein, J., and Hammerberg, C. (1977). *Immunol. Rev.* **33**, 70–104.
Kleinsmith, L. J., and Pierce, G. B. (1964). *Cancer Res.* **24**, 1544–1551.
Köhler, G., and Milstein, C. (1975). *Nature (London)* **256**, 495–497.
Kornfeld, R., and Kornfeld, S. (1976). *Annu. Rev. Biochem.* **45**, 217–237.
Kuhns, W. J., Weinbaum, G., Turner, R., and Burger, M. M. (1974). *Ann. N.Y. Acad. Sci.* **234**, 58–74.
Lehman, J. M., Speers, W. C., Swartzendruber, D. E., and Pierce, G. B. (1974). *J. Cell. Physiol.* **84**, 13–28.

Manasek, F. J. (1976). *In* "The Cell Surface in Animal Embryogenesis and Development" (G. Poste and G. L. Nicolson, eds.), Vol. 1, pp. 545–598. North-Holland Publ., Amsterdam.

Martin, G. (1975). *Cell* **5**, 229–243.

Martin, G. R., and Evans, M. J. (1975). *Proc. Natl. Acad. Sci. U.S.A.* **72**, 1441–1445.

McLaren, A., and Smith, R. (1977). *Nature (London)* **267**, 351–353.

Mintz, B., and Illmensee, U. (1975). *Proc. Natl. Acad. Sci. U.S.A.* **72**, 3585–3589.

Morello, D., Gachelin, G., Dubois, P., Tanigaki, N., Pressman, D., and Jacob, F. (1978). *Transplantation* **26**, 119–125.

Muramatsu, T., Atkinson, P. H., Nathenson, S. G., and Ceccarini, C. (1973). *J. Mol. Biol.* **80**, 781–799.

Muramatsu, T., Gachelin, G., and Jacob, F. (1979). *Biochim. Biophys. Acta* (in press).

Muramatsu, T., Cachelin, G., Nicolas, J. F., Condamine, H., and Jakob, H. (1978). *Proc. Natl. Acad. Sci. U.S.A.* **75**, 2315–2319.

Nicolas, J. F., Dubois, P., Jakob, H., Gaillard, J., and Jacob, F. (1975). *Ann. Microbiol. (Inst. Pasteur)* **126A**, 3–22.

Nicolas, J. F., Jakob, H., and Jacob, F. (1978). *Proc. Natl. Acad. Sci. U.S.A.* **75**, 3292–3296.

Ogata, S., Muramatsu, T., and Kobata, A. (1976). *Nature (London)* **259**, 580–582.

Papaioannou, V. E., McBurney, M. W., and Gardner, R. L. (1975). *Nature (London)* **258**, 70–73.

Pierce, G. B. (1967). *In* "Current Topics in Developmental Biology" (A. A. Moscona and A. Monroy, eds.), Vol. 2, pp. 223–246. Academic Press, New York.

Pierce, G. B., and Beals, T. F. (1964). *Cancer Res.* **24**, 1553–1558.

Reisner, Y., Gachelin, G., Dubois, P., Nicolas, J. F., Sharon, N., and Jacob, F. (1977). *Dev. Biol.* **61**, 20–27.

Rosenthal, M. D., Wishnow, R. M., and Sato, G. (1970). *J. Natl. Cancer Inst.* **44**, 1001–1009.

Sherman, M. I., and Solter, D., eds. (1975). "Teratomas and Differentiation." Academic Press, New York.

Solter, D., and Knowles, B. B. (1975). *Proc. Natl. Acad. Sci. U.S.A.* **72**, 5099–5102.

Stern, P. L., Martin, G. R., and Evans, M. J. (1975). *Cell* **6**, 455–465.

Stevens, L. C. (1967a). *Adv. Morphogen.* **6**, 1–31.

Stevens, L. C. (1967b). *J. Natl. Cancer Inst.* **38**, 549–552.

Stevens, L. C. (1970). *Dev. Biol.* **21**, 364–382.

Stevens, L. C., and Little, C. C. (1954). *Proc. Natl. Acad. Sci. U.S.A.* **40**, 1080–1087.

Stevens, L. C., and Varnum, D. S. (1974). *Dev. Biol.* **37**, 369–380.

Vitteta, E. S., Artzt, K., Bennett, D., Boyse, E. A., and Jacob, F. (1975). *Proc. Natl. Acad. Sci. U.S.A.* **72**, 3215–3219.

CHAPTER 7

DEVELOPMENTAL STAGE-SPECIFIC ANTIGENS DURING MOUSE EMBRYOGENESIS

Davor Solter and Barbara B. Knowles

THE WISTAR INSTITUTE OF ANATOMY AND BIOLOGY
PHILADELPHIA, PENNSYLVANIA

I. Introduction	139
II. Immunological Analysis of Stage-Specific Antigens	141
A. Antigens Detected on Mouse Embryos by Xenogeneic Antisera	142
B. Antigens Detected on Mouse Embryos by Alloantisera	147
C. Antigens Detected on Mouse Embryos by Syngeneic Antisera	147
D. Embryonic Antigens Detected by Monoclonal Antibodies	153
III. Conclusions and Perspectives	160
IV. Concluding Remarks	162
References	163

I. Introduction

To study development and differentiation in a multicellular organism we must make certain assumptions on the basis of current biological knowledge. Although the details of these processes are practically unknown, a few basic premises are probably correct and can serve as the starting point for any experimental approach.

Each one of the hundreds of different cell types found in adult organisms is precisely defined by its biochemical and morphological characteristics and is quite distinct from any other cell type in the same organism. All cell types, however, are derived from a single cell, the fertilized egg or zygote. We assume that differentiation proceeds in a stepwise fashion and that between the zygote and the adult cell type there exist an unspecified number of intermediary cell types which, again, differ from each other, from their ancestors, and from their final progeny. Intermediary cell types exist for a limited time only and then disappear by changing into one or more other intermediary or final cell types. We can also assume that although developmental strategies differ and the development of an oak tree, a frog, and a mouse are guided by different control mechanisms, they probably all boil down to essentially the same principle, namely, expressing different gene sets in different cell types (differential gene expression). Whether differential gene expression is achieved by activation—deactivation or by irreversible gene repression, as recently suggested (Caplan and Ordahl, 1978), is irrelevant for the present discussion (otherwise this distinction is of enormous importance as it bears on the question of reversibility of the differentiated state and of transition from one pathway of differentiation to another). Differential gene expression implies that each intermediary cell type is characterized by its own set of gene products. Finding those products and determining what they do during differentiation and how they do it has been, and still is, the major goal of developmental biology.

In subsequent pages we will discuss the use of immunological methods in the search for developmental gene products, considering briefly other methods for studying developmental genes and the advantages of using immunological techniques. Before that, however, we should examine briefly the particular problems and advantages of using the mouse embryo as an experimental model.

The main disadvantage in studying mouse development is the relative scarcity of material and the long, tedious process of collecting it. Miniaturization of most biochemical techniques partially offsets such a disadvantage and makes it possible to exploit several unique features of mouse development. Murine genetics has been extensively studied; the existence of numerous inbred strains makes it possible to work on a stable genetic background, against which developmental genes and their products will not disappear, as they would in the allelic noise of a random bred population.

Activation of the embryonic genome occurs very early in mouse development, possibly immediately after fertilization and certainly after two divisions. It is therefore possible to study the activity of developmental genes in a very simple structure that is composed of only a few cells and in which gene activity directs basic events such as the separation of the embryo into its embryonic part (giving rise to the adult organism) and its extraembryonic part (giving rise to fetal membranes). Early mouse embryos can be easily manipulated; even very rough treatment can be absorbed and embryos will continue to develop normally, which gives us an opportunity to test the effects of various mechanical and chemical manipulations on development.

The ideal approach to the study of developmental genes would be to study the genes themselves, but in most cases this is beyond our technical ability. Analysis of primary gene products, namely messenger RNAs, is just starting, and we know, for example, that synthesis of polyadenylated RNA occurs during early mouse development (Levey *et al.*, 1978). The newly developed techniques of genetic engineering, reverse transcriptase, and plasmid technology offer the exciting likelihood that in the near future developmental genes will be amplified and become accessible for much more detailed analysis than we believed possible just a few years ago.

Two-dimensional gel electrophoresis (O'Farell, 1975) has facilitated analysis of protein patterns and their changes during development as a secondary reflection of changes in patterns of activity of developmental gene sets. Several reports described protein patterns characteristic for some (Van Blerkom *et al.*, 1976; Dewey *et al.*, 1978) or all (Levinson *et al.*, 1978; Howe and Solter, 1979) stages of mouse preimplantation embryo development. As expected, several polypeptides appear and several others disappear in each developmental stage. Although illustrative, two-dimensional gel electrophoresis has some disadvantages, the major problem, in our opinion, being the impossibility of determining the importance of observed differences since none of the stage-specific polypeptides have been identified and it is not known whether newly observed proteins are the result of gene activation or some kind of posttranscriptional modification. Although the method is very sensitive (detecting protein synthesized at rates as low as 0.05% of the total synthesis), it has limitations. For example, Levinson *et al.* (1978) did not observe any difference in protein patterns of embryos from several different inbred strains although these strains differ in several isozymes that are synthesized during the period examined (Brinster, 1973).

To supplement biochemical methods in the study of developmental genes, immunological approaches which have potential preparative as well as analytical possibilities have repeatedly been used. The remainder of this work will describe several different immunological techniques applied to the study of mouse development and the results obtained.

II. Immunological Analysis of Stage-Specific Antigens

To systematize the results obtained thus far, the antisera used have been divided on the basis of the immunization procedure into xenogeneic, allogeneic, or syngeneic. Tables I–III summarize the antisera described so far and the distribution of the detected activity through different stages of development. All of the antisera are to some extent embryo specific and do not react with the majority of adult cell types. Results dealing with antigens shared with all adult cells (minor histocompatibility antigens) and with the absence in embryos of some membrane constituents that are normally found on all adult cells, such as the major histocompatibility antigens, will be reported only briefly here.

A. ANTIGENS DETECTED ON MOUSE EMBRYOS BY XENOGENEIC ANTISERA

Several antisera produced by xenoimmunization with whole or part of the mouse embryo have been described (Table I). One of the most detailed descriptions is that of the egg antigen (EA). This antigen is detected by antiserum produced by injection of guinea pigs with unfertilized mouse eggs in complete Freund adjuvant (Baranska et al., 1970; Koprowski et al., 1971; Moskalewski and Koprowski, 1972). As detected by complement-mediated cytotoxicity and indirect immunofluorescence, the *unabsorbed* antiserum reacted with all stages of mouse preimplantation development including trophoblasts but not with inner cell masses of the late blastocyst. Postimplantation stages were also nonreactive. The unabsorbed serum reacted with simian virus 40 (SV40)-transformed mouse cells which absorbed antiserum activity when retested on embryos. The serum did not react with other mouse cells, whether normal or transformed with other viruses, and it also did not react with human or monkey cells transformed with SV40 or with rat preimplantation embryos. It would appear that this antiserum detects antigen(s) common to mouse preimplantation embryos and SV40-transformed mouse cell lines. No further identification of the antigen was attempted, and this work was not independently repeated. One puzzling feature (in view of subsequent reports about anti-embryo xenoantisera) is that antibodies against species-specific antigens were not detected.

Glass and Hanson (1974) describe three antisera raised in rabbits against mouse cumulus oocyte masses, morulae, and blastocysts. Embryos with intact zonae pellucidae were injected in complete Freund adjuvant. Although the activity of the sera differed to some extent, in general, unabsorbed sera reacted to various degrees with embryonic cytoplasm, follicular fluid, uterine and oviductal epithelium, and zonae pellucidae when mounted sections of ovaries, oviducts, and uteri were examined by indirect immunofluorescence. Absorption with whole mouse serum removed all reactivity except that with zonae pellucidae. Anti-cummulus oocyte mass and anti-blastocyst serum, even after absorption with mouse serum, caused the immediate death of cultivated mouse embryos. The antisera described apparently detect two antigens: one that is the same or similar to mouse serum antigens and another that is specific for the zona pellucida.

Rabbit anti-mouse blastocyst serum was also produced and described by Wiley and Calarco (1975), although the immunization procedure was somewhat different (see also Wiley, this volume). Blastocysts were freed from zonae pellucidae by mouth pipette and injected together with complete Freund adjuvant. Antiserum was absorbed

TABLE I

REACTIVITY OF XENOANTISERA WITH MOUSE EMBRYOS

Immunogen	Unfertilized mouse eggs[b]	Blastocyst[c]	Placenta[c]	Placenta[d]	Ectoplacental cone[e]	Cerebellum[f]	Sperm[g]	Sperm[g,h]	Teratocarcinoma cells[i]	Teratocarcinoma cells[i,j]
Immunized animal	Guinea pig	Rabbit	Rabbit	Rabbit	Rabbit	Rabbit	Rabbit	Rabbit	Rabbit	Rabbit
Methods of detection[a]	CML, IF	CML, IF	CML, IF	IF	IP	CML	IF	IF	IF	IF
Embryonic stages										
Unfertilized egg	+	−	+	−	−	+	+	−	+	+/−
Zygote	+	+/−	+	+/−	−	+	+	−	+	−
2-cell	+	+/−	+	+	−	+	+	−	+	−
4- to 8-cell	+	+	+	+	+/−	+	+	+	+	−
Morula	+	+/−	+	+	+	+	+	+	−	−
Trophoblast	−	+/−	NT	NT	−	+	+	NT	+	+
Inner cell mass		NT[k]	NT	NT		+	NT	NT	+	+

[a] CML, complement-mediated lysis; IF, indirect immunofluorescence; IP, immunoperoxidase labeling.
[b] Baranska et al., 1970; Koprowski et al., 1971; Moskalewski and Koprowski, 1972.
[c] Wiley and Calarco, 1975.
[d] Kometani et al., 1973.
[e] Searle and Jenkinson, 1978.
[f] Solter and Schachner, 1976.
[g] Menge and Fleming, 1978.
[h] Absorbed with mouse ovaries.
[i] Gooding and Edidin, 1974; Gooding et al., 1976.
[j] Absorbed with Cl 1D (mouse fibroblast cell line).
[k] NT, not tested.

with a crude membrane preparation of mouse kidney, liver, and spleen and tested in indirect immunofluorescence and complement-mediated cytotoxicity. It did not react with unfertilized eggs and had weak activity on zygotes and 2-cell stage embryos; activity increased and peaked at the 8- to 16-cell stage and diminished thereafter. The serum was cytotoxic (in a very low dilution, 1 : 2) for 8-cell stage embryos but not for 2-cell stage embryos. When added to culture medium, dialyzed anti-blastocyst serum (but not normal rabbit serum) completely prevented development of 2-cell stage embryos in concentrations of 4% or higher.

Comparison of the results obtained by Wiley and Calarco (1975) with those of Glass and Hanson (1974) brings to light several problems related to the use of xenogeneic antisera in the study of embryo-specific antigens. The main activity in the sera described by Glass and Hanson is directed against the zona pellucida. It appears that the zona is a very strong immunogen; activity against it is always very strong after xenogeneic immunization (see Solter, 1977) and therefore can mask weaker activity against other embryonic antigens. It is also possible that in the course of immune reaction weak antigens are ignored in the presence of strong ones or that the response of individual animals may differ. Although the immune response to zona-specific antigens might play an important role in fertility control (Yanagimachi et al., 1976), it is hardly relevant to the study of stage-specific embryonic antigens. Therefore, it is advisable to remove zonae from embryos before using them as immunogens. On the other hand, the use of proteolytic enzymes, such as Pronase, to digest zonae might remove just those stage-specific antigens we wish to study. Mechanical removal or use of low pH (Brun and Psychoyos, 1972) can circumvent such problems.

Another drawback to the use of xenogeneic immunization with embryos is the presence of serum or serum-like antigens in embryos (Glass and Hanson, 1974). It is possible, though unlikely, that antiserum activities reported by Wiley and Calarco (1975) are due to the presence of such antigens both in the embryos used as immunogen and those used as target. Syngeneic immunization would obviously obviate such problems (see Section II, C).

Another series of xenoantisera cross-reacting with embryos was produced by immunization with adult cells or cultured cell lines. Several antisera were produced by immunization with placenta (Kometani et al., 1973; Wiley and Calarco, 1975) or ectoplacental cone trophoblast (Searle and Jenkinson, 1978). All these antisera reacted with mouse preimplantation embryos, trophoblasts, and trophoblastic outgrowths to various degrees of specificity. The antiserum described by Wiley and

Calarco (1975) reacted with unfertilized and fertilized eggs, whereas the other two antisera did not, but reactivity was detectable from the 2-cell stage (Kometani et al., 1973) and from the 8-cell stage (Searle and Jenkinson, 1978) onward. The different gestational ages of the placentas used as immunogen and variations in absorption of antisera might explain these differences. It is probable that each of the antisera detects different, or only partially overlapping, sets of antigens. Without further characterization of the antigens in question not much more can be learned from such an approach.

Several organ-specific antisera have been found to cross-react with early mouse embryos. Antiserum produced in rabbits against the cerebellum of 4-day-old mice and defining nervous system antigen 4 (NS-4) (Schachner et al., 1975) cross-react with all preimplantation stage embryos including the trophoblast and inner cell mass (Solter and Schachner, 1976). This activity is specifically absorbed with mouse brain and sperm but not with kidney, liver, and spleen. [Recent results (Goridis et al., 1978) suggest that activity against mouse teratocarcinoma cells can be removed from some NS-4 antisera with kidney absorption, so it is unclear whether NS-4 is expressed by kidney or not.] Similar antiserum raised in rabbits against reaggregates of 6- to 8-day-old mouse cerebellar cells (Seeds, 1975) also reacted with preimplantation mouse embryos (Solter and Schachner, 1976). Our preliminary results also indicate that several other antisera raised by xenogeneic or syngeneic immunization against cerebellar cells or neural tumors also react with mouse embryos in cytotoxicity assays. These include antisera defining NS-5 (Zimmerman and Schachner, 1976), NS-6 (Chaffee and Schachner, 1978a), and NS-7 (Chaffee and Schachner, 1978b). These results indicate that preimplantation mouse embryos express the wide variety of antigens expressed on various neural cells (see chapters by Akeson, Fields, and Schachner, this volume). These antigens are also expressed on sperm and some of them are expressed on kidneys; thus, it was no surprise that xenogeneic immunization of rabbits with mouse sperm also produced antisera reacting with mouse embryos (Menge and Fleming, 1978). However, these antisera are directed against sets of antigens different from those detected by anti-NS sera. Absorption with kidney and brain did not affect activity of anti-sperm sera on sperm or preimplantation embryos. It is interesting that absorption with ovaries removed activity against unfertilized, fertilized, and 2-cell embryos but not activity against later stages.

The existence of brain-specific and sperm-specific antisera that cross-react with early embryos has evoked interesting speculations. It has been suggested (Solter and Schachner, 1976; Solter, 1977) that the mouse embryo, or any embryo for that matter, possesses a complete set of organ-specific antigens though they might be present in minute quantities. These antigens would persist in totipotent cells of the embryo and later be sequestered to germ cells only. As the other cells differentiate into adult tissues they would retain only the appropriate organ-specific antigens and stop expressing others. Expression of organ-specific antigens might be a consequence of differentiation, and it also might play a regulatory role in differentiation. It has also been suggested that sperm can contribute its own antigens (Jacob, 1975; Solter and Schachner, 1976) which would complement antigens present in the oocyte, thus forming a complete set. Sperm antigens would serve as the blueprint for renewed synthesis of the same antigens and represent some sort of extragenomic inheritance carried by the membrane. The contribution of sperm antigens not detected in unfertilized eggs but present on fertilized eggs has been described (O'Rand, 1977) and gives credence to such a hypothesis. Additional circumstantial evidence that the sperm membrane is necessary for successful development can be found in the fact that although mammals are not capable of parthenogenetic development, fertilized eggs from which the male pronucleus is subsequently removed can develop normally (Hoppe and Illmensee, 1978).

Edidin and co-workers (reviewed in Edidin and Gooding 1975; Edidin, 1976) have so far presented the most detailed analysis of antigens detected on mouse embryos with xenogeneic antiserum. Antiserum was raised in rabbits against a teratocarcinoma stem cell line derived from transplantable mouse teratocarcinoma 402AX. Antiserum preabsorbed with lymph node, spleen, thymus, and kidney cells of adult mice detected three different antigens or antigenic classes. Antigen I, which is present on mouse L cells and numerous mouse tumor cell lines (Gooding and Edidin, 1974), is expressed on unfertilized eggs, cleavage-stage embryos, and inner cell masses but not on trophoblasts of hatched blastocysts (Gooding et al., 1976). Antigen II, present on teratoma and hepatoma cells, reacted only with trophoblast cells prior to implantation and no other embryonic stages. Antigen III is teratoma-specific and absent from embryonic cell surfaces. Antigen I is physically associated with H-2 antigens in the plasma membrane of live mouse L cells. It was suggested that in embryonic cells antigen I can serve as an anchoring point for newly synthesized H-2 antigens and, at the same time, can block some of the antigenic specificities of the H-2 molecule, thus preventing its detection by conventional methods. Recent reports indicate that antigen I is either glycolipid or associated with glycolipid. Specific immunoprecipitates can be extracted with chloroform–methanol and resolved into multiple peaks on thin-layer gels (Larraga and Edidin, 1979).

B. ANTIGENS DETECTED ON MOUSE EMBRYOS BY ALLOANTISERA

The alloantisera were mainly used to confirm expression of minor and major histocompatibility antigens and H-Y antigen on mouse embryos and also to determine the onset of synthesis of antigens coded by the paternal genome. Minor histocompatibility antigens (H-3, H-6, and probably others) have been repeatedly detected on preimplantation mouse embryos (Palm et al., 1971; Muggleton-Harris and Johnson, 1976; Searle et al., 1976). The only major controversy is related to the expression of antigens of the H-2 complex. It was surmised that H-2 antigens are not present on cleavage-stage embryos (Muggleton-Harris and Johnson, 1976) but are detected for the first time on trophoblastic cells of the blastocyst (Searle et al., 1976), especially during experimental delay of implantation (Hakansson et al., 1975).

Recent results obtained by immunoprecipitation techniques challenge this view and suggest that H-2 antigens are synthesized in the inner cell mass of the blastocyst and not in the trophoblast (Webb et al., 1977). It is possible that the amount of H-2 antigen on trophoblastic cells is too small to be detected by immunoprecipitation but can be detected by immunolabeling and electron microscopy (Searle et al., 1976). The results (Krco and Goldberg, 1977) suggesting the detection of H-2 antigen on cleavage-stage embryos by complement-mediated cytotoxicity await further confirmation. The possibility that anti-H-2 sera are contaminated with anti-Ia antibodies and antiviral antibodies (Milner et al., 1976) should be kept in mind, and direct cytotoxicity data should be confirmed by specific absorption experiments. The question of whether Ia antigens are expressed on mouse embryos is still open. Our results (Solter and Goetze, unpublished data) and those reported by Hammerling (1976) indicate that Ia antigen cannot be detected by indirect immunofluorescence and complement-mediated cytotoxicity.

C. ANTIGENS DETECTED ON MOUSE EMBRYOS BY SYNGENEIC ANTISERA

Antisera that were raised by syngeneic immunization with cells and embryos and that react with all or some stages of mouse preimplantation development are described in Table II. The so-called F9 antigen has, so far, been studied in greater detail than any other embryonic antigen. This antigen was described by Artzt et al. (1973, 1974) and reviewed in detail by Jacob (1977, 1978) and Gachelin

TABLE II

REACTIVITY OF SYNGENEIC SERA WITH MOUSE EMBRYOS

Immunogen	Terato-carcinoma cells F9[b]	Terato-carcinoma cells SIKR[c]	Embryoid bodies OTT6050[d]	Terato-carcinoma cells PCC4[e]	Mouse–human hybrids SV40-transformed[f,g]	8-Cell stage embryos[h]
Strain of mice immunized	129	129	129	129	BALB/c	BALB/c
Methods of detection[a]	IF, IP, CML	IF	IF	IF	IF, CML, RIA	IF, CML
Embryonic stages						
Unfertilized egg	−	NT[h]	−	−	+	−
Zygote	+	NT	NT	NT	+	+/−
2-cell stage	+	NT	+	NT	+	+
4- to 8-cell stage	+	+	+	−	+	+
Morula	+	+	+	−	−	−
Trophoblast	+	NT	+	−		NT
Inner cell mass	+	NT	+	+	+	

[a] IF, indirect immunofluorescence; IP, immunoperoxidase labeling; CML, complement-mediated lysis; RIA, indirect binding of iodinated antimouse immunoglobulin sera.
[b] reviewed in Jacob, 1977, 1978.
[c] Stern et al., 1975.
[d] Dewey et al., 1977.
[e] Gachelin et al., 1977.
[f] Hybrid between BALB/c mouse peritoneal macrophage and SV40-transformed human cells (Croce et al., 1973).
[g] Solter, Shevinsky, Aden, and Knowles, unpublished data.
[h] NT, not tested.

(1978). F9 antigen is present on preimplantation embryos and on the surface of embryonic cells isolated from 7- and 8-day-old (but not 9-day-old) postimplantation mouse embryos (Buc-Caron *et al.*, 1978). It is also present on mature sperm and on all stages of spermatogenesis. It is not species-specific and can be found on other mammalian embryos, mammalian sperm, and mouse and human teratocarcinoma cells (Holden *et al.*, 1977; Hogan *et al.*, 1977). It is possibly related to the T locus in mice (Artzt *et al.*, 1974), though its genetic control is still unclear (Kemler *et al.*, 1976; Marticorena *et al.*, 1978; Erickson and Lewis, 1979). The role of F9 antigen in maintaining embryonic structure and development has been documented (Kemler *et al.*, 1977). For a more detailed description of F9 antigen see Jacob (this volume). Several other antisera have been raised by immunizing mice with syngeneic teratocarcinoma cells (Stern *et al.*, 1975; Dewey *et al.*, 1977; Gachelin *et al.*, 1977), and though they differ from anti-F9 serum in the range of activities (detected by direct testing and absorptions), all of these polyspecific sera probably overlap to a large extent.

Another type of syngeneic antiserum was recently analyzed in our laboratory (D. Solter, L. Shevinsky, D. P. Aden, and B. B. Knowles, unpublished results). The antiserum was raised against mouse–human hybrids, containing human chromosome 7 with the SV40 genome integrated on it as the only human chromosome (Croce *et al.*, 1973). This serum was used to define the cell surface antigen coded by human chromosome 7 (Aden and Knowles 1976; Knowles *et al.*, 1977; Ford *et al.*, 1978), but it also reacted with early mouse embryos. It was later found that the same type of activity against embryos can be found in the sera of mice immunized with syngeneic SV40-transformed cells. Since activity against mouse embryos cannot be absorbed with human cells, it was concluded that the activity originally observed is related to SV40 transformation. The antiserum detects antigen(s) present on unfertilized mouse eggs, zygotes, and cleavage stages and on inner cell masses but not on trophoblasts, as shown by complement-mediated cytotoxicity (Fig. 1), indirect immunofluorescence (Fig. 2), or indirect binding of iodinated sera against mouse immunoglobulin (Figs. 3, 4). The antibody(s) belong to the IgM class. Absorption analysis indicates that antigen(s) found on early embryos are also expressed on mouse teratocarcinoma cells, the parietal yolk sac cell line PYS-2 (Lehman *et al.*, 1974), and some, but not all, mouse cell lines transformed with SV40 (Figs. 1, 4). Human and monkey cell lines transformed with SV40, unfertilized monkey eggs, and mouse cell lines either normal or transformed spontaneously or with other viruses do not express the antigen in question. In summary, it appears that the antigen is em-

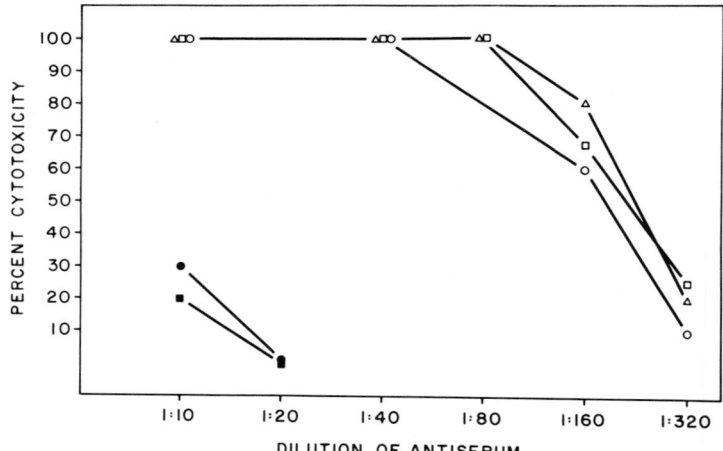

FIG. 1. Cytotoxic activity of syngeneic antiserum raised against mouse–human somatic cell hybrids containing chromosome 7, with the integrated SV40 genome, as the only human chromosome. Antiserum was preabsorbed with mouse spleen cells and then absorbed with F9 mouse teratocarcinoma cells (■——■); SV40-transformed mouse fibroblasts KCSV (●——●); another SV40-transformed mouse fibroblast SSSV (○——○); polyoma-transformed mouse fibroblasts Py3T3 (△——△); and chemically transformed mouse fibroblasts MC57G (□——□). Absorbed sera were tested in a two-step complement-mediated cytotoxicity assay on 8-cell stage mouse embryos.

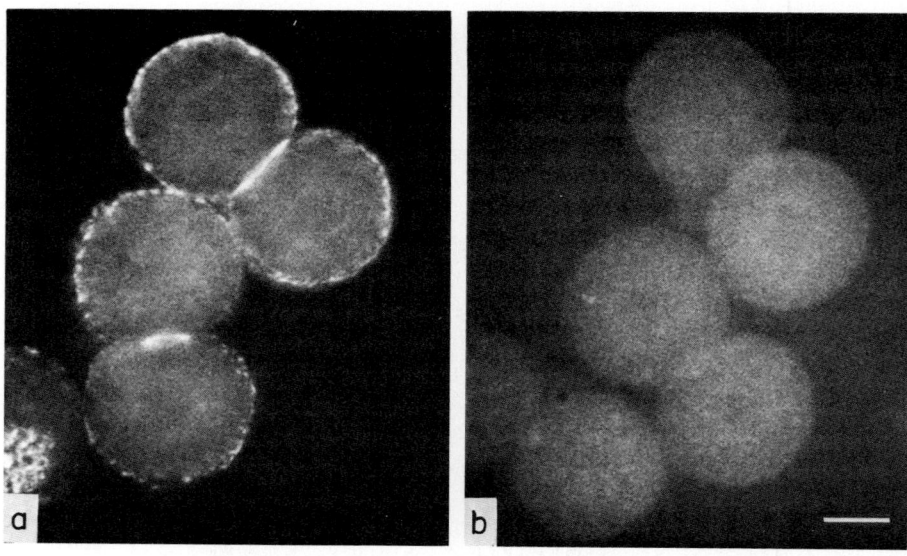

FIG. 2. The same antiserum as described in Fig. 1 was tested in indirect immunofluorescence on 2-cell stage embryos. Second reagent was goat anti-mouse IgM serum (μ-chain specific) labeled with fluorescein isothiocyanate (FITC). Preabsorbed serum was absorbed with MC57G cells (a) or F9 cells (b). Bar represents 20 μm.

7. ANTIGENS DURING MOUSE EMBRYOGENESIS 151

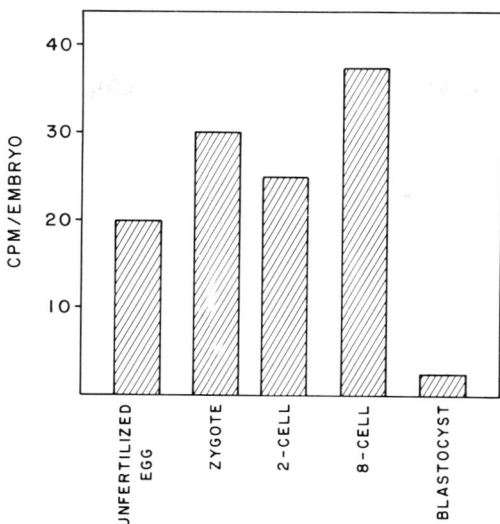

FIG. 3. The same preabsorbed antiserum as described in Fig. 1 tested on all preimplantation stages in indirect antibody-binding radioimmunoassay. The second reagent was goat anti-mouse IgM serum (μ-chain specific) labeled with ^{125}I. The values obtained by using normal mouse serum have been substracted.

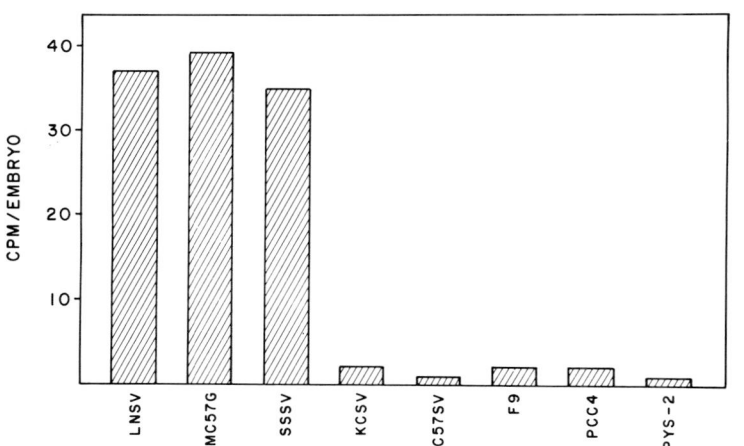

FIG. 4. The same preabsorbed antiserum as described in Fig. 1 tested in radioimmunoassay on 8-cell stage embryos. Antiserum was absorbed with SV40-transformed human cells (LNSV); chemically transformed mouse cells (MC57G); SV40-transformed mouse cells (SSSV, KCSV, C57SV); mouse teratocarcinoma cells (F9, PCC4); and mouse parietal yolk sac cells (PYS-2).

bryonic in nature and can be induced in adult cells as a by-product of transformation with SV40. However, since not all SV40-transformed mouse cells express this antigen, its expression is not a necessary, but rather a possible consequence of SV40 transformation.

All syngeneic antisera so far described have been raised by immunization with cells and then tested on embryos in a search for cross-reacting antigens. Though this approach is certainly valid and produced interesting results, it is obvious that antigens that are really embryo-specific and not expressed on other cell types will be missed by such analysis. We therefore decided to attempt syngeneic immunization with specific embryonic stages. The main problem here is the scarcity of embryonic material that can be used as an immunogen. Nevertheless, our preliminary results indicate that repeated syngeneic immunization with unfertilized eggs and cleavage stage embryos can produce antisera with detectable cytotoxic titers (Fig. 5). When one such antiserum was absorbed with mouse teratocarcinoma cells, its activity decreased but did not disappear, indicating that the serum might be reacting with some antigens shared with teratocarcinoma cells and some specific for embryos only.

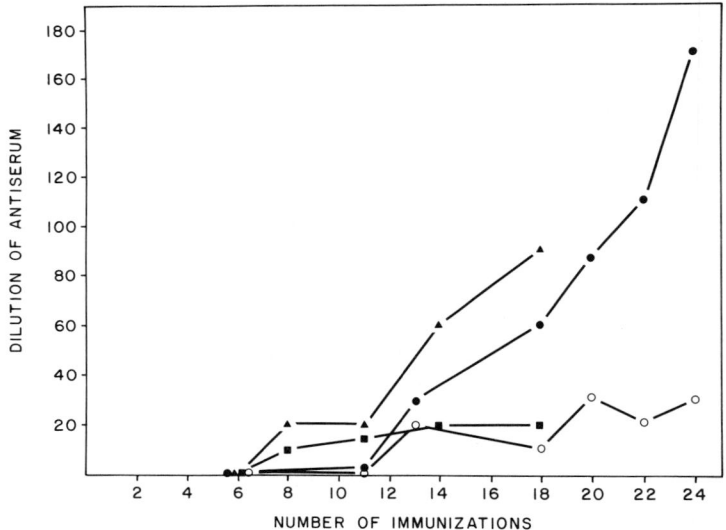

FIG. 5. Changes in cytotoxic titer of sera from four BALB/c mice injected subcutaneously with BALB/c 8-cell stage embryos weekly. For each injection 100–400 embryos were used. The antiserum was tested in a two-step complement-mediated cytotoxicity assay on 8-cell stage embryos and the dilution of antiserum that gave 50% lysis was determined. (▲, ●, ○, ■: each symbol represents serum from an individual mouse).

Production of syngeneic anti-embryo sera gave encouraging preliminary results, although the scarcity of antiserum and its low titer make detailed analysis somewhat difficult. Fortunately, recently developed techniques (Köhler and Milstein, 1975, 1976) that make possible production of monoclonal antibodies of desired specificities in essentially unlimited quantities might prove an elegant way of circumventing our problem with anti-embryo sera.

D. Embryonic Antigens Detected by Monoclonal Antibodies

The use of xenogeneic and even syngeneic sera for defining unknown antigenic determinants is beset with various technical and theoretical pitfalls. Xenogeneic antisera have to be extensively absorbed to remove antibodies to species-specific antigens. Only absorption of presumed anti-embryo serum *in vivo* will remove all the antibodies shared by embryos and adult tissues. However, after such absorption the antiserum is usually greatly diluted, and the decrease in titer is quite considerable. Additionally, most if not all normal mouse sera react with mouse embryos in low dilution so it is possible that antiserum recovered after absorption *in vivo* is contaminated with "naturally" occurring mouse antibodies. Syngeneic sera present fewer problems since we can assume that only antibodies against molecules present on immunizing cells or embryos and absent from adult organisms will be produced. That this is not always the case is suggested by the existence of different organ-specific antigens (Zimmermann and Schachner, 1976) produced by syngeneic immunization. The presence of such antigens necessitates explanations like breakdown of tolerance or existence of immunologically privileged sites. In addition, all xenogeneic and syngeneic sera are probably polyspecific. Their analyses have to be completed by a complex series of absorptions and eventually by immunoprecipitation which is difficult, time consuming, and rarely done.

It is therefore not surprising that a new immunological technique, namely hybridization of antibody-producing lymphocytes with permanent plasmacytoma cell lines resulting in hybrids producing monoclonal antibodies (Köhler and Milstein, 1975, 1976; Galfre *et al.*, 1977), found immediate appreciation in analysis of stage-specific embryonic antigens (Knowles *et al.*, 1978; Stern *et al.*, 1978; Willison and Stern, 1978; Solter and Knowles, 1978).

Monoclonal antibodies that react with mouse preimplantation embryos and teratocarcinoma stem cells have been produced in a variety of ways, and a few so far available will be described in some detail. Monoclonal antibody against stage-specific embryonic antigen-1

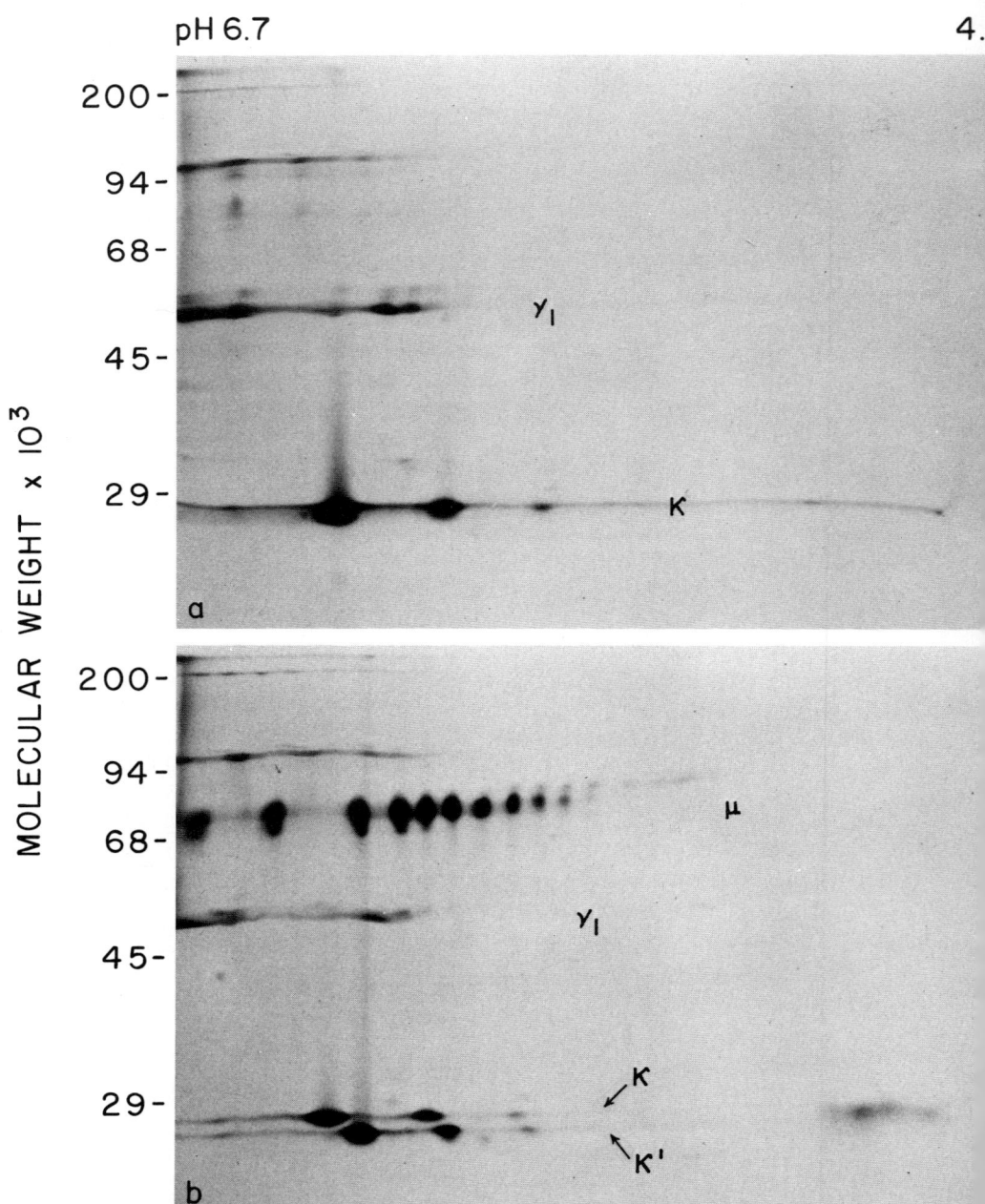

TABLE III

REACTIVITY OF MONOCLONAL ANTIBODIES WITH MOUSE EMBRYOS

Immunogen	Teratocarcinoma cells F9[b]	Mouse spleen cells[c]
Immunized animal	BALB/c mice	Rat
Method of detection[a]	CML, IF, RIA	IF
Embryonic stages		
Unfertilized egg	−	−
Zygote	−	−
2-cell stage	−	−
4- to 8-cell stage	+/−	−
Morula	+	+/−
Trophoblast	+/−	+
Late trophoblast	−	−
Inner cell mass	+	+

[a] IF, indirect immunofluorescence; CML, complement-mediated lysis; RIA, indirect binding of iodinated anti-immunoglobulin sera.
[b] Knowles et al., 1978; Solter and Knowles, 1978.
[c] Stern et al., 1978; Willison and Stern, 1978.

(SSEA-1) has been produced (Knowles et al., 1978; Solter and Knowles, 1978) by immunizing BALB/c mouse with the nullipotent embryonic carcinoma cell line F9 (Bernstine et al., 1973). This antibody (anti-SSEA-1) belongs to the IgM class (Fig. 6) and reacts with all mouse teratocarcinoma cell lines regardless of the strain of mice from which the cells were isolated as shown by complement-mediated cytotoxicity, indirect immunofluorescence, and indirect antibody-binding radioimmunoassay (RIA). It reacts with human teratocarcinoma lines (Solter and Knowles, 1978, and unpublished observations) but does not react with any other mouse or human cell line tested. It also reacts with mouse and human sperm and possibly with mouse brain and kidney, but not with liver or spleen. Anti-SSEA-1 reacts with preimplantation mouse embryos in a way not previously observed with conventional antisera. It does not react with unfertilized eggs and cleavage stages until the 8-cell stage (Table III). Using RIA, maximal binding of anti-

FIG. 6. Supernatants from cultures of the mouse myeloma line P3-X63 Ag8 (a) and of the cloned hybrid between P3-X63 Ag8 cells and spleen cells from a mouse immunized with F9 teratocarcinoma cells (b) were separated in two-dimensional gels. Cells were labeled with [^{35}S]methionine for 3 hours before supernatants were harvested. The P3-X63 Ag8 line secretes γ- and κ-chains, whereas the hybrid secretes these two chains as well as μ-chain and a slightly different κ'-chain.

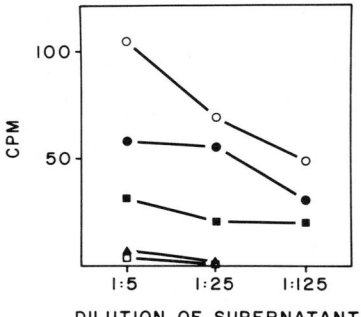

FIG. 7. Supernatant from the culture of the hybridoma producing anti-SSEA-1 monoclonal antibody (see Fig. 6b) was tested in indirect antibody-binding radioimmunoassay. The second reagent was goat antimouse IgM serum (μ-chain specific) labeled with ^{125}I. Triplicates of 10 embryos per group were used for each dilution and each developmental stage. (□——□), 2-cell stage; (▲——▲), 4-cell stage; (■——■), 8-cell stage; (○——○), morula; (●——●), early blastocyst.

SSEA-1 is detected at the morula stage and decreases on trophoblast cells but remains high on inner cell masses (Fig. 7). Using immunofluorescence and cell-mediated cytotoxicity, we can detect unequal distribution of SSEA-1 in embryos of the same developmental stage and even in blastomeres of the same embryo (Fig. 8). Although the majority of 8-cell stage embryos are negative, some are completely positive and some have a few positive blastomeres (Fig. 8a,b). Morulae are predominantly positive, but completely negative morulae can be found and negative blastomeres can be found in otherwise positive embryos (Fig. 8c,d). Trophoblastic cells are mostly negative although patches of positive cells can be observed (Fig. 8c). So far, examination of a large number of embryos of the same developmental stage has not detected any consistent pattern in the distribution of positive and negative blastomeres. Inner cell masses isolated by immunosurgery (Solter and Knowles, 1975) are predominantly positive with some negative cells. When inner cell masses are grown *in vitro* and prevented from attaching, an outside layer of entodermal cells is formed

---

FIG. 8. Indirect immunofluorescence using anti-SSEA-1 serum (from a hybridoma-bearing mouse) diluted 1:500 and goat antimouse IgM serum (μ-chain specific) labeled with FITC diluted 1:10. (a) 8-cell stage, two positive blastomeres. (b) 8-cell stage, all but one blastomere positive, focus on the upper surface of the embryo. (c) Completely negative morula. (d) Partially positive morula from the same sample as (c). (e) Blastocyst, few trophoblastic cells positive. (f) Inner cell mass grown *in vitro* for 3 days, patches of positive and negative entodermal cells, focus on the upper surface of the embryo. Bar represents 20 μm.

7. ANTIGENS DURING MOUSE EMBRYOGENESIS

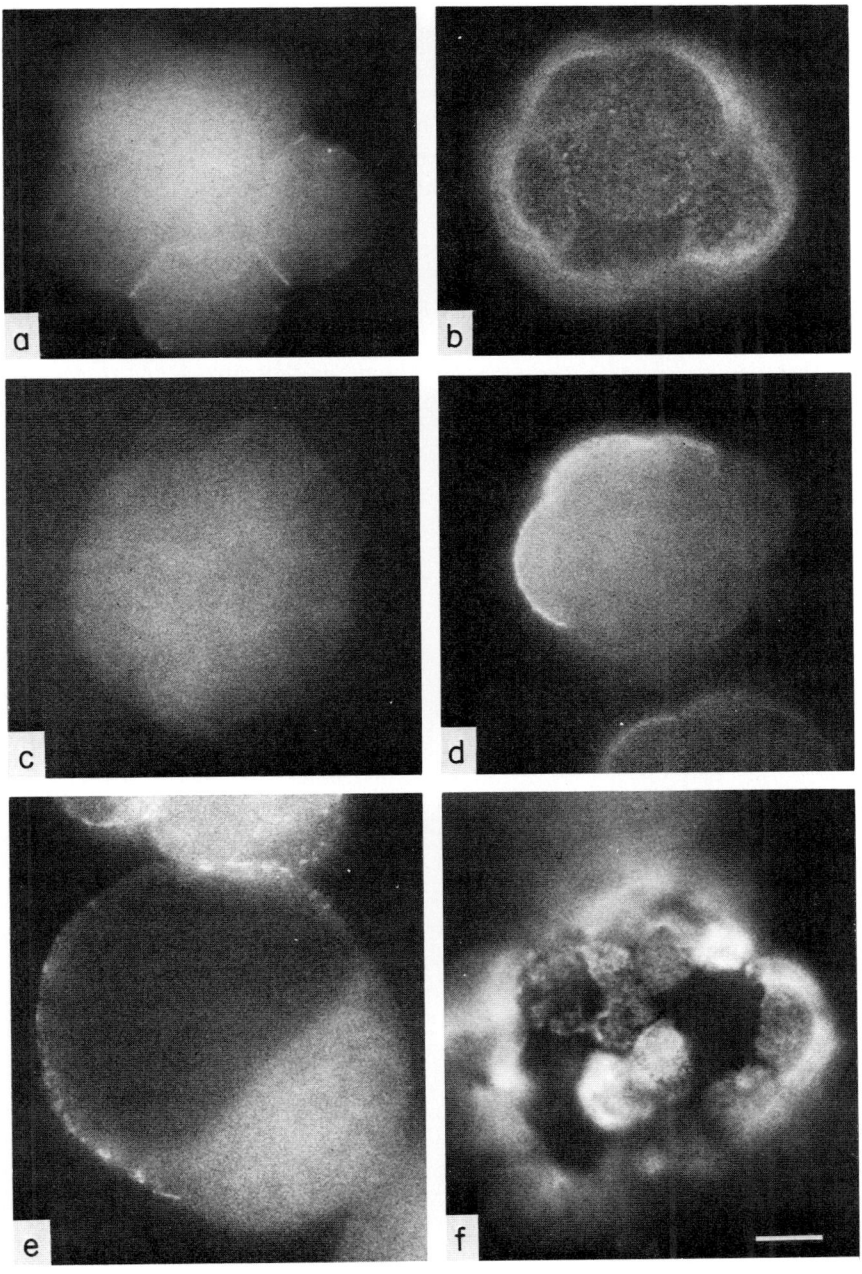

(Solter and Knowles, 1975; Hogan and Tilly, 1978). These entodermal cells are, again, partly positive and partly negative (Fig. 8b), without distinguishable pattern. The entodermal layer can be removed by repeated immunosurgery (Strickland et al., 1976), and the resulting ectodermal cells are predominantly positive, although the presence of cellular debris makes analysis of these structures difficult.

In summary, SSEA-1 is present on all embryonal carcinoma cells tested and disappears as teratocarcinoma stem cells differentiate *in vitro* (Solter et al., 1979). The reason for its inconsistent distribution on blastomeres of preimplantation embryos is not clear and cannot easily be explained by unequal expression during the cell cycle or related to any hypothesis of cell lineage in early mouse embryos. Our preliminary results (B. B. Knowles, P. W. Andrews, and D. Solter, unpublished observations) indicate that SSEA-1 is a glycolipid since its specific immunoprecipitate runs in front of tracking dye in sodium dodecyl sufate (SDS)-polyacrylamide gels and SSEA-1 can be extracted by the chloroform–methanol procedure. Radiolabeled anti-SSEA-1 and whole body external scintigraphy can be used for *in vivo* localization of teratocarcinoma (Ballou et al., 1979).

Stern et al. (1978) described a monoclonal antibody produced by hybrids between a mouse myeloma cell line and spleen cells from rats immunized with mouse spleen. This antibody reacts with a small subpopulation of cells in the mouse spleen, bone marrow, lymph node, brain, kidney, and testis but not in liver and thymus. The antibody also reacts very strongly with mouse teratocarcinoma cell lines and sheep red blood cells. The distribution of antigen suggests that this monoclonal antibody reacts with an antigeneic determinant of the Forssman antigen. This suggestion is corroborated by the facts that the antibody can be absorbed by autoclaved guinea pig kidney and that its binding to sheep red blood cells is inhibited by rabbit anti-Forssman serum. This antibody belongs to the IgM class, and the antigen is probably a small glycopeptide or glycolipid (Stern et al., 1978).

Forssman-like antigen was also detected with the same monoclonal antibody on preimplantation mouse embryos by indirect immunofluorescence (Willison and Stern, 1978). This antigen is first detectable on morulae and is expressed in trophoblastic cells of early and expanded blastocysts (Table III). The antigen decreases and disappears from trophoblastic cells of hatched blastocysts but persists on cells of inner cell masses. Some blastocysts were completely positive, whereas in others only a few trophoblastic cells were stained. Freshly isolated inner cell masses and those grown *in vitro* for a few days were uniformly positive.

Though the SSEA-1 antigen described by us and the Forssman antigen described by Stern et al. (1978) are somewhat similar in distribution, several lines of evidence indicate that they are not the same. SSEA-1 is not expressed on mouse spleen and lymph node cells and is not present on sheep red blood cells. In addition, its time of appearance in and distribution on mouse embryos is consistently different from those of the Forssman antigenic specificity.

So far, only a few studies have used monoclonal antibodies but the results are promising and, considering the problems in dealing with conventional antisera, monoclonal antibodies are definitely a step forward. It is, however, possible that use of monoclonal antibodies will bring its own problems, and some of them can be deduced from the results described in the previous pages. It appears that all antibodies described so far as reacting with preimplantation mouse embryos belong to the IgM class, and the same was also true for conventional antibodies. It is possible that this is the only class of antibodies that react with embryonic antigens or that the methods of detection were geared to detect antibodies of the IgM class more easily than those of other classes. In addition to the hybrid-producing antibody against SSEA-1, we screened hybrids from several other fusions between mouse myeloma cells and spleen cells from mice immunized with embryo, teratocarcinoma, or SV40-transformed cells (Solter and Knowles, unpublished). Several hybrids were detected whose supernatants reacted with teratocarcinoma cells but since they also reacted with the majority of other mouse cell lines tested, they were considered nonspecific and of no interest in a study of stage-specific antigens. However, all of these monoclonal antibodies also belonged to the IgM class.

The fact that all monoclonal antibodies reacting more or less specifically with teratocarcinoma cells belong to the IgM class is probably not accidental and raises some interesting possibilities. It is conceivable that teratocarcinoma stem cells and embryos possess Fc receptors specific for IgM molecules, and this notion is currently under intensive investigation in our laboratory. Two IgM produced by the plasmacytoma lines MOPC 104 E and TEPC 183 do indeed bind to both ECC and embryos. However, when the binding of IgM molecules produced by MOPC 104 E and TEPC 183 is compared with the binding of anti-SSEA-1 on a strictly quantitative basis, the latter binds 100-fold more efficiently (D. Solter, B. B. Knowles, and D. P. Aden, unpublished results). This seems to indicate that in the case of anti-SSEA-1 the binding was the consequence of antigen recognition, but does not exclude the possibility that other IgM molecules can bind through Fc receptors. These findings suggest that it is necessary for us to determine the quantity of IgM antibodies and to compare presumably positive monoclonal antibodies with equal amounts of unrelated IgM as a control in order to exclude "positive" binding due to the large amount of IgM. It has recently been reported that in nonimmunized mice exist many lymphocytes producing autoantibodies of IgM class and that the number of such cells can be increased by nonspecific means (Steele and Cunningham, 1978; Dresser, 1978). This would explain low levels of reactivity of normal or unrelated immune sera with mouse embryos. It also raises the somewhat disturbing possibility (if the same situation also exists in rats) that the monoclonal antibodies described here are not really related to the immunization but are products of already existing IgM-secreting cells that were rescued by fusion. Our preliminary results (W. Gerhard, F. Melchers, B. B. Knowles, and D. Solter) indicate that numerous monoclonal antibodies of IgM class secreted by hybridoma which were produced by fusion of lipopolysaccharide (LPS)-stimulated fetal lymphocytes and P3-X63-Ag8 myeloma reacted with teratocarcinoma cells. Some of those antibodies, but not all, also reacted with other mouse cell lines. Although all these reservations do not deny the enormous potential use of monoclonal antibodies in the study of embryonic antigens, they certainly suggest that headlong enthusiasm should be somewhat tempered.

III. Conclusions and Perspectives

A description of the results so far obtained using immunological methods to search for stage-specific embryonic antigens leaves one with the feeling that not much has been accomplished. Despite the existence of a score or more antisera that are more or less specific for some stages in embryonic development, no stage-specific embryonic antigen (SSEA) has been even preliminarily characterized, with the possible exception of the F9 antigen (Jacob, 1978). However, all this work has certainly not been in vain. On the basis of partial results and recently developed techniques, we can, with reasonable confidence, chart future experiments that should lead to the identification and isolation of SSEA. Several experimental approaches can be suggested, each reflecting the particular bias and beliefs of its author. Our approach is based on the assumption that identification and assignment of function to antigens that are present for a finite period in development is the most promising way of researching genetic control of differentiation. To accomplish this, several theoretically simple, but practically difficult, experimental sequences have been formulated.

Syngeneic immunization with embryos of a given stage should produce an antibody response specific for antigens expressed in embryos but not in adult animals. As stated before, the small quantity of immunogen available and the problem of the best immunization route and schedule are some of the pitfalls of this approach. However, our successful production of several anti-embryo sera indicate that this approach is feasible. The amount of antisera produced by immunization with embryos will always be small, but hybridization of myeloma cells with spleen cells from immunized mice and isolation of hybrids secreting monoclonal antibodies against embryonic antigens should solve that problem.

Before embarking on the difficult process of isolation of SSEA, it would be wise to somehow screen the embryonic antigens as to their importance. The idea that embryonic antigens, especially those on surface membranes, play an important role in embryonic organization is not a new one, but experimental evidence has been scarce. Recently, however, there have been several reports, dealing with widely different experimental systems, that indicate that a specific cell surface molecule directs aggregation and subsequent organization in developing systems. Antibodies against such molecules prevent aggregation of *Dictyostelium* cells (Müller and Gerisch, 1978; Gerisch, this series, Vol. 14), aggregation of neural retina cells of chick embryos (Rutishauser *et al.*, 1978), organization of mouse and rat testes (Ohno *et al.*, 1978; Zenzes *et al.*, 1978), and compaction and blastulation of mouse embryos (Kemler *et al.*, 1977). Prevention of compaction of the mouse morula was accomplished by the use of Fab fragments prepared from rabbit anti-F9 serum but not by the whole antibody molecule. Indeed, the action of the Fab fragments was competitively inhibited by simultaneous exposure to intact IgG molecules from anti-F9 sera. No inhibition was seen when embryos were simultaneously exposed to rabbit antibody recognizing other cell surface molecules expressed on the F9 cell and Fab fragments from anti-F9 sera (Jacob, 1978). Therefore, it appears that for successful compaction to occur, specific cell surface molecules must come into contact with each other, either directly or through the action of divalent antibody molecules; it is this contact that is inhibited when these molecules are covered with Fab fragments. The described results clearly indicate that it is possible to detect the function of antigens that in some way affect development and that the use of whole antibody and Fab fragments should be contemplated.

The next step, probably essential if we are to find why and how stage-specific embryonic antigens direct development, is their isolation. It is unlikely that such isolation and purification could be accomplished using embryos as the starting material; thus we have two alternatives. If the antigen is also present on teratocarcinoma stem cells (as has always been the case so far), isolation will be a straightforward, though time-consuming, procedure. If the antigen is unique to embryos, we can attempt to isolate it by using the rapidly developing technology of genetic engineering and gene amplification. This approach is still in the planning stage and will work only in cases in which the antigenic determinant is a primary gene product residing within the protein molecule. If the antigenic determinant is the result of some specific posttranscriptional modification that occurs only in intact embryonic cells, its isolation would pose a considerable problem. Knowledge of the chemical structure of the antigenic determinant and/or the molecule with which it is associated should give us insight into the molecular control of the cell–cell interactions that are necessary to trigger normal embryonic development.

Establishment of a firm intercellular contact is the first prerequisite for successful development, at least in the mouse embryo, and it is not surprising that the first antibodies described as affecting mammalian development act by interfering with compaction. The nature of the relationship between the F9 antigen and molecules of the junctional complexes that appear at the time of compaction and effectively seal mouse blastocysts (Ducibella et al., 1975) is open to speculation. One can surmise that antibodies interfering with intercellular communication on a more complex level will also be found and that these antibodies will act even on properly compacted embryos. The binding of such antibodies to the antigen and the consequence of this binding in terms of affecting development might be temporarily close or widely separated. The search for molecules that direct development and the elucidation of their action are areas of developmental biology where immunological approaches are the most appropriate.

IV. Concluding Remarks

The important role of cell surface molecules in regulating developmental processes has been assumed but experimental evidence has been scarce. Immunological methods offer, perhaps, the best approach to the study of such molecules and in this chapter we have summarized the results obtained thus far and offer some suggestions for future experimentation. Several xenogeneic, allogeneic, and syngeneic antisera, with the common property that they all react with some or all stages of early mouse development, have been described. These antisera cross-react most notably with teratocarcinoma stem cells and sperm and with various other cell types as well. Probably all antisera described so far are polyspecific and their range of specificities partially overlap. None of the described antisera react with embryos alone, with the possible exception of some raised by syngeneic immunization with embryos. The search for stage-specific embryonic antigens (SSEA) necessitates the production of monospecific antisera; here the newly developed techniques of antibody cloning can be used to considerable advantage in embryonic systems, as shown in several recent reports. Advances in membrane biology, a better understanding of basic questions in developmental biology, and new immunological techniques bode well for future analysis of stage-specific embryonic antigens and their role in development.

ACKNOWLEDGMENTS

Our studies reviewed in this chapter were supported by USPHS Research grants CA-18470, CA-17546, CA-10815, and CA-21069 from the National Cancer Institute; HD-12487 from the NICHHD; PDT-26 and IM-88 from the American Cancer Society; PCM-78-16177 from The National Science Foundation; I-301 from The National Foundation–March of Dimes; and by NIAID Research Career Award AI-00053 to B.B.K.

The excellent technical assistance of Lynne Shevinsky and John Barnes is gratefully acknowledged. We thank Robert Erickson and Karen Artzt for preprints of their work.

REFERENCES

Aden, D. P., and Knowles, B. B. (1976). *Immunogenetics* 3, 209–221.
Artzt, K., Dubois, P., Bennett, D., Condamine, H., Babinet, C., and Jacob, F. (1973). *Proc. Natl. Acad. Sci. U.S.A.* 70, 2988–2992.
Artzt, K., Bennett, D., and Jacob, F. (1974). *Proc. Natl. Acad. Sci. U.S.A.* 71, 811–814.
Ballou, B., Levine, G., Hakala, T. R., and Solter, D. (1979). *Science* (in press).
Baranska, W., Koldovsky, P., and Koprowski, H. (1970). *Proc. Natl. Acad. Sci. U.S.A.* 67, 193–199.
Bernstine, E. G., Hooper, M. L., Grandchamp, S., and Ephrussi, B. (1973). *Proc. Natl. Acad. Sci. U.S.A.* 70, 3899–3902.
Brinster, R. L. (1973). *Biochem. Genet.* 9, 187–191.
Brun, J. L., and Psychoyos, A. (1972). *J. Reprod. Fertil.* 30, 489–491.
Buc-Caron, M. H., Condamine, H., and Jacob, F. (1978). *J. Embryol. Exp. Morph.* 47, 149–160.
Caplan, A. I., and Ordahl, C. P. (1978). *Science* 201, 120–130.
Chaffee, J. K., and Schachner, M. (1978a). *Dev. Biol.* 62, 173–184.
Chaffee, J. K., and Schachner, M. (1978b). *Dev. Biol.* 62, 185–192.
Croce, C. M., Girardi, A. J., and Koprowski, H. (1973). *Proc. Natl. Acad. Sci. U.S.A.* 70, 3617–3620.
Dewey, M. J., Gearhart, J. D., and Mintz, B. (1977). *Dev. Biol.* 55, 359–374.
Dewey, M. J., Filler, R., and Mintz, B. (1978). *Dev. Biol.* 65, 171–182.
Dresser, D. W. (1978). *Nature (London)* 274, 480–482.
Ducibella, T., Albertini, D. F., Anderson, E., and Biggers, J. D. (1975). *Dev. Biol.* 45, 231–250.
Edidin, M. (1976). *In* "Embryogenesis in Mammals" (Ciba Foundation Symposium 40), pp. 177–197. Elsevier, Amsterdam.

Edidin, M., and Gooding, L. R. (1975). *In* "Teratomas and Differentiation" (M. I. Sherman and D. Solter, eds), pp. 109–121. Academic Press, New York.
Erickson, R. P., and Lewis, S. E. (1979). *Nature (London)* (in press).
Ford, S. R., Aden, D. P., Mausner, R., Trinchieri, G., and Knowles, B. B. (1978). *Immunogenetics* **6,** 293–300.
Gachelin, G. (1978). *Biochim. Biophys. Acta* **516,** 27–60.
Gachelin, G., Kemler, R., Kelly, F., and Jacob, F. (1977). *Dev. Biol.* **57,** 199–209.
Galfre, G., Howe, S. C., Milstein, C., Butcher, G. W., and Howard, J. C. (1977). *Nature (London)* **266,** 550–552.
Glass, L. E., and Hanson, J. E. (1974). *Fertil. Steril.* **25,** 484–493.
Gooding, L. R., and Edidin, M. (1974). *J. Exp. Med.* **140,** 61–78.
Gooding, L. R., Hsu, Y.-C., and Edidin, M. (1976). *Dev. Biol.* **49,** 479–486.
Goridis, C., Artzt, K., Wortham, K. A., and Schachner, M. (1978). *Dev. Biol.* **65,** 238–243.
Hakansson, S., Heyer, S., Sundquist, K.-G., and Bergstrom, S. (1975). *Int. J. Fertil.* **20,** 137–140.
Hammerling, G. J. (1976). *Transplant. Rev.* **30,** 64–82.
Hogan, B., and Tilly, R. (1978). *J. Embryol. Exp. Morphol.* **45,** 93–105.
Hogan, B., Fellous, M., Avner, P., and Jacob, F. (1977). *Nature (London)* **270,** 515–518.
Holden, S., Bernard, O., Artzt, K., Whitmore, W. F. Jr., and Bennett, D. (1977). *Nature (London)* **270,** 518–520.
Hoppe, P. C., and Illmensee, K. (1978). *Proc. Natl. Acad. Sci. U.S.A.* **74,** 5657–5661.
Howe, C. C., and Solter, D. (1979). *J. Embryol. Exp. Morph.* (in press).
Jacob, F. (1975). *In* "The Early Development of Mammals" (M. Balls and A. E. Wild, eds.), pp. 233–241. Cambridge Univ. Press, London and New York.
Jacob, F. (1977). *Immunol. Rev.* **33,** 3–32.
Jacob, F. (1978). *Proc. R. Soc. B* **201,** 249–270.
Kemler, R., Babinet, C., Condamine, H., Gachelin, G., Guenet, J. L., and Jacob, F. (1976). *Proc. Natl. Acad. Sci U.S.A.* **73,** 4080–4084.
Kemler, R., Babinet, C., Eisen, H., and Jacob, F. (1977). *Proc. Natl. Acad. Sci. U.S.A.* **74,** 4449–4452.
Knowles, B. B., Solter, D., Trinchieri, G., Maloney, K., Ford, S. R., and Aden, D. P. (1977). *J. Exp. Med.* **145,** 314–326.
Knowles, B. B., Aden, D. P., and Solter, D. (1978). *Curr. Top. Microbiol. Immunol.* **81,** 51–53.
Köhler, G., and Milstein, C. (1975). *Nature (London)* **256,** 495–497.
Köhler, G., and Milstein, C. (1976). *Eur. J. Immunol.* **6,** 511–519.
Kometani, K., Paine, P., Cossman, J., and Behrman, S. J. (1973). *Am. J. Obstet. Gynecol.* **116,** 351–357.
Koprowski, H., Sawicki, W., and Koldovsky, P. (1971). *J. Natl. Cancer Inst.* **46,** 1317–1323.
Krco, C. J., and Goldberg, E. M. (1977). *Transplant Proc.* **9,** 1367–1370.
Larraga, V., and Edidin, M. (1979). *Proc. Natl. Acad. Sci. U.S.A.* **76,** 2912–2916.
Lehman, J. M., Speers, W. C., Swartzendruber, D. E., and Pierce, G. B. (1974). *J. Cell. Physiol.* **84,** 13–28.
Levey, I. L., Stall, G. B., and Brinster, R. L. (1978). *Dev. Biol.* **64,** 140–148.
Levinson, J., Goodfellow, P., Vadeboncoeur, M., and McDevitt, H. (1978). *Proc. Natl. Acad. Sci. U.S.A.* **75,** 3332–3336.
Marticorena, P., Artzt, K., and Bennett, D. (1978). *Immunogenetics* **7,** 337–347.
Menge, A. C., and Fleming, C. H. (1978). *Dev. Biol.* **63,** 111–117.

Milner, R. J., Henning, R., and Edelman, G. M. (1976). *Eur. J. Immunol.* **6**, 603–607.
Moskalewski, S., and Koprowski, H. (1972). *Nature (London)* **237**, 167–168.
Muggleton-Harris, A. L., and Johnson, M. H. (1976). *J. Embryol. Exp. Morphol.* **35**, 59–72.
Müller, K., and Gerisch, G. (1978). *Nature (London)* **274**, 445–449.
O'Farrell, P. H. (1975). *J. Biol. Chem.* **250**, 4007–4021.
Ohno, S., Nagai, Y., and Ciccarese, S. (1978). *Cytogenet. Cell Genet.* **20**, 351–364.
O'Rand, M. G. (1977). *J. Exp. Zool.* **202**, 267–273.
Palm, J., Heyner, S., and Brinster, R. L. (1971). *J. Exp. Med.* **133**, 1282–1293.
Rutishauser, H., Blackenbury, R., Thiery, J.-P., and Edelman, G. M. (1978). *In* "The Molecular Basis of Cell-Cell Interaction" (R. A. Lerner and D. Bergsma, eds.), pp. 305–316. Liss, New York.
Schachner, M., Wortham, K. A., Carter, L. D., and Chaffee, J. K. (1975). *Dev. Biol.* **44**, 313–325.
Searle, R. F., and Jenkinson, E. J. (1978). *J. Embryol. Exp. Morphol.* **43**, 147–156.
Searle, R. F., Sellers, M. H., Elson, J., Jenkinson, E. J., and Billington, W. D. (1976). *J. Exp. Med.* **143**, 348–359.
Seeds, N. W. (1975). *Proc. Natl. Acad. Sci. U.S.A.* **72**, 4110–4114.
Solter, D. (1977). *In* "Immunobiology of Gametes" (M. Edidin and M. H. Johnson, eds.), pp. 207–226. Cambridge Univ. Press, London and New York.
Solter, D., and Knowles, B. B. (1975). *Proc. Natl. Acad. Sci. U.S.A.* **72**, 5099–5102.
Solter, D., and Knowles, B. B. (1978). *Proc. Natl. Acad. Sci. U.S.A.* **75**, 5565–5569.
Solter, D., and Schachner, M. (1976). *Dev. Biol.* **52**, 98–104.
Solter, D., Shevinsky, L., Knowles, B. B., and Strickland, S. (1979). *Dev. Biol.* **70**, 515–521.
Steele, D., and Cunningham, A. J. (1978). *Nature (London)* **274**, 483–484.
Stern, P. L., Martin, G. R., and Evans, M. J. (1975). *Cell* **6**, 455–465.
Stern, P. L., Willison, K. R., Lennox, E., Galfre, J., Milstein, C., Secher, D., Ziegler, A., and Springer, T. (1978). *Cell* **14**, 775–783.
Strickland, S., Reich, E., and Sherman, M. I. (1976). *Cell* **9**, 231–240.
Van Blerkom, J., Barton, S. C., and Johnson, M. H. (1976). *Nature (London)* **259**, 319–321.
Webb, C. G., Gall, W. E., and Edelman, G. M. (1977). *J. Exp. Med.* **146**, 923–932.
Wiley, L. M., and Calarco, P. G. (1975). *Dev. Biol.* **47**, 407–418.
Willison, K. R., and Stern, P. L. (1978). *Cell* **14**, 785–793.
Yanagimachi, R., Winkelhake, J., and Nicolson, G. L. (1976). *Proc. Natl. Acad. Sci. U.S.A.* **73**, 2405–2408.
Zenzes, M. T., Wolf, U., Gunther, E., and Engel, W. (1978). *Cytogenet. Cell Genet.* **20**, 365–372.
Zimmerman, A., and Schachner, M. (1976). *Brain Res.* **115**, 297–310.

CHAPTER 8

EARLY EMBRYONIC CELL SURFACE ANTIGENS AS DEVELOPMENTAL PROBES

Lynn M. Wiley

DEPARTMENT OF ANATOMY
UNIVERSITY OF VIRGINIA
CHARLOTTESVILLE, VIRGINIA

I. Introduction	167
II. Cell Surface Antigens of Gametes and Preimplantation Embryos	169
A. Antisera to Histocompatibility Antigens	169
B. Antisera to Teratocarcinoma Cells	172
C. Antisera to Gametes and Preimplantation Embryos	175
D. Antisera to Placentas or to Other Parts of the Postimplantation Embryo	177
E. Antisera to Human Chorionic Gonadotropin (hCG)	182
F. Antisera to Mouse L Cells and to LETS Protein	185
III. Future Studies on Embryonic Cell Surface Antigens as Developmental Probes	187
IV. Concluding Remarks	192
References	193

I. Introduction

One currently popular idea on developmental control is that events are timed and coordinated by reciprocal relationships between cell surface properties and intracellular processes (Edelman, 1976; Nicolson, 1976). Such relationships could, for example, provide a way for sequential embryonic gene activity to be responsive to and coordinated with sequential modifications of extracellular conditions (Brunner, 1977). In this way, morphogenetic processes and cytodifferentiation could be timed and coordinated appropriately to produce an embryo consisting of various cell types arranged in specific ways.

During preimplantation development the mouse embryo undergoes four consecutive overlapping morphogenetic processes: cleavage, compaction, peripheral tight junction formation, and cavitation. Provided that the embryo also attains a minimum cell number, (1) differentiation of the trophectoderm and the inner cell mass (ICM) and (2) formation of the blastocele then take place, resulting in a blastocyst. The first three morphogenetic processes transform the zygote into a collection of approximately 8 pluripotent blastomeres; these increase to 16 or more cells to become a morula consisting of an outer cell layer of presumptive trophectoderm that encloses a presumptive ICM. It appears highly probable that only trophectoderm will form if cleavage has resulted in a morula with insufficient cell numbers to enclose one or more blastomeres (Tarkowski and Wroblewska, 1967). Furthermore, for those enclosed blastomeres to form ICM, compaction together with peripheral tight junction formation (Ducibella and Anderson, 1975; Ducibella *et al.*, 1975) and retaining an "inside" position in the morula (Tarkowski, 1959; Hillman *et al.*, 1972) all appear essential for ICM formation and for suppression of trophectoderm differentiation. It seems logical, therefore, to propose that trophectoderm/ICM cytodifferentiation is cued by information that is relayed by cell surface properties sensitive to cell contact, relative cell position, and/or cellular microenvironment (Mintz, 1965).

The morula-to-blastocyst transformation occurs by the process of cavitation and appears to involve the secretion of intracellular vesicles (Melissinos, 1907; Calarco and Brown, 1969; Wiley and Eglitis, 1978). To be effective in blastocele formation, vesicle deployment—and subsequent discharge—is polarized, tending to occur at those cell surfaces that are within the peripheral tight junctions and face the interior of the embryo. Assuming that secretion of these vesicles is necessary for cavitation, one might ask what is the mechanism(s) by which vesicle deployment is (1) polarized toward those cell surfaces that are within the peripheral tight junctions and (2) timed to follow cleavage, compaction, and peripheral junction formation. As is the case for trophectoderm/ICM cytodifferentiation, it seems logical to propose that vesicle deployment and the timing of cavitation are cued by extracellular information acting through the cell surface.

The cell surface properties of preimplantation embryos are only now being characterized and none have yet been associated causally with developmental control. Immunological methods have been used extensively in these studies, the aim being to identify cell surface antigens with potential as developmental probes. The intention of this chapter, therefore, is to review the progress made in producing antisera suitable for examining cell surface properties of mouse preimplantation embryos in relation to the control of morphogenesis and cytodifferentiation. The equally fascinating topic of maternal–fetal immunological relationships has been discussed in detail elsewhere and will not be discussed here (see Simmons and Russell, 1962, 1966; Kirby *et al.*, 1964, 1966; Kirby, 1968; Hellström *et al.*, 1969; James, 1969; Fauve *et al.*, 1974; Voisin and Chaouat, 1974; Bagshawe and Lawler, 1975; Billington, 1975; Johnson, 1975).

II. Cell Surface Antigens of Gametes and Preimplantation Embryos

The behavior of embryonic cells in certain instances more closely resembles that of tumor cells than that of normal adult cells, especially with regard to cell surface properties or to behavior attributed to cell surface function. Consequently, one of the collective aims of previous studies has been to determine how the expression of cell surface antigens by embryonic cells compares with that by normal adult and tumor cells. The major findings of these studies can be summarized as follows. First, gametes and preimplantation embryos express antigens that are not detectable on adult cells, and vice versa. Second, some "embryonic" antigens are also found on certain tumor cells. Third, some of the embryo-specific antigens are only transiently expressed during a specific stage of development; these antigens are called phase-specific (Coggin and Anderson, 1974) or stage-specific embryonic antigens. Fourth, and most critical for our purposes, antisera to certain stage-specific embryonic antigens can cause reversible, stage-specific arrest of the development of cultured preimplantation mouse embryos.

Most of the antisera that will be discussed here can be grouped according to their immunogens: (1) cells carrying specific histocompatibility antigens, (2) teratocarcinoma cells, (3) gametes and preimplantation embryos, (4) placentas or other parts of the postimplantation conceptus, and (5) hormones such as human chorionic gonadotropin (hCG). In addition, antisera to LETS protein and to mouse L cells will be discussed. Attention will be limited to those antisera of which the specificities were tested on preimplantation embryos.

A. Antisera to Histocompatibility Antigens

Products of the major histocompatibility complex (MHC) in the adult are known largely for their association with tissue graft rejection and with the immune lysis of foreign cells by specific antibodies. Embryos were originally tested for the presence of H-2 antigens because of their ability to survive *in utero* even though they are technically allografts (see references listed at the end of Section I). An interest in postcoital fertility control provided another reason for studying the cell surface antigens expressed on gametes (Voisin et al., 1975; Glass and Hanson, 1974) and on preimplantation embryos (Glass and Hanson, 1974). More recently, additional incentive has been provided by the working hypothesis that the MHC complex codes for cell surface properties that control essential cell–cell interactions during embryo development (Edidin, 1976a,b; Ohno, 1977). As a result considerable work has been done on examining the expression of H-2 antigens (and other histocompatibility antigens as well) during mouse development with the aim of determining (1) when H-2 antigens are first expressed and (2) if the expression of H-2 antigens is restricted to certain cell types (i.e., ICM versus trophoblast).

There have been various reports of H-2 antigens on mouse sperm as detected by immunofluorescence (Voitiskova et al., 1969; Erickson, 1972) and by cytotoxicity (Goldberg et al., 1970; Johnson and Edidin, 1972). However, the H-2 gene complex consists of several closely linked regions coding for classical H-2 products and for additional ones. All of these gene products seem to be associated with cell surface properties, some being concerned with various immunological responses, others being concerned with metabolic responses (see Edidin, 1976a,b). As a result, "complex" antisera to H-2 antigens can contain antibodies to classical H-2 antigens (H-2D, H-2K) and to some of the other H-2 complex products such as Ia antigens (Edidin, 1976a,b). More recent reports suggest that many antisera to H-2 antigens are, in fact, detecting Ia antigens when tested on sperm (Hammerling et al., 1975; Fellous et al., 1976). However, further testing of sperm with antisera specific for H-2 antigens (H-2D, H-2K) suggest that sperm express both Ia and H-2 antigens (Lemonnier et al., 1975; Erickson, 1972).

In contrast to the case with sperm, most studies have failed to detect products of the H-2 complex on mouse eggs. Early studies, which were done with complex H-2 antisera, suggested that mouse eggs expressed H-2 antigens (Olds, 1968). However, in later studies in which narrowly defined antisera were used no H-2 antigens were detected on mouse eggs or on preimplantation embryos. Indeed, preimplantation embryos developed into normal blastocysts when cultured in dilutions of H-2 antisera plus complement (Heyner et al., 1969; Palm et al., 1971). After implantation, H-2 antigens appeared to be limited to ICM derivatives (Heyner, 1973; Palm et al., 1971; Gardner et al., 1973; Muggleton-Harris and Johnson, 1976). Failure of neuraminidase to "unmask" H-2 antigens on trophectoderm (Simmons, 1971) and the ability to detect H-2 antigen synthesis by the ICM but not by the trophectoderm (Webb et al., 1977) made it appear as though the blastocyst trophectoderm failed to express H-2 antigens.

Just when there appeared to be a consensus on trophectoderm lacking (serologically detectable) H-2 antigens, an electron microscope study using immunoperoxidase labeling techniques revealed low levels of (apparent) H-2 antigen expression on trophectoderm of the blastocyst (Searle et al., 1976). This revelation was attributed to the greater sensitivity of the techniques and cannot be ascribed to non-H-2 antigens detected by complex antisera because specific antisera to certain H-2 antigens were used. Such low levels of H-2 antigen expression may be related to the reported increased (or initial?) expression of H-2 antigens on mouse blastocysts that is induced by experimental delay of implantation, only to disappear after the onset of blastocyst activation to implant (Håkansson et al., 1975; Håkansson and Sundqvist, 1975). In addition to the study by Searle et al., (1976), one other laboratory recently reported detecting H-2 antigens on preimplantation embryos beginning with the 8-cell stage (Krco and Goldberg, 1977), further confusing the issue of when and by which cells are H-2 antigens first expressed.

However, some non-H-2 histocompatibility antigens (H-3, H-6) have been unequivocally detected on all stages of preimplantation development (Palm et al., 1969, 1971). With embryos from F_1 crosses of inbred strains differing only in non-H-2 loci, maternal alloantigens were detectable throughout preimplantation development whereas paternal alloantigens first became evident at the 6- to 8-cell stage (Muggleton-Harris and Johnson, 1976). In contrast to the case reported for H-2 antigens (excepting Krco and Goldberg, 1977), non-H-2 antigens detected on these F_1 embryos were present on both ICM and trophectoderm, with the expression of non-H-2 antigens (on trophoblast outgrowths) being somewhat strain-specific (Sellens, 1977). These findings are in agreement with those from a study of the effects of H-2 and non-H-2 alloantigens on blastocyst survival in ectopic sites. The non-H-2 mismatches—but not the H-2 mismatches—caused blastocyst rejection (Patthey and Edidin, 1973; Searle et al., 1974).

Two other histocompatibility-associated antigens have been found on preimplantation embryos. The first one is the H-Y (male) antigen, which is serologically detectable on sperm (Goldberg, 1971) and on 50% of preimplantation mouse embryos beginning with the 6- to 8-cell stage (Krco and Goldberg, 1976). The second one is β_2-microglobulin (β_2-M), which is a small protein that is physically associated with several (and possibly all) MHC-coded cell surface glycoproteins including H-2D and H-2K (Rask et al., 1974; Silver and Hood, 1974). Such β_2-M/MHC antigen complexes have been proposed as being ubiquitous plasma membrane anchorage sites of organogenesis-directing antigens (Ohno, 1977). However, β_2-M is expressed on blastocysts on which H-2 antigens are undetectable (Håkansson and Peterson, 1976). β_2-M expression on blastocysts should be looked at more carefully because it could lead, for example, to the identity of embryonic β_2-M/MHC organogenesis-directing antigens (H-2 antigen precursor? see following discussion) that direct blastocyst morphogenesis.

Do any of these histocompatibility antigens have potential as developmental probes in experiments on blastocyst morphogenesis? With two exceptions (Olds, 1968; Krco and Goldberg, 1977), the data indicate that normal preimplantation embryos do not appear to express significant levels of serologically accessible H-2 antigens except during experimental delay. Thus, adult-type H-2 antigens per se may not hold much promise as developmental probes during preimplantation development. Because some non-H-2 antigens (H-3, H-6, H-Y, β_2-M, others?) are serologically accessible on preimplantation embryos, they merit further examination for potential use as probes.

B. ANTISERA TO TERATOCARCINOMA CELLS

The stem cells of teratocarcinomas exhibit an apparent developmental equivalence to normal pluripotent embryo cells. When injected into mouse blastocysts that are subsequently transferred to foster mothers, these cells can colonize the resultant offspring and differentiate into a variety of cell types (Brinster, 1974, 1976), including fertile gametes (Mintz and Illmensee, 1975). Several teratocarcinoma-derived cell lines have thus been used as alternative models for mouse embryogenesis (see Martin et al., 1977; Damjanov and Solter, 1974; Martin, 1975; Graham, 1977). Several antisera have been generated against these cells to investigate their antigenicity in comparison with that of normal embryo and adult cells and with other tumor cells (Edidin et al., 1971; Artzt et al., 1973; Stern et al., 1975; Dewey et al., 1977; Banka and Calarco, 1978; see also Edidin, 1976a,b; Graham, 1977). For the most part, however, the following discussion will be limited to two of these antisera that have been most intensively studied (anti-402AX teratoma, Edidin et al., 1971; anti-F9, Artzt et al., 1973).

The first such antiserum was prepared by injecting rabbits with an ascites subline (402AX; Stevens, 1958) of strain 129 teratoma, which forms typical embryoid bodies (core of stem cells covered by endoderm-like cells). When this antiserum was rendered "teratoma-specific" by absorption with normal mouse tissues, it remained positive (immunofluorescence) for the teratoma and cross-reacted with antigens on unfertilized mouse eggs. These antigens became restricted to the ICM after implantation and further restricted to the embryonic endoderm (Edidin et al., 1971). These teratoma antigens were also found on SV40-transformed mouse 3T3 fibroblasts and on several mouse tumor cell lines, but were absent from normal mouse cells (Gooding and Edidin, 1974), including brain and probably sperm (Edidin et al., 1971). Subsequent absorption studies showed that this antiserum was detecting at least three different antigens. One of these (antigen III) was specific for the teratoma (Gooding and Edidin, 1974), while the other two (antigens I and II) were responsible for the antiserum's reaction with mouse eggs and embryos (Gooding et al., 1976). Antigen I, which was found to be first expressed on unfertilized eggs, became restricted to the ICM in blastocysts prior to implantation. Antigen II, which was found to be first expressed on the mural trophectoderm of the blastocyst, spread over to include the polar trophectoderm and later also became restricted to the ICM of postimplantation embryos.

In the laboratory of François Jacob a syngeneic antiserum to mouse teratocarcinoma (Artzt et al., 1973; Jacob, 1977) was prepared against primitive (nullipotent F9) cells derived from tumor OTT6050 (Stevens, 1970). The resultant anti-F9 serum was reported negative for unfertilized eggs and to become increasingly positive from the 1-cell stage to the morula stage. Both the ICM and the trophectoderm were labeled by anti-F9 serum. After implantation anti-F9 labeling became restricted to the embryonic ectoderm, finally disappearing altogether by day 10 of development (Jacob, 1977), when H-2 antigens are definitely detectable on ICM derivatives (see Section II, A). The activity of anti-F9 serum could not be ascribed to products of the (adult) H-2 locus because F9 cells lack H-2 antigens (Artzt and Jacob, 1974; Forman and Vitetta, 1975). In contrast to the case with the antiserum to 402AX teratoma, anti-F9 serum did not react with a variety of transformed mouse cells, including SV40-transformed mouse 3T3 fibroblasts (Artzt et al., 1973).

The results obtained with both the anti-402AX teratoma serum and the anti-F9 serum have led to the conclusion that teratoma antigens are not generally coexpressed with H-2 antigens and disappear with development as the H-2 antigens become detectable. This conclusion has in turn led to the intriguing hypothesis that teratoma antigens are the embryonic precursors to adult-type H-2 antigens (Edidin et al., 1971; Edidin, 1976a,b; Bennett, 1975; Jacob, 1977). Subsequently, it was found that (1) under certain conditions teratoma antigens, when coexpressed with H-2s, will co-cap with the H-2 antigens on a transformed subline of the C3H L-cell fibroblast, clone ID (Gooding and Edidin, 1974) and (2) H-2 antigens isolated from splenocytes and F9 antigens isolated from F9 cells have identical molecular weights and subunit structure, including the presence of a subunit that resembles β_2-M (Vitetta et al., 1975) but which differs from the β_2-M present on adult cells (Dubois et al., 1976). These observations suggest that teratoma antigens are structurally similar to H-2 antigens, and, when coexpressed, that these two antigens are physically related.

In addition to mouse early embryos and teratocarcinoma stem cells, the F9 antigen or cross-reacting species is/are present on human teratomas (Hogan *et al.*, 1977; Holden *et al.*, 1977; Ostrand-Rosenberg *et al.*, 1977), mouse sperm (Artzt *et al.*, 1973, 1974), and even on human sperm (Buc-Caron *et al.*, 1974; Gachelin *et al.*, 1976). Absorption studies have since found that F9 antigen may be a product of (or associated with?) the T complex in the mouse; mutations in this region are associated with the death of homozygous embryos at different stages of development (see Sherman and Wudl, 1977). Consequently, there has been speculation on whether cell surface antigens coded for by the T complex and the MHC complex are developmentally related to the F9 antigen. One working hypothesis is that F9 antigen is a cell surface protein specified by the wild type allele of t^{12} (recessive mutation in the T complex; Artzt and Bennett, 1975) and that the T complex in general codes for embryonic precursors to adult-type H-2 antigens, all of which are concerned with cell–cell interactions (Bennett *et al.*, 1972).

However, there are several problems with this hypothesis. First, studies with a rabbit antiserum to nullipotent F9-41 cells suggested that the F9 genetic determinant was not the "wild-type" allele of a particular t^x gene (Kemler *et al.*, 1976). Second, a more recent study with a syngeneic antiserum to OTT6050 teratoma-derived embryoid bodies (same source of cells that can colonize blastocysts and differentiate into sperm; Mintz and Illmensee, 1975) failed to produce evidence supporting the hypothesis that the wild-type allele of t^{12} coded for surface components required for preimplantation development (Dewey *et al.*, 1977). Third, large genetic variability in H-2 antigens have no effect on cell-cell recognition events as measured by an *in vitro* quantitative aggregation assay (McClay and Gooding, 1978). Finally, it is not clear whether the T complex codes for cell surface properties and whether t^x homozygotes die because of (stage-specific) cell surface defects or because of general additive metabolic defects (see Sherman and Wudl, 1977).

Putting aside, for the moment, the validity of this hypothesis, is there any information on the effects of anti-teratoma sera on preimplantation development? Some intriguing results in this regard have been obtained with an antiserum to F9 antigen (Kemler *et al.*, 1977) prepared by injecting rabbits with nullipotent EC cells (F9-41; Bernstine *et al.*, 1973). The Fab fragments from this antiserum had no effect on cell cleavage of 2-cell mouse embryos but prevented compaction and reversed compaction already underway. Treated embryos that were rinsed free of Fab and transferred to control culture medium, subsequently compacted, formed blastocysts that, when transferred to foster mothers, developed into normal offspring. Divalent anti-F9 antibodies, however, had no obvious effect on preimplantation development *in vitro* (Kemler *et al.*, 1977). One other anti-teratoma serum has also been investigated for effects on preimplantation development, this one also being a rabbit antiserum, which was prepared using the clonally derived nondifferentiating N_1 cell line (Martin and Evans, 1975) as the immunogen (Calarco and Banka, 1979). Antigens detected by this antiserum were expressed by all stages of preimplantation development with a marked increase on the ICM and the trophectoderm of blastocysts, by the female germ line but not by the male germ line. The development of cultured 2-cell embryos was progressively inhibited by this antiserum in a concentration-dependent manner.

C. ANTISERA TO GAMETES AND PREIMPLANTATION EMBRYOS

There are several studies using sperm or sperm components to make antibodies to sperm, mostly with the aim of controlling fertility. In one such study (Menge and Fleming, 1977) serum IgG and colostrum IgA to mouse sperm were produced in rabbits. After absorption with adult mouse tissues including brain, both types of antibodies reacted with sperm (immunofluorescence). In addition, the anti-sperm IgG reacted with unfertilized mouse eggs, zygotes, and all preimplantation stages including blastocysts. The anti-sperm IgA, however, reacted only with later cleavage stages beginning with the 4- to 8-cell stage. Further absorption of the anti-sperm IgG with ovarian tissue removed its reactivity with unfertilized eggs, zygotes, and 2-cell embryos so that its specificities resembled those of the anti-sperm IgA. These results led to the conclusion that these antibodies were detecting two cell surface antigens expressed by mouse preimplantation embryos, one already expressed on unfertilized eggs and the other first being expressed after fertilization (Menge and Fleming, 1977).

In contrast to the case with sperm, there are very few reported attempts of using isolated mouse eggs as immunogens, due mainly to the difficulty of obtaining large quantities of mouse eggs. In the first such report an anti-egg serum was produced by injecting female guinea pigs with mouse unfertilized eggs that had been treated with hyaluronidase (to remove follicle cells) and with Pronase (to remove the zona pellucida). This antiserum was cytotoxic to Pronased unfertilized eggs and to (non-Pronased) SV40-transformed fibroblasts from mice but not from other species (Baranska *et al.*, 1970; Koprowski *et al.*, 1971). It was also cytotoxic to mouse preimplantation embryos, the ICM being somewhat less susceptible to lysis than the trophectoderm. No portion of the 6- to 7-day-old mouse egg cylinder or adult mouse blood cells, cumulus oophorus cells, or sperm were lysed by this antiserum (Moskalewski and Koprowski, 1972). The problem with interpreting results obtained with this antiserum is the use of Pronase on eggs used for immunization together with the failure to test the antiserum on Pronased target cells. Although tests showed that the anti-egg serum could lyse eggs with the zonae pellucidae intact, lysis was greatly reduced and might have been nonspecific since the antiserum was not absorbed in any way (Moskalewski and Koprowski, 1972). Pronase can profoundly alter cell surface properties (Wallach, 1972) and the antigenicity of mouse eggs and preimplantation embryos in particular (Wiley, 1975; Wiley and Calarco, 1975). However, it is necessary to remove the zonae from material used as immunogens because of the high immunogenicity of mouse zonae (Glass and Hanson, 1974; Tsunoda, 1977) would probably eliminate any reasonable chance of eliciting an immune response from weaker embryonic cell surface antigens due to immunological competition. Nevertheless, this particular antiserum to mouse unfertilized eggs has some intriguing specificities and certainly merits additional attention.

In an effort to overcome the potential problems inflicted by Pronase, we attempted to generate an anti-egg serum by injecting male rabbits with mouse unfertilized eggs from which the zonae were removed by mouth pipet (Wiley, 1975). Repeated immunizations with a total of 3000 eggs per rabbit failed to produce an antiserum that reacted with mouse unfertilized eggs after absorption with adult mouse tissues. This observation raised the question of whether the cell surface of the unfertilized egg expressed egg-specific antigens that were normally immunogenic.

There is only one reported success at obtaining an antiserum using mouse preimplantation embryos as immunogen. The antiserum was produced by injecting male rabbits with blastocysts freed of their zonae by mouth pipet (Wiley, 1975; Wiley and Calarco, 1975). After this antiserum was absorbed with adult mouse tissues (liver, kidney and spleen), it reacted weakly (immunofluorescence) with unfertilized mouse eggs and zygotes. The reaction increased with each cleavage to peak at the 8- to 12-cell stage, then decreased with successive stages, being absent on all portions of blastocyst outgrowths (Fig. 1). This antiserum lysed preimplantation embryos in a manner that paralleled the immunofluorescence results (Table I). Anti-blastocyst serum did not react with brain or sperm (Wiley, unpublished observations), or with mouse ovary cells (L. V. Johnson, 1978), or with mouse-transformed cells so far tested, including some teratocarcinoma cell lines (Wiley, unpublished observations). Thus, insofar as its specificities are concerned, anti-blastocyst serum differs from all other antisera so far described. However, most pertinent to our purpose is that this antiserum

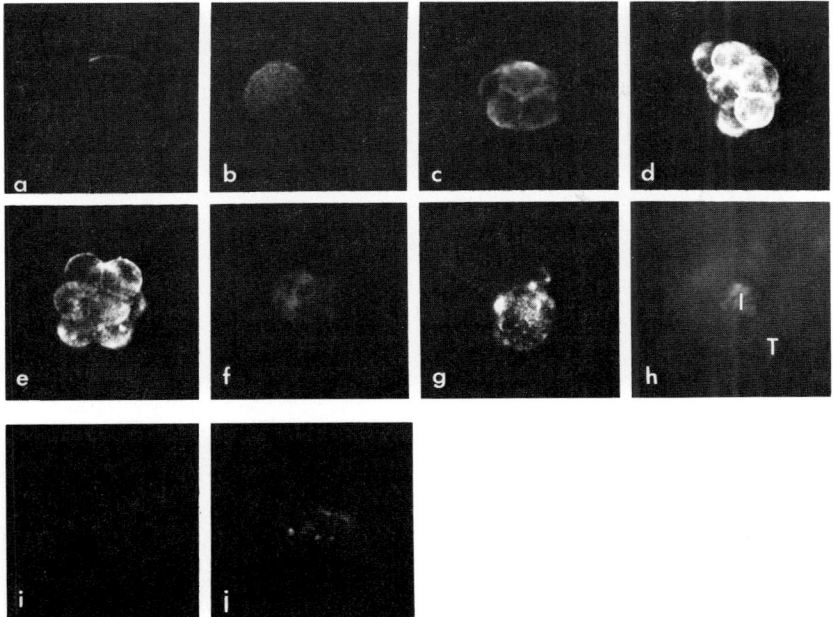

FIG. 1. Indirect immunofluorescence. Anti-blastocyst serum localization on zona-free mouse preimplantation embryos. What little membrane fluorescence is seen on 1-cell (a) and 2-cell (b) embryos increases on the 4-cell (c) and peaks on the 8-cell (d) to 12-cell stage (e). Thereafter membrane fluorescence diminishes on the morula (f) and on the blastocyst (g), which in this instance has transported fluorescein conjugate into the blastocele and has a "filled," fluorescent (therefore dead) blastomere. No fluorescence is seen on the trophoblast (T) or on the ICM (I) of a blastocyst outgrowth (h). Normal rabbit serum control embryos; 2-cell (i) and 8-cell (j) stages. Magnification: a–g, i, j, ×365; h, ×219. From Wiley and Calarco, 1975.

caused a concentration-dependent inhibition of preimplantation development *in vitro* when cultures were begun at the 2-cell stage (Fig. 2; Wiley, 1975; Wiley and Calarco, 1975).

D. Antisera to Placentas or to Other Parts of the Postimplantation Embryo

There are two different antisera to mouse placentas that have been tested on preimplantation embryos. Both were produced in rabbits. The first (anti-placenta-A) was produced with a view toward fertility control (Kometani and Bherman, 1971; Kometani *et al.*, 1973). After absorption with mouse liver, kidney, and spleen anti-placenta-A serum reacted with placenta and early embryos starting with the 2-cell stage,

TABLE I
Cytotoxicity of Antisera to Preimplantation Embryos[a]

Embryonic stage	A-PL serum No./total (%)	A-BL serum No./total (%)	Normal rabbit serum No./total (%)
2-Cell embryos Lysed or blue embryos per total embryos	23/25 (92%) 6/18[b] (33%)	0/23 (0%)	0/21 (0%)
8-Cell embryos Lysed or blue blastomeres per total blastomeres	27/80[b] (34%)	60/80 (75%)	5/40 (12.5%)

[a] Viability was judged by the ability of cells to exclude Trypan Blue.
[b] A different lot of antiserum was used in these assays. Reprinted from Wiley and Calarco, 1975.

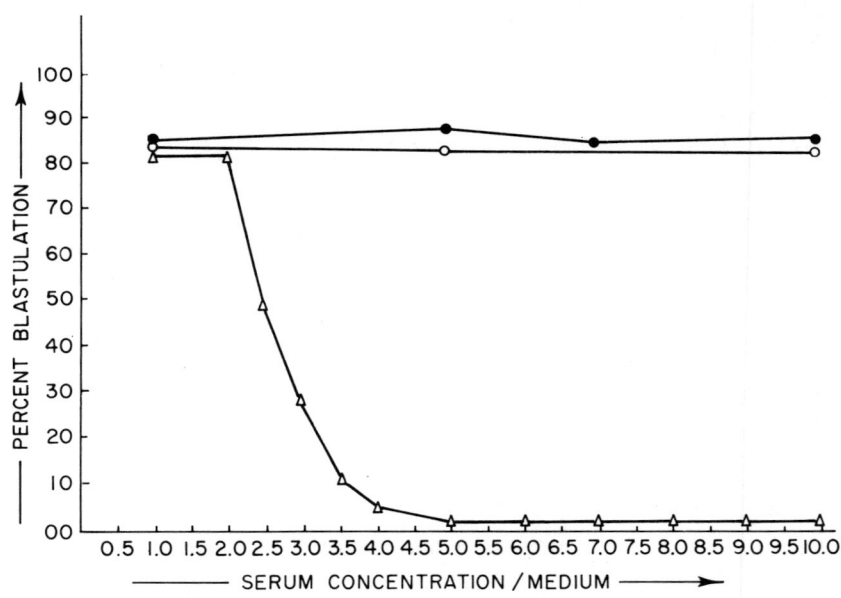

FIG. 2. *In vitro* development of mouse preimplantation embryos in immune antisera. All cultures were started at the 2-cell stage. Each point represents the response of a minimum of 100 embryos. (Δ), antiblastocyst serum; (O), antiplacenta serum; (●), normal rabbit serum. From Wiley and Calarco, 1975.

unfertilized eggs and zygotes being negative (immunofluorescence). Labeling persisted to the blastocyst stage and occurred on both ICM and trophoblastic portions of early postimplantation embryos. When injected into pregnant mice anti-placenta-A serum caused embryo loss by the fourth day of gestation, before implantation.

The second rabbit anti-placenta serum (anti-placenta-B) was produced specifically for use in studies on preimplantation development (Wiley, 1975; Wiley and Calarco, 1975). Although the same protocol was used to produce both antisera, anti-placenta-B serum had different specificities than did anti-placenta-A serum. After absorption with adult mouse liver, kidney, and spleen anti-placenta-B serum reacted with (immunofluorescence) and lysed (cytotoxicity) unfertilized mouse eggs and all preimplantation stages of development (Fig. 3; Table I). The positive reaction with unfertilized eggs was startling in light of the

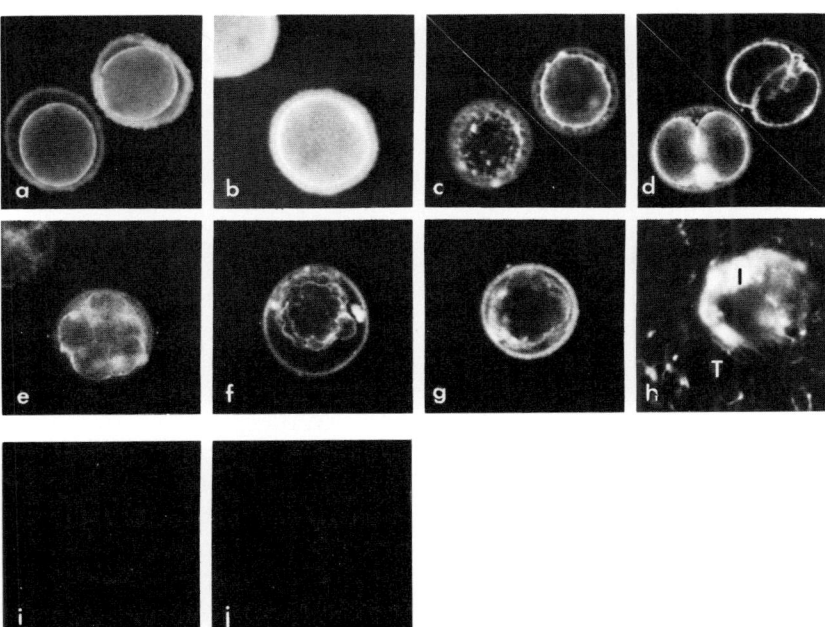

FIG. 3. Indirect immunofluorescence. Anti-placenta serum localization on mouse preimplantation embryos. The intensity of membrane fluorescence is constant on (a) ovarian oocytes, (b) ovulated ova, and on all of the preimplantation stages [(c) zygote; (d), 2-cell; (e), 8-cell; (f), morula; and (g), blastocyst]. Membrane fluorescence is also present on the trophoblast (T) and on the ICM (I) of blastocyst outgrowths (h). Normal rabbit serum control embryos: blastocyst (i), and 2-cell embryo (j). Magnification: a–g,i,j, ×365; h, ×219. From Wiley and Calarco (1975).

negative reaction with unfertilized eggs reported for anti-placenta-A serum (Kometani et al., 1973) and the failure to obtain an antiserum to unfertilized mouse eggs injected into rabbits (Wiley, 1975). Both the ICM and the trophoblast of early blastocyst outgrowths (Wiley and Calarco, 1975), and the yolk sac endoderm but not the embryonic ectoderm of cultured isolated ICMs (Wiley, unpublished observations), labeled with anti-placenta-B serum. When tested on sections of mouse placenta, labeling was confined to the trophoblastic components, whereas the fetal capillaries and maternal blood cells were negative. No labeling was observed on any adult mouse tissues including brain, testes, vas deferens, and sperm therein (Wiley and Calarco, unpublished observations).

Interestingly, anti-placenta-B serum cross-reacted with some transformed mouse cell lines, raising the possibility that some of the antigens present on the trophoblast and the transformed cells were associated with the invasive properties characteristic of trophoblastic and malignant cells (Calarco and Wiley, unpublished observations).

Besides these specificities the pattern of the membrane fluorescence produced by anti-placenta-B serum changed with the first cleavage (Fig. 4; Wiley, 1975; Wiley and Calarco, 1975). With a 30-minute incubation at 37°C, anti-placenta-B serum produced a patchy membrane fluorescence on zygotes. Under identical assay conditions this fluorescence was smooth on 2-cell and older preimplantation stages. If the zygotes were prefixed with paraformaldehyde, subsequent immunolabeling with anti-placenta-B serum was smooth, resembling that observed with unfixed 2-cell embryos (Wiley, 1975). This observation suggested that anti-placenta antigens become fixed when cleavage begins (i.e., less mobile and consequently less susceptible to cross-linking into patches by divalent antibody).

Despite these interesting reactions with preimplantation embryos, anti-placenta-B serum had no effect on mouse preimplantation development *in vitro* (Fig. 2; Wiley and Calarco, 1975). However, preliminary results suggested that the antiserum impaired trophoblast outgrowth (Wiley, unpublished observations) when blastocysts were cultured for 3 days in dilutions of the antiserum in the appropriate culture medium (Spindle and Pedersen 1973). The culturing data together with the immunofluorescence observations led to the speculation that these antigens were concerned with trophoblast differentiation, function, and/or maintenance (Wiley and Calarco, 1975).

Another interesting antiserum that may be detecting "trophoblast"

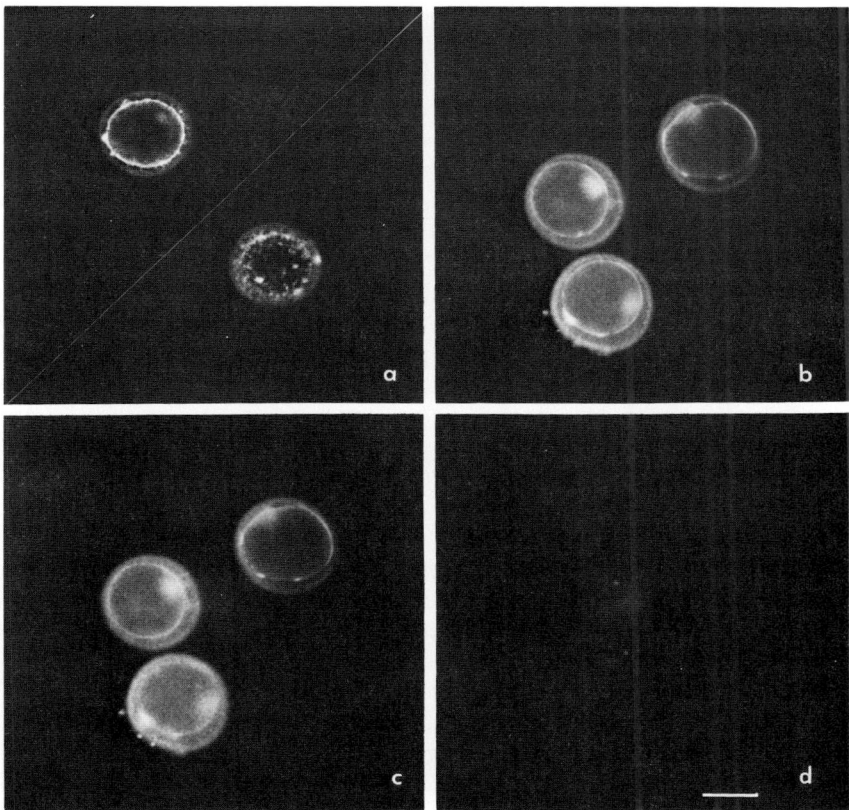

FIG. 4. Indirect immunofluorescence. Anti-placenta serum localization on mouse preimplantation embryos before and after paraformaldehyde fixation. Membrane fluorescence on unfixed 1-cell embryos is patchy (a; the perimeter of the embryo is in focus in the upper view and the surface of the same embryo is in focus in the lower view). After paraformaldehyde fixation, subsequent immunolabeling produces a smooth membrane fluorescence (b and c; b is with the perimeter of the embryo in focus, c is with the surfaces in focus). Normal rabbit serum control (d). The bar in (d) represents 25 μm.

antigens was produced by injecting rabbits with ectoplacental cone cells obtained from 7.5-day gestation mouse egg cylinders (anti-EPC serum; Searle and Jenkinson, 1978). Unlike the results obtained with anti-placenta-B serum, anti-EPC serum (preabsorbed with mouse spleen cells) did not react with unfertilized eggs or zygotes. However, it did react with 8-cell embryos, morulae and blastocysts with the trophectoderm reacting more strongly than the ICM. With development, reactivity became confined to the trophectodermal derivatives, the ectoplacental cone, and the extraembryonic ectoderm (Searle and Jenkinson, 1978).

One final point that should be mentioned here is that antisera to normal mouse neural tissues or to neural tumor cell lines can cross-react with mouse sperm (Chaffee and Schachner, 1978a,b) and with preimplantation embryos (Solter and Schachner, 1976). This observation raises the following question: Why did not the aforementioned antisera to mouse blastocysts and to placenta also react with mouse brain and/or sperm since both antisera reacted with preimplantation embryos? Possibly, this question could be explained by multiple specificities caused by different immunization procedures and/or by varying degrees of immunogenicity of shared antigens dependent on which cell type those antigens were expressed.

E. ANTISERA TO HUMAN CHORIONIC GONADOTROPIN (hCG)

Human chorionic gonadotropin coats the maternal side of the syncytiotrophoblast (Martin et al., 1974). A protein that cross-reacts immunologically with hCG is secreted into the rabbit blastocele (Haour and Saxena, 1974; Fugimoto et al., 1973). These observations suggested that perhaps mammalian trophoblast cells in general might synthesize and secrete hCG-like substances, and consequently the mouse blastocyst was examined for the presence of hCG-like material (Wiley, 1974).

When tested (immunofluorescence) on a panel of adult mouse tissues and cells and on live mouse preimplantation embryos, an unabsorbed rabbit antiserum to hCG (Miles Laboratories) proved negative for adult mouse tissues and cells including brain, sperm, and unfertilized eggs (Wiley and Calarco, unpublished). However, with preimplantation embryos, a low level of cell surface labeling was observed with cleavage-stage embryos, which increased to a maximum on late morulae and early blastocysts (Fig. 5; Wiley, 1974). Thereafter, as blastocysts matured and escaped from their zonae, surface labeling diminished to undetectable levels. Similar cell surface labeling of mouse morulae and blastocysts has also been reported using immunoperoxidase techniques (Sengupta et al., 1978).

Recently, these studies have been extended using a different rabbit antiserum to hCG (Cappel Laboratories). To ensure that this antiserum contained no nonspecific, xenogeneic anti-mouse activity, it was absorbed with intact liver cells (1469 mouse embryo liver cell line; 10^4 cells per microliter of serum). When tested on preimplantation embryos, this antiserum gave similar results as was reported earlier, with the additional observation that certain cells of late morulae labeled

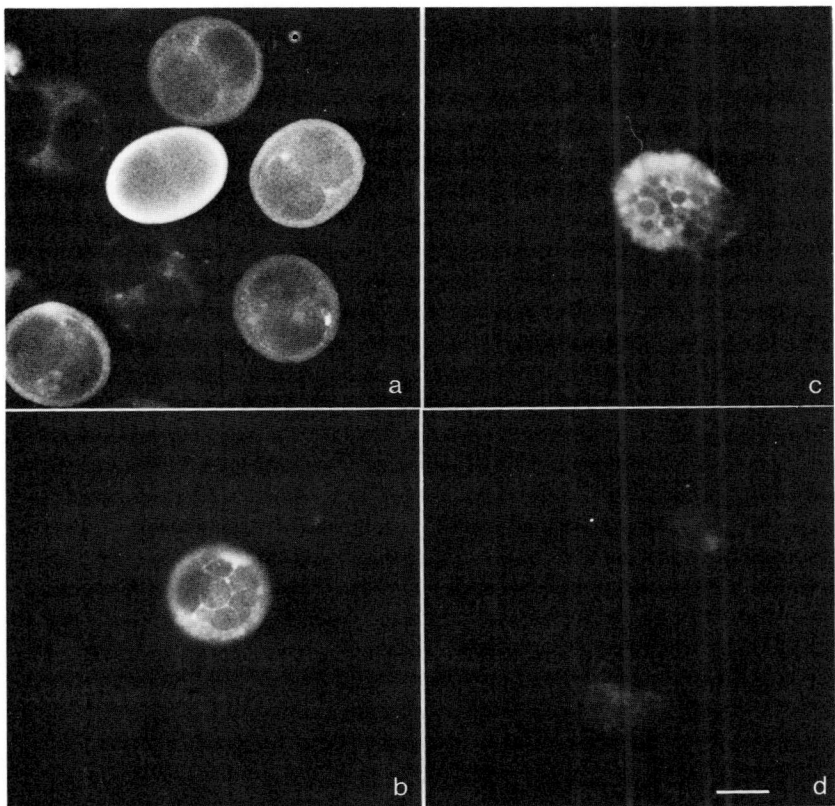

FIG. 5. Indirect immunofluorescence. Anti-hCG serum localization on mouse preimplantation embryos. (a) Two-cell embryos show very little membrane fluorescence; staining of zonae is variable and disappears by the 8-cell stage. (b) Morula showing membrane fluorescence outlining individual blastomeres. (c) "Hatched" blastocyst showing membrane fluorescence outlining individual trophectodermal cells. (d) Normal rabbit serum controls; upper embryo is a blastocyst, lower embryo is a morula. The bar in (d) represents 25 μm.

more intensely than did neighboring cells (Fig. 6). Late blastocysts and cultured early postimplantation embryos were then examined to determine the fate of the hCG-like antigen after implantation (Fig. 6). As was discovered earlier (Wiley, 1974), labeling diminished on the trophectoderm of maturing blastocysts. However, fluorescent labeling was still intense on the inner cell mass (isolated from blastocysts by immunosurgery; Solter and Knowles, 1975). When inner cell masses were cultured further and allowed to differentiate into endoderm and

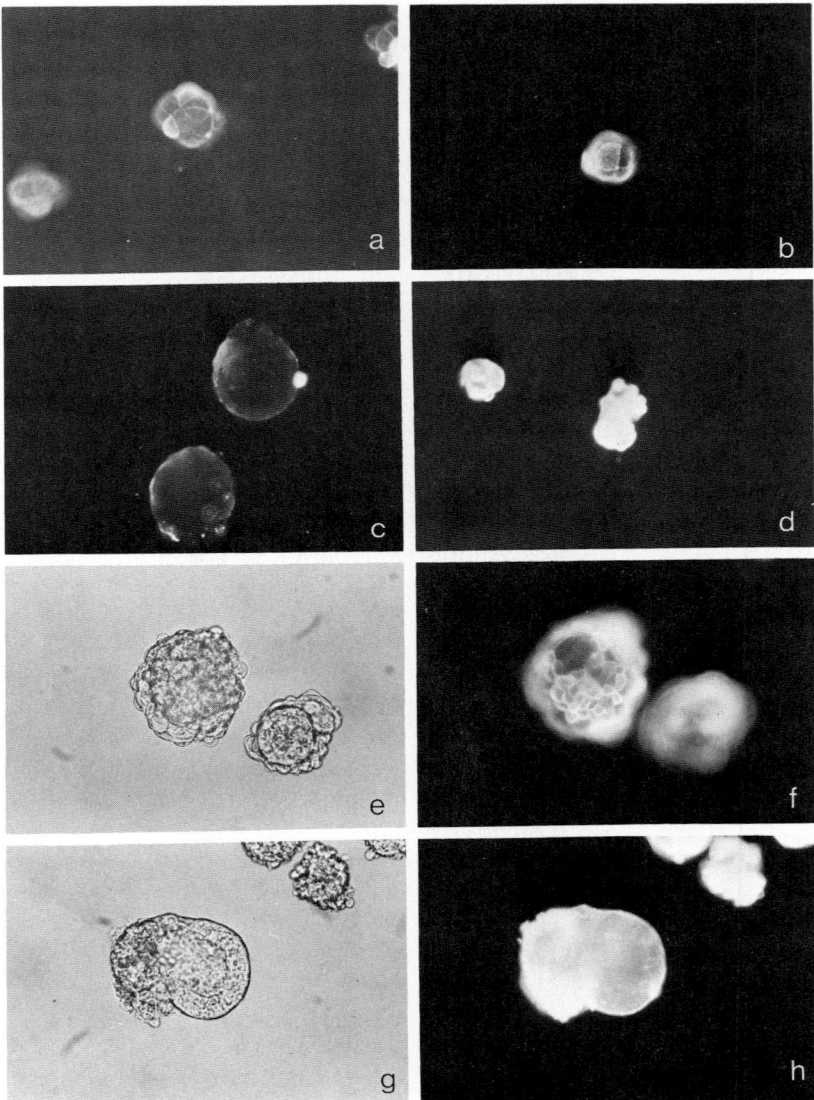

FIG. 6. Indirect immunofluorescence. Anti-hCG serum localization on late mouse preimplantation embryos and on cultured isolated inner cell masses. (a,b) Morulae, showing that some cells label more intensely than do adjacent cells. (c) Mature, hatched blastocyst, showing diminished labeling of the trophectoderm. (d) Freshly isolated inner cell masses from hatched blastocysts; immunolabeling is intense. (e,f) Isolated inner cell masses after 5d in culture: (e) phase contrast image showing the inner ectodermal and

ectoderm, this hCG-like surface antigen continued to be expressed on both germ layers (Fig. 6). These observations suggest that this hCG-like antigen that is initially expressed on morulae becomes restricted to the inner cell mass with blastocyst maturation and implantation. The antigen then remains serologically accessible on the cell surfaces of both the primitive endoderm and the embryonic ectoderm. Whether the antigen continues to be expressed during subsequent primitive streak formation and mesoderm differentiation has not yet been examined.

Unfortunately, the molecular identity of this hCG-like antigen remains obscure. The antigen does, however, cross-react with antisera to LH (Sengupta *et al.*, 1978), suggesting that the antigen may be similar to the alpha subunit that is common to LH and hCG. Compatible with this suggestion is the observation that the antigen does not cross-react with antiserum specific for the beta subunit of hCG (Sengupta *et al.*, 1978; Wiley and Calarco, unpublished).

F. Antisera to Mouse L Cells and to LETS Protein

The anti-L serum to be described also begins reacting with preimplantation embryos during the morula-to-blastocyst transformation (Wiley, unpublished). It was produced by injecting male rabbits with mouse (NCTC 929) L-cell fibroblasts with the original intent of making a "mesoderm-specific" cell surface marker for other purposes. To render the antiserum "mesoderm-specific" it was absorbed with an "endodermal" cell line (1469 liver cells), the rationale being that this absorption would remove generalized anti-mouse activity, leaving only that activity specific for the L cell (i.e., mesoderm). Of course this approach rests on the unproved assumption that cells with a common embryonic derivation express antigens specific for cells of that derivation (in the adult as well as in the embryo).

After being so absorbed, the anti-L serum, which still reacted strongly with L cells (immunofluorescence), was also positive for preimplantation embryos beginning with the morula stage (Fig. 7). With blastocysts, both the ICM and the trophectoderm were positive. When tested on a panel of sectioned near-term fetal and adult mouse tissues, the absorbed anti-L serum was positive only

---

outer endodermal cell layers; (f) the same inner cell mass under UV illumination, showing that the endodermal layer is positive for the hCG-like antigen. (g,h) Ectodermal cell layer of an isolated inner cell mass that was cultured 5d: (g) phase contrast image; (h) fluorescent image. The ectodermal layer is also positive for the hCG-like antigen. The bar in h represents 50 μm.

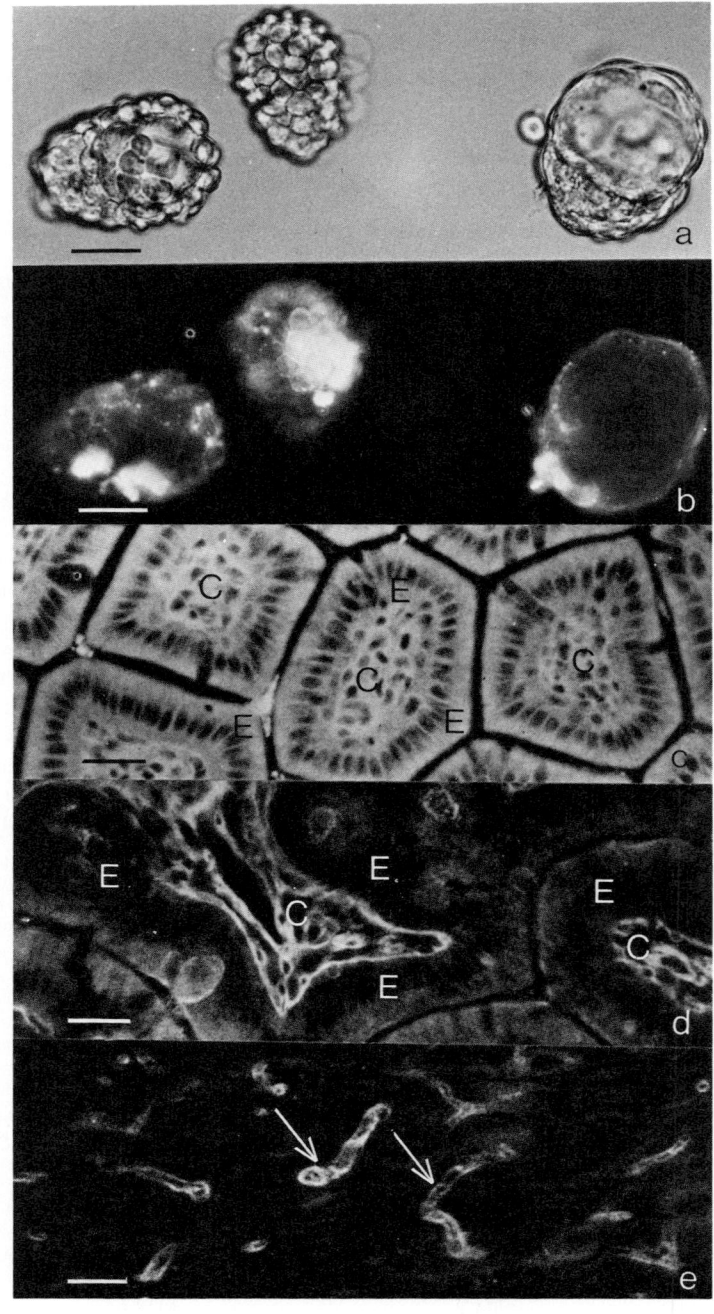

for the connective tissue components and for classical mesodermal derivatives such as skeletal and smooth muscle, cartilage, and bone.

Basement membranes were also positive as were the cell surfaces of these mesodermal derivatives. The epithelial (endoderm and ectoderm) components such as the mucosa of the intestine and the parenchyma of the brain were completely negative (Fig. 7).

One possibility that came to mind was that these results were due to LETS (large external transformation-sensitive) protein. LETS protein is a major component of basement membranes (Vaheri et al., 1976) and of the cell surfaces of many cell types, notably fibroblasts (Hynes, 1973) but also some epithelial cells (Chen et al., 1977). Experimental evidence suggests that LETS protein functions in cell–substratum and cell–cell interactions (see Hynes, 1976). Rabbit antiserum to LETS protein has been reported as being positive (immunofluorescence) for mouse blastocysts (but not for earlier stages), with the reaction being restricted to the ICM (Zetter and Martin, 1978). In contrast, recall that anti-L serum is positive for both the ICM and the trophectoderm of blastocysts. Therefore, it seems unlikely that all of the antigens that are detected by anti-L serum and shared by L cells, mesodermal derivatives in adult tissues, and by mouse preimplantation embryos can be ascribed to LETS protein. It would be interesting to absorb anti-L serum further with LETS protein and then reexamine its activity with preimplantation embryos and adult tissues. In addition, it might also prove interesting to absorb anti-L serum with different types of collagen and then repeat the immunofluorescence experiments.

III. Future Studies on Embryonic Cell Surface Antigens as Developmental Probes

There are available a respectable number of antisera whose specificities have provided some intriguing information on the cell surface antigens expressed during preimplantation development (Table II). However, there have been few direct comparisons made between these antisera, the exception being antisera to histocompatibility anti-

---

FIG. 7. Indirect immunofluorescence. Anti-L serum localization on mouse blastocysts, small intestine, and brain. (a) Phase-contrast micrograph of zona-free blastocysts. (b) Same blastocysts treated with absorbed anti-L serum. (c) Adult mouse small intestine treated with unabsorbed anti-L serum. Both the epithelium (E) and underlying connective tissue (C) are positive. (d) Adult mouse small intestine treated with absorbed anti-L serum. Only the connective tissue is positive. (e) Mouse fetal (16-day-gestational age) brain treated with absorbed anti-L serum. The capillaries (arrows) are positive while the neural material is negative. The bars represent 25 μm.

TABLE II

SUMMARY OF THE SPECIFICITIES OF ANTISERA TESTED ON PREIMPLANTATION EMBRYOS

Antiserum to	Brain	Sperm	Unfertilized egg	Zygote	2-Cell	4- to 8-Cell	Morula	Blastocyst[a] ICM	Blastocyst[a] TB	Affect development in vitro?	References
Antisera to histocompatibility antigens											
1. H-2 antigens[b]	+	+	–	–	–	–	–	–	–	No	Heyner (1973), Heyner et al. (1969), and Lemonnier et al. (1975)
2. H-3, H-6 antigens[c]	NT	NT	+	+	+	+	+	+	–	NT[a]	Palm et al. (1969) and Muggleton-Harris and Johnson (1976)
3. H-Y antigen[e]	+	+	–	–	–	+	+	+	–	NT	Goldberg et al. (1971) and Krco and Goldberg (1976)
4. β_2-M	+	+	NT	NT	NT	NT	NT	+	–	NT	Håkansson and Peterson (1976) and Dubois et al. (1976)
Antisera to teratocarcinoma cells											
5. 402AX teratoma[f,g,k]											
Antigen I	–	–	+	+	+	+	+	NT	–	NT	Edidin et al. (1971)
Antigen II	–	–	+	–	–	–	–	NT	+	NT	Gooding et al. (1976)
Antigen III	–	–	–	–	–	–	–	–	–	NT	
6. F9 nullipotent teratocarcinoma[i]	+	+	–	±	±	+	+	+	+	Yes	Artzt et al. (1973, 1974) and Kemler et al. (1977)
Antisera to gametes and preimplantation embryos											
7. Sperm											
IgG	NT	+	+	+	+	+	+	+	+	NT	Menge and Fleming (1978)
IgA	NT	+	–	–	–	+	+	+	+	NT	

									References
8. Unfertilized egg[a]	NT	–	–	+	+	+	+	NT	Baranska et al. (1970), Koprowski et al. (1971), and Moskalewski and Koprowski (1972)
9. Blastocysts[j]	–	–	±	+	±	+	+	Yes	Wiley (1975) and Wiley and Calarco (1975)
Antisera to placenta or to other parts of the postimplantation embryo									
10. Placenta	NT	NT	–	–	+	+	+	NT	Kometani and Behrman (1971) and Kometani et al. (1973)
11. Placenta[j]	–	–	+	+	+	+	+	No	Wiley (1975) and Wiley and Calarco (1975)
12. Ectoplacental cone (EPC)	NT	NT	–	NT	+	±	+	NT	Searle and Jenkinson (1978)
Antisera to human chorionic gonadotropin and to mouse L cells									
13. Human chorionic[j] gonadotropin (hCG)	–	–	–	±	+	+	+	Yes	Wiley (1974, 1975) and Sengupta et al. (1978)
14. Mouse L cell	–	–	–	–	+	–	+	NT	Wiley (unpublished)

[a] A plus sign or minus sign centered (mid-column) means that no distinction was made between ICM and trophoblast (TB).
[b] H-2 antigens may be present at very low levels on blastocysts (Searle et al., 1976) and are definitely present during implantation delay (Håkansson et al., 1975; Håkansson and Sunqvist, 1975).
[c] The paternal non H-2s are not expressed until the 6- to 8-cell stage (Muggleton-Harris and Johnson, 1976).
[d] NT, not tested.
[e] Present on 50% of embryos.
[f] Negative for brain and sperm inferred from absorption studies.
[g] Positive for SV40-transformed 3T3 fibroblasts.
[h] Negative for SV40-transformed 3T3 fibroblasts.
[i] Only Fab, and not IgG affects development.
[j] Culturing done with antiserum.
[k] Blastocyst data are for late (hatched) blastocysts.

gens, thus making it impossible to draw conclusions as to whether certain pairs or groups of antisera overlap in the antigens they detect on preimplantation embryos. For example, the specificities of anti-402AX teratoma (No. 5; antigens I and II), anti-egg (No. 8), anti-blastocyst (No. 9), anti-placenta (No. 11), anti-EPC (No. 12), and anti-hCG (No. 13) sera overlap in that the antisera are all positive for mouse morulae and blastocysts but, where tested, are all negative for brain and sperm.

Of these 14 antisera only 5 have been examined for effects on the development of preimplantation embryos *in vitro,* and these experiments were preliminary ones that gave no data on the underlying cause of the observed effects, the most blatant effect being the arrest of development. However, it is very easy to arrest development by nonspecific, uninformative ways that may result from generalized toxicity inflicted by the investigator. This is the main drawback with the studies done with anti-blastocyst (No. 9) and anti-hCG (No. 13) sera, both of which were obtained by culturing embryos in antisera without monitoring any parameters for assessing overall viability, a difficult task in itself.

Aside from the problem of assessing viability, there is the problem of decreasing viability for reasons that stem from cell surface antigens being bound by divalent antibodies per se. Divalent antibodies can cause cell surface antigens to become redistributed (patching and capping), which in itself may alter development for reasons completely unrelated to the function of the antigens so affected. In addition, antigens cross-linked into patches and/or caps may be removed from the surface or may result in permanent resynthesis of the antigens. Finally, intact antibodies to one set of antigens can indirectly alter the mobility and/or distribution of other unrelated antigens by (cytoskeleton-mediated) anchorage modulation. This last point is especially relevant here because H-2 and β_2-M are among the many known antigens showing anchorage modulation (see Edelman, 1976). Thus, if these particular antigens, or, perhaps, their embryonic counterparts (F9?) are indeed essential to morphogenetic control, then results obtained from the use of antibodies will have to be interpreted very carefully.

One solution to this problem is to use Fab prepared from the antibodies because univalent ligands (like Fab) do not cause anchorage modulation or the other aforementioned problems brought on by using antibodies. Indeed, certain divalent antibodies have been shown to produce different effects than the corresponding Fab. In one such case with an antiserum to *Arbacia* eggs, the immune antibodies caused a variety of morphological changes in target *Arbacia* eggs, while the anti-egg Fab had no such effects (see Metz, 1972). Here, the effects observed with antibodies were attributed to secondary complications resulting from using multivalent antibodies. Conversely, recall that antibodies to F9 had no effect on preimplantation development while Fab prepared from anti-F9 antibodies prevented compaction in a reversible manner (Kemler *et al.*, 1977). In this case, the ineffectiveness of the F9 antibodies was attributed to permanent resynthesis of the antigen (Kemler *et al.*, 1977). However, an alternative explanation of these results is that due to steric reasons less F9 antigen was bound by antibody than by Fab, resulting in more of the F9 antigen being rendered "nonfunctional" (assuming that F9 antigen performs functions essential to compaction) by Fab than by antibody. Nevertheless, the culturing experiments previously done with antiplacenta (No. 11, anti-blastocyst [No. 9], and with anti-hCG [No. 13]), for example, should all be repeated using the corresponding Fab before any conclusions are drawn regarding their effects on preimplantation development.

However, substituting Fab for antibodies will not completely eliminate the problem of nonspecific toxicity, for Fab could, for example, still be toxic due to simple steric interference of unrelated cell surface function. It would be better if antigens of interest could be selectively removed so that the effects of their absence on development might reveal their function. However, this is probably not yet feasible without additional information on antigen synthesis and expression. Hence the need, for example, for information obtained with inhibitors to determine if the synthesis of a given antigen is under transcriptional or posttranscriptional control, or whether synthesis is pulsed so that it could be inhibited by pulsed treatment with the appropriate inhibitor, thereby eliminating the antigen from the cell surface in as selective a way as possible. Alternatively, it might be possible to selectively "eliminate" antigens genetically by examining preimplantation embryos expressing embryo-lethal genes (t^x alleles of the T complex?) for the expression of the antigens in comparison with that of normal embryos.

However, because of their ability to cross-link antigens, multivalent antibodies can be exploited to study the topography of cell surface antigens and their relationships with cytoplasmic components. As discussed in Section I, such relationships may be central to developmental regulation. Thus, the combined use of antibodies together with Fab is essential to realizing the potential of these antisera and the antigens they detect as developmental probes.

IV. Concluding Remarks

This chapter has reviewed the properties of several antisera, which have been used for examining the cell surface of the mouse preimplantation embryo, with the objective of assessing their use as developmental probes. The major conclusions drawn from the specificities of these antisera are that certain cell surface antigens are expressed only during embryonic life, some of which are further restricted temporally to specific embryonic stages or spatially to specific tissues. Some embryonic antigens may also be expressed by tumor cells. The bases for these conclusions are summarized below according to the immunogen used to elicit the antisera.

1. *Antisera to histocompatibility antigens.* Normal preimplantation embryos may not express significant levels of adult-type H-2 antigens, which first become serologically detectable shortly after implantation. However, there are several non-H-2 histocompatibility antigens that are serologically accessible on preimplantation embryos, although none of these has yet been studied directly for developmental importance.

2. *Antisera to teratocarcinoma cells.* There are some embryonic antigens that are shared by teratoma cells, sperm, and by certain stages of preimplantation embryos, but which disappear from the embryo after implantation. In general, these teratoma antigens are not coexpressed with H-2 antigens and their disappearance overlaps the initial appearance of the H-2 antigens. Teratoma antigens, it has been proposed, may be embryonic precursors to adult-type H-2 antigens. Since these antisera (or their Fab fragments) inhibit the development of preimplantation embryos *in vitro,* these teratoma antigens may be developmentally important.

3. *Antisera to gametes and preimplantation embryos.* Several antisera to sperm detect embryonic antigens shared by sperm, preimplantation embryos, and, in some cases, also by adult brain cells. Only one antiserum to mouse eggs has been reported, and it cross-reacts with some transformed cells, preimplantation embryos, but not with normal adult cells. Only one antiserum to mouse preimplantation embryos has been reported, which was prepared to blastocysts. This anti-blastocyst serum detects embryonic antigens that have a peak expression on 8- to 12-cell-stage embryos. Because this antiserum inhibits the development of cultured mouse preimplantation embryos, it may be detecting cell surface antigens that are developmentally important.

4. *Antisera to placentas or to other parts of the postimplantation embryo.* Antisera developed to mouse term placenta and to ectoplacen-

tal cone detect embryonic antigens that are expressed on preimplantation embryos and which later segregate to the trophectodermal derivatives after implantation. Some of these antisera also cross-react with unfertilized mouse eggs and with some transformed cell lines.

5. *Antisera to human chorionic gonadotropin (hCG).* These antisera detect embryonic antigens whose expression peaks on morulae and blastocysts and which becomes restricted to the ICM prior to implantation. The antisera can cause a reversible inhibition of mouse preimplantation development *in vitro,* suggesting that these antigens may be developmentally important.

6. *Antisera to mouse L cells and to LETS protein.* An antiserum to mouse L-cell fibroblasts detects embryonic antigens that are first expressed on morulae and which persist on both the ICM and the trophectoderm of the blastocyst. In the adult, these antigens are restricted to cells of mesodermal origin and to basement membranes. An antiserum to LETS protein cross-reacts with antigens that are first detectable on mouse blastocysts and later become restricted to the ICM.

ACKNOWLEDGMENTS

The author wishes to express her sincere gratitude to Drs. M. I. Sherman and P. G. Calarco and to L. V. Johnson, C. L. Banka, and M. A. Eglitis for their critical review of the manuscript, and to M. C. Siebert for her excellent technical assistance during portions of the work described for results obtained with anti-hCG sera.

REFERENCES

Artzt, K., and Jacob, F. (1974). *Transplantation* **17,** 633–634.
Artzt, K., and Bennett, D. (1975). *Nature (London)* **256,** 545–547.
Artzt, K., Dubois, P., Bennett, D., Condamine, H., Babinet, C., and Jacob, F. (1973). *Proc. Natl. Acad. Sci. U.S.A.* **70,** 2988–2992.
Artzt, K., Bennett, D., and Jacob, F. (1974). *Proc. Natl. Acad. Sci. U.S.A.* **71,** 811–814.
Bagshawe, K. D., and Lawler, S. (1975). *In* "Immunobiology of Trophoblast" (R. G. Edwards, C. W. S. Howe, and M. H. Johnson, eds.), pp. 67–86. Cambridge Univ. Press, London and New York.
Banka, C. L., and Calarco, P. G. (1978). *Anat. Rec.* **190,** 331a.
Baranska, W., Koldovsky, P., and Koprowski, H. (1970). *Proc. Natl. Acad. Sci. U.S.A.* **67,** 193–199.
Bennett, D. (1975). *Nature (London)* **272,** 539.
Bennett, D., Boyse, E. A., and Old, L. J. (1972). *In* "Cell Interactions" (L. G. Silvestri, ed.), pp. 247–263. North-Holland Publ., Amsterdam.
Bernstine, E. G., Hooper, M. L., Grandchamp, S., and Ephrussi, B. (1973). *Proc. Natl. Acad. Sci. U.S.A.* **70,** 3899–3903.
Billington, W. D. (1975). *In* "Immunobiology of Trophoblast" (R. G. Edwards, C. W. S. Howe, and M. H. Johnson, eds.), pp. 67–86. Cambridge Univ. Press, London and New York.

Brinster, R. L. (1974). *J. Exp. Med.* **140**, 1049–1056.
Brinster, R. L. (1976). *Cancer Res.* **36**, 3412–3414.
Brunner, G. (1977). *Differentiation* **8**, 123–132.
Buc-Caron, M., Gachelin, G., Hofnung, M., and Jacob, F. (1974). *Proc. Natl. Acad. Sci. U.S.A.* **71**, 1730–1733.
Calarco, P. G., and Banka, C. L. (1979). *Biol. Reprod.* **20**, 699–704.
Calarco, P. G., and Brown, E. H. (1969). *J. Exp. Zool.* **171**, 253–284.
Chaffee, J. K., and Schachner, M. (1978a). *Dev. Biol.* **62**, 173–184.
Chaffee, J. K., and Schachner, M. (1978b). *Dev. Biol.* **62**, 185–192.
Chen, L. B., Maitland, N., gallimore, P. H., and McDougall, J. K. (1977). *Exp. Cell Res.* **106**, 39–46.
Coggin, J. H., and Anderson, N. G. (1974). *Adv. Cancer Res.* **19**, 105–165.
Damjanov, I., and Solter, D. (1974). *Curr. Top. Pathol.* **59**, 69–130.
Dewey, M. J., Gearhart, J. D., and Mintz, B. (1977). *Dev. Biol.* **55**, 359–374.
Dubois, P., Fellous, M., Gachelin, G., Jacob, F., Kemler, D., Pressman, D., and Tanigaki, N. (1976). *Transplantation* **22**, 467–473.
Ducibella, T., and Anderson, E. (1975). *Dev. Biol.* **47**, 45–58.
Ducibella, T., Albertini, D. F., Anderson, E., and Biggers, J. D. (1975). *Dev. Biol.* **45**, 231–250.
Edelman, G. M. (1976). *Science* **192**, 218–226.
Edidin, M. (1976a). *In* "The Cell Surface in Animal Embryogenesis and Development" (G. Poste and G. L. Nicolson, eds.), pp. 127–143. Elsevier/North-Holland, Amsterdam.
Edidin, M. (1976b). *In* "Embryogenesis in Mammals" (CIBA Foundation Symposium 40, New Series, pp. 177–197. Elsevier-Excerpta Medica, Amsterdam.
Edidin, M., Patthey, H. L., McGuire, E. J., and Sheffield, W. D. (1971). *In* "Conference Workshop on Embryonic and Fetal Antigens in Cancer" (N. G. Anderson and J. H. Coggin, Jr., eds.), pp. 239–248. Oak Ridge National Laboratory, Oak Ridge, Tenn.
Erickson, R. P. (1972). *In* "Genetics of the Spermatozoan" (R. A. Beatty and S. Gluecksohn-Waelsch, eds.), pp. 191–202. Edinburgh Univ. Press, Edinburgh and New York.
Fauve, R. M., Hevin, B., Jacob, H., Gaillard, J. A., and Jacob, F. (1974). *Proc. Natl. Acad. Sci. U.S.A.* **71**, 4052–4056.
Fellous, M., Erickson, R. P., Gachelin, G., and Jacob, F. (1976). *Transplantation* **22**, 440–444.
Forman, J., and Vitetta, E. S. (1975). *Proc. Natl. Acad. Sci. U.S.A.* **72**, 3661–3665.
Fugimoto, S., Woody, H. D., and Dukelow, W. R. (1973). *Pediat. Proc.* **32**, 214.
Gachelin, G., Fellous, M., Guenet, J. L., and Jacob, F. (1976). *Dev. Biol.* **50**, 310–320.
Gardner, R. L., Johnson, M. H., and Edwards, R. G. (1973). *In* "Immunology of Reproduction" (K. Bratanov, ed.), pp. 480–484. Bulgarian Academy of Science Press, Sofia.
Glass, L. E., and Hanson, J. E. (1974). *Fertil. Steril.* **25**, 484–493.
Goldberg, E. H., Aoki, T., Boyse, E. J., and Bennett, D. (1970). *Nature (London)* **228**, 570–572.
Goldberg, E. H., Boyse, E. A., Bennett, D., Scheid, M., and Carswell, E. A. (1971). *Nature (London)* **232**, 478–480.
Gooding, L. R., and Edidin, M. (1974). *J. Exp. Med.* **140**, 61–78.
Gooding, L. R., Hsu, Y.-C., and Edidin, M. (1976). *Dev. Biol.* **49**, 479–486.
Graham, C. F. (1977). *In* "Concepts in Mammalian Embryogenesis" (M. I. Sherman, ed.), pp. 315–394. MIT Press, Cambridge, Massachusetts.

Håkansson, S., and Peterson, P. A. (1976). *Transplantation* **21**, 358–360.
Håkansson, S., and Sundquist, K.-G. (1975). *Transplantation* **19**, 479–484.
Håkansson, S., Heyner, S., Sundquist, K.-G., and Bergstrom, S. (1975). *Int. J. Fertil.* **20**, 137–140.
Hammerling, G. J., Mauve, G., Goldberg, E., and McDevitt, H. O. (1975). *Immunogenetics* **1**, 428–437.
Haour, F., and Saxena, B. B. (1974). *Science* **185**, 444–446.
Hellström, K. E., Hellstrom, I. G., and Grawn, J. (1969). *Nature (London)* **224**, 914–915.
Heyner, S. (1973). *Transplantation* **16**, 675–677.
Heyner, S., Brinster, R. L., and Palm, J. (1969). *Nature (London)* **222**, 783–784.
Hillman, N., Sherman, M. I., and Graham, C. (1972). *J. Embryol. Exp. Morphol.* **28**, 263–278.
Hogan, B., Fellous, M., Avner, P., and Jacob, F. (1977). *Nature (London)* **290**, 515–518.
Holden, S., Bernard, O., Artzt, K., Whittmore, W. F., Jr., and Bennett, D. (1977). *Nature (London)* **270**, 518–520.
Hynes, R. O. (1973). *Proc. Natl. Acad. Sci. U.S.A.* **70**, 3170–3174.
Hynes, R. O. (1976). *Biochim. Biophys. Acta* **458**, 73–107.
Jacob, F. (1977) *Immunol. Rev.* **33**, 3–32.
James, D. A. (1969). *Transplantation* **8**, 846–851.
Johnson, L. V. (1978). Ph.D. Thesis, University of California, San Francisco.
Johnson, M. H. (1975). *In* "Immunobiology of Trophoblast" (R. G. Edwards, C. W. S. Howe, and M. H. Johnson, eds.), pp. 87–112. Cambridge Univ. Press, London and New York.
Johnson, M. H., and Edidin, M. (1972). *Transplantation* **14**, 781–786.
Kemler, R., Babinet, C., Condamine, H., Gachelin, J. L., Guenet, L., and Jacob, F. (1976). *Proc. Natl. Acad. Sci., U.S.A.* **73**, 4080–4084.
Kemler, R., Babinet, C., Eisen, H., and Jacob, F. (1977). *Proc. Natl. Acad. Sci. U.S.A.* **74**, 4449–4452.
Kirby, D. R. S. (1968). *Transplantation* **6**, 1005–1009.
Kirby, D. R. S., Billington, W. D., Bradbury, S., and Goldstein, D. J. (1964). *Nature (London)* **204**, 548–549.
Kirby, D. R. S., Billington, W. D., and James, D. A. (1966). *Transplantation* **4**, 713–718.
Kometani, K., and Behrman, S. J. (1971). *Int. J. Fertil.* **16**, 139–143.
Kometani, K., Paine, P., Cossman, J., and Behrman, S. J. (1973). *Am. J. Obstet. Gynecol.* **116**, 351–357.
Koprowski, H., Koldovsky, P., Sawicki, W., and Baranska, W. (1971). *In* "Conference and Workshop on Embryonic and Fetal Antigens in Cancer" (N. G. Anderson and J. H. Coggin, Jr., eds.), pp. 291–301. Oak Ridge National Laboratory, Oak Ridge, Tenn.
Krco, C. J., and Goldberg, E. H. (1976). *Science* **193**, 1134–1135.
Krco, C. J., and Goldberg, E. H. (1977). *Transpl. Proc.* **9**, 1367–1370.
Lemonnier, F., Neauport-Sautes, C., and Kourilsky, F. M. (1975). *Immunogenetics* **2**, 517.
Martin, G. R., and Evans, M. J. (1975). *In* "Teratomas and Differentiation" (M. I. Sherman and D. Solter, eds.), pp. 237–250. Academic Press, New York.
Martin, B. J., Spicer, S. S., and Smyth, N. M. (1974). *Anat. Rec.* **178**, 769–786.
Martin, G. R. (1975). *Cell* **5**, 229–243.
Martin, G. R., Wiley, L. M., and Damjanov, I. (1977). *Dev. Biol.* **61**, 230–244.
McClay, D. R., and Gooding, L. R. (1978). *Nature (London)* **274**, 367–368.
Melissinos, K. (1907). *Mikr. Anat.* **70**, 577–628.
Menge, A. C., and Fleming, C. H. (1978). *Dev. Biol.* **63**, 111–117.

Metz, C. B. (1972). *Biol. Reprod.* **6,** 358–383.
Mintz, B. (1965). *In* "Preimplantation Stages of Pregnancy" (G. E. W. Wolsten-Holme and M. O'Connor, eds.), pp. 194–207. Churchill, London.
Mintz, B., and Illmensee, K. (1975). *Proc. Natl. Acad. Sci. U.S.A.* **72,** 3585–3589.
Moskalewski, S., and Koprowski, H. (1972). *Nature (London)* **237,** 167–168.
Muggleton-Harris, A. L., and Johnson, M. H. (1976). *J. Embryol. Exp. Morphol.* **35,** 59–72.
Nicolson, G. L. (1976). *Biochim. Biophys. Acta* **457,** 57–108.
Ohno, S. (1977). *Immunol. Rev.* **33,** 59–69.
Olds, P. (1968). *Transplantation* **6,** 478–479.
Ostrand-Rosenberg, S., Edidin, M., and Jewett, M. A. S. (1977). *Dev. Biol.* **61,** 11–19.
Palm, J., Heyner, S., and Brinster, R. L. (1969). *Fed. Proc.* **28,** 379.
Palm, J., Heyner, S., and Brinster, R. L. (1971). *J. Exp. Med.* **133,** 1282–1293.
Patthey, H. L., and Edidin, M. (1973). *Transplantation* **15,** 211–214.
Rask, L., Lindblom, B., and Peterson, P. A. (1974). *Nature (London)* **249,** 833.
Searle, R. F., and Jenkinson, E. J. (1978). *J. Embryol. Exp. Morphol.* **43,** 147–156.
Searle, R. F., Johnson, M. H., Billington, W. D., Elson, J., and Clutterbuck-Jackson, S. (1974). *Transplantation* **18,** 136–141.
Searle, R. F., Sellens, M. H., Elson, J., Jenkinson, E. J., and Billington, W. D. (1976). *J. Exp. Med.* **143,** 348–359.
Sellens, M. H. (1977). *Nature (London)* **269,** 60–61.
Sengupta, J., Gupta, P. D., Manchanda, S. K., and Talwar, G. P. (1978). *J. Reprod. Fertil.* **52,** 163–165.
Sherman, M. I., and Wudl, L. R. (1977). *In* "Concepts in Mammalian Embryogenesis" (M. I. Sherman, ed.), pp. 136–234. M.I.T. Press, Cambridge, Massachusetts.
Silver, J., and Hood, L. (1974). *Nature (London)* **249,** 764.
Simmons, R. L. (1971). *Nature (London)* **231,** 111–112.
Simmons, R. L., and Russell, P. S. (1962). *Ann. N.Y. Acad. Sci.* **99,** 717–732.
Simmons, R. L., and Russell, P. S. (1966). *Ann. N.Y. Acad. Sci.* **129,** 35–45.
Solter, D., and Knowles, B. B. (1975). *Proc. Natl. Acad. Sci., U.S.A.* **72,** 5099–5102.
Solter, D., and Schachner, M. (1976). *Dev. Biol.* **52,** 98–104.
Spindle, A. I., and Pedersen, R. A. (1973). *J. Exp. Zool.* **186,** 305–318.
Stern, P. L., Martin, G. R., and Evans, M. J. (1975). *Cell* **6,** 455–465.
Stevens, L. C. (1958). *J. Natl. Cancer Inst.* **20,** 1257–1270.
Stevens, L. C. (1970). *Dev. Biol.* **21,** 364–382.
Tarkowski, A. K. (1959). *Acta Theriol.* **3,** 191–267.
Tarkowski, A. K., and Wroblewska, J. (1967). *J. Embryol. Exp. Morphol.* **18,** 155–180.
Tsunoda, Y. (1977). *J. Reprod. Fertil.* **50,** 353–355.
Vaheri, A., Ruoslahti, E., Linder, E., Wartiovaara, J., Keski-Oja, J., Kuusela, P., and Saksela, O. (1976). *J. Supramol. Struct.* **4,** 63–70.
Vitetta, E. S., Artzt, K., Bennett, D., Boyse, E. A., and Jacob, F. (1975). *Proc. Natl. Acad. Sci. U.S.A.* **72,** 3215–3219.
Voisin, G. A., and Chaouat, G. (1974). *J. Reprod. Fertil. Suppl.* **21,** 89–103.
Voisin, G. A., Toullet, F., and D'Almeida, M. (1975). *Acta Endocrinol.* **78,** Suppl. **194,** 173–222.
Voitiskova, M., Polackova, M., and Pokorna, Z. (1969). *Folia. Biol. (Prague)* **15,** 322–332.
Wallach, D. F. H. (1972). *Biochim. Biophys. Acta* **265,** 61–83.

Webb, C. G., Gall, W. E., and Edelman, G. M. (1977). *J. Exp. Med.* **146,** 923–932.
Wiley, L. D. (1974). *Nature (London)* **252,** 715–716.
Wiley, L. M. (1975). Ph.D. Thesis, University of California, San Francisco.
Wiley, L. M., and Calarco, P. G. (1975). *Dev. Biol.* **47,** 407–418.
Wiley, L. M., and Eglitis, M. A. (1978). *J. Cell Biol.* **79,** 159a.
Zetter, B. R., and Martin, G. R. (1978). *Proc. Natl. Acad. Sci. U.S.A.* **75,** 2324–2328.

CHAPTER 9

SURFACE ANTIGENS INVOLVED IN INTERACTIONS OF EMBRYONIC SEA URCHIN CELLS

David R. McClay

DEPARTMENT OF ZOOLOGY
DUKE UNIVERSITY
DURHAM, NORTH CAROLINA

I.	Introduction	199
II.	Adhesive Specificity: Demonstrations of Recognition Capabilities	200
III.	Immunochemical Specificity: Changes Associated with Gastrulation	203
IV.	Relationship between Cell Surface Antigens and Cell–Cell Interactions	209
V.	Synthesis and Conclusions	211
VI.	Concluding Remarks	213
	References	213

I. Introduction

Throughout development cells undergo a series of movements and an exchange of contacts. Nowhere is this more evident than with the reconstruction that takes place at gastrulation. At the blastula stage an embryo consists of a single layer of cells. A short time later, as a result of cell movements and an exchange of cell contacts, the triploblastic structure of the gastrula is established. Many cellular phenomena contribute to the process of gastrulation. These include invagination of sheets of cells accompanied by cell shape changes (Holtfreter, 1943a,b; Burnside, 1971, 1973), shifts and movement of cells within layers accompanied by an increase or decrease in exposed cell surface area (Gustafson and Wolpert, 1962, 1963, 1967; Gibbins et al., 1969), movement by pseudopodial (Holtfreter, 1948; Gustafson, 1963) or filipodial extension (Trinkaus, 1967; Trelstad et al., 1967), and an exchange of adhesive contacts (Townes and Holtfreter, 1955). Most models attempting to explain aspects of gastrulation require a cell recognition mechanism to account for the observed rearrangement of cells (Holtfreter, 1939; Weiss, 1947; Hood et al., 1977). Usually recognition is thought of in terms of the formation of cell adhesive interactions, although there are many examples where recognition results in the release of cells or tissues from one another. For example, determination of the primary mesenchyme in the sea urchin at the beginning of gastrulation appears to be a recognition event leading to the release of an adhesion. Conceptually, recognition has been thought of as an "affinity" between cells (Holtfreter, 1939), and in a classic model recognition was proposed to be a "lock and key" type mechanism (Weiss, 1947; Tyler, 1947). Of importance to this chapter, the cell surface was thought to contain some sort of molecular specificity to govern the recognition processes of morphogenesis.

Demonstrations of cell adhesive specificity have often been taken as support for the notion that embryogenesis is governed by a molecular recognition mechanism. The validity of such a molecular specificity hypothesis remains to be fully established, however, because a number of questions remain unanswered. As yet not a single molecular component has been isolated and demonstrated to be a participant in morphogenesis. The questions as to how cell surfaces might govern morphogenesis and the extent to which molecular specificities contribute to adhesive interactions therefore will continue to be unanswered until molecular components for these phenomena are described.

This chapter explores the specificity hypothesis, the intent being to show how the cell surface might contribute to the gastrulation process. Two lines of research will be discussed: First, data on the formation of cell–cell contacts will be presented to demonstrate an array of age-specific and tissue-specific recognition capabilities available to embryonic cells; second, immunochemical data will show age-specific and tissue-specific differences in cell surface antigens during early development of the sea urchin.

II. Adhesive Specificity: Demonstrations of Recognition Capabilities

The pioneering work of Giudice (Giudice, 1962; Giudice and Mutolo, 1970) and the more recent studies by Spiegel and Spiegel (1975) showed that embryonic sea urchin cells had an impressive capacity to readhere and to reorganize following complete dissociation of the embryo. The recognition pattern was species-specific and cells of an aggregate demonstrated a germ layer recognition capacity by sorting out into ectoderm, mesoderm, and endoderm layers.

In studies where sorting out occurred, the cells initially associated into clumps nonspecifically. It could be argued, therefore, that the aggregation was not specific and that recognition was a secondary event. Also, although the aggregates clearly had the capacity to reorganize, it was not clear from an experiment by Giudice et al. (1969) whether cells had an age- or tissue-dependent recognition capacity, or whether sorting out was actually a demonstration of dedifferentiation followed by redifferentiation appropriate to the new location of cells in the aggregate. The likelihood of the latter possibility was reduced by the demonstration of Spiegel and Spiegel (1975) that isolated macromeres, mesomeres, or micromeres aggregate and then differentiate into structures appropriate only to the normal fate of the blastomeres tested. The classic studies of Holtfreter had shown that dissociated amphibian gastrula cells aggregated and then sorted out into the three germ layers by differential "affinities" (Holtfreter, 1939; Townes and Holtfreter, 1955). This experiment has not been repeated on sea urchin embryos because of an inability to distinguish cells of the three germ layers (thus not eliminating the dedifferentiation–redifferentiation possibility). In any case, the question of cell surface specificity changes remained open since it was still not possible to eliminate hypotheses stating that cell sorting could be explained without involving specific membrane changes (e.g., Steinberg, 1970).

Although sorting out served as a demonstration of some kind of specificity, it was a difficult process to work with experimentally. Therefore, before specificity could be examined in any detail, a method of analysis was needed that would permit a direct examination of the phenomenon. A method, developed originally by Roth and Weston (1967), demonstrated a remarkable array of age- and tissue-specific recognition capacities in the chick (Roth, 1968). The method measured the ability of cell aggregates to collect cells at the aggregate surface. It was found that cells adhered to aggregates of their own specificity in greater numbers than to aggregates of different tissues. This approach was adapted to show a species-specific recognition pattern in sponges (McClay, 1971) and in sea urchin embryos (McClay and Hausman, 1975). From the sorting-out studies, it had been hypothesized that species-specific and germ layer-specific recognition molecules existed at the cell surface and governed morphogenesis (Moscona, 1976). This hypothesis was supported by Roth and co-workers using the projection of retina neurons to the optic tectum as a model system. Though the system was a complicated one, it was possible to suggest dorsoventral gradients of cell surface molecules that could account for the recognition patterns observed (Marchase et al., 1975; Marchase, 1977).

Gastrulation in the sea urchin has offered another, perhaps simpler, system for testing the membrane specificity hypothesis. At gastrulation a number of new cell–cell contacts are established (for review, see Gustafson and Wolpert, 1967). The specificity hypothesis would predict that as the germ layers are organized, new cell surface specificities govern the cell behavior. Guidice et al. (1969) attempted to demonstrate new adhesive specificities at gastrulation but were unsuccessful, either because there were no changes or because the changes were too subtle to be detected by their aggregation methods. The aggregate collection methods of Roth were applied to this problem (McClay and Baker, 1975) and at first only species-specific differences could be detected (Fig. 1) (McClay and Hausman, 1975). Again, changes at gastrulation were not detected. The question was asked again, this time taking advantage of a genetic probe. Cells from *Lytechinus variegatus* and from *Tripneustes esculentus* were shown to collect to aggregates species-specifically (Fig. 1). Viable hybrid crosses between these species then permitted a maximization of adhesive differences at gastrulation. *Lytechinus* eggs (L) could be fertilized by *Tripneustes* sperm (T) to form LT hybrids and the reciprocal cross formed TL hybrids. The hybrid embryos were fully viable, at least to the pluteus stage (rearing the larvae was not attempted), and at the pluteus stage hybrids expressed both maternal and paternal genes (McClay and Hausman, 1975; McClay, 1976). It was reasoned that if the hybrid cells expressed both genomes they should express cell surface molecules as well. If that were so and if those molecules participated in adhesive recognition events, then the aggregate collection system might pick up hybrid cell changes. Figure 2A and 2B shows the results of an aggregate collection test with hybrid cells isolated before the gastrula stage, and with cells isolated after the embryos had reached the gastrula stage (McClay *et al.*, 1977). The collection pattern

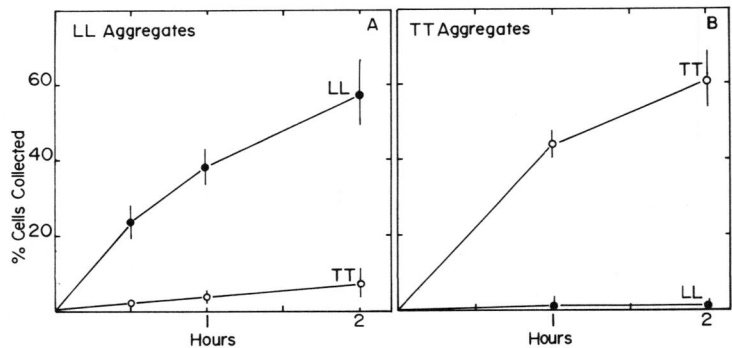

FIG. 1. Species-specific adhesion of cells. The procedures of McClay and Hausman (1975) were used to test the affinity of LL and TT gastrula cells for LL and TT collecting aggregates. (A) Flasks of LL collecting aggregates collected either LL or TT cells for periods up to 2 hours. Each point represents the percentage single cells in suspension collected to aggregate surfaces per unit time with three replicates per time point. (B) The reciprocal experiment using TT aggregates. The cells were labeled with [^{3}H]leucine, and the aggregates were labeled with ^{14}C as an internal standard. Vertical bars indicate standard error of the mean.

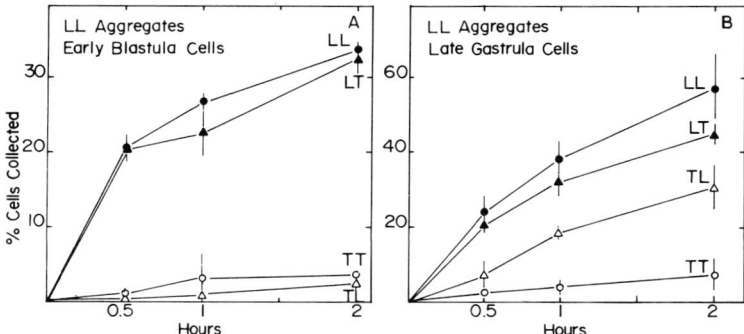

FIG. 2. Age-dependent specificity of collection of hybrid cells. (A) Specificity of collection of early blastula stage cells to blastula stage cell aggregates of LL. TL cells do not adhere to the LL aggregates. (B) By contrast, TL gastrula cells collect to LL gastrula cell aggregates (methods as described in Fig. 1). Vertical bars indicate standard error of the mean.

demonstrated an increase in the rate of collection of TL cells to LL aggregates at a time corresponding to the early stages of gastrulation (there was a similar increase in the affinity of LT cells to TT aggregates at gastrulation). The best explanation for these observations was that hybrid cells had begun to express cell surface components of the paternal specificity, this expression first being detected at gastrulation, and that the cell surface components participated in cell recognition phenomena. Thus, this study supported the concept that cell surface changes occur at gastrulation and that some of those changes influence cellular recognition. The data were insufficient, however, to permit the conclusion that the changes observed were related necessarily to separation of germ layers.

Given this defined, genetically correlated change in adhesive specificity, an immunochemical study was undertaken to approach the germ layer question. Cell surfaces were examined to determine if germ layer-specific antigens could be detected.

III. Immunochemical Specificity: Changes Associated with Gastrulation

The demonstration of adhesive recognition of TL cells for LL aggregates suggested that L-specific molecules were expressed on the TL cells. (To simplify this series of experiments only one of the two hybrids will be discussed; the LT to TT data show the same pattern in all of the studies below providing a reciprocal test as confirmation in each case.) The TL cells had a T-specific background since they were derived from

Tripneustes eggs. Therefore, most of the membrane molecules would be expected to be derived from T-specific maternal and embryonic genes. L-Specific molecules on TL cells could be present only through expression of the embryonic genome. This system therefore had some unique advantages for an immunological approach. Antiserum made against LL membranes could be absorbed with TT cells to assure its specificity for L antigens and a non-cross-reactivity with T antigens. Then, even though the serum would be polyspecific, it should identify only those antigens appearing on TL cells as a result of L gene activation. Table I shows the results of an agglutination study demonstrating the presence of L-specific antigens on TL cells. The antigens were first detected during the early phases of gastrulation (McClay and Chambers, 1978). [As an example, the reciprocal experiment on antiserum directed against TT cell membranes identified the presence of T antigens on LT hybrid cells surfaces by agglutination and by indirect immunofluores-

TABLE I

AGGLUTINATION OF TL CELLS OF DIFFERENT EMBRYONIC AGES[a]

CELLS	SERUM	RECIPROCAL SERUM DILUTION 4 8 16 32 64 128 256 512
TL BLASTULA CELLS	PREIMMUNE SERUM	N.A.
	LL ANTISERUM UNABS.	▨
	LL ANTISERUM ABSORBED WITH TT GASTRULA CELLS	N.A.
TL GASTRULA CELLS	LL ANTISERUM ABSORBED WITH TT GASTRULA CELLS	▨▨
TL PLUTEUS CELLS	LL ANTISERUM ABSORBED WITH	
	TT GASTRULA CELLS	▨▨▨▨▨▨
	LL BLASTULA CELLS	▨▨▨
	LL PLUTEUS CELLS	N.A.
	LL ECTODERM CELLS	▨▨▨▨
	LL ENDODERM CELLS	▨▨▨▨
	LL ECTODERM CELLS THEN ENDODERM CELLS	N.A.

[a] TL blastula cells (top), gastrula cells (middle), or pluteus cells (bottom) were incubated in antiserum to LL gastrula membranes. The antiserum was preabsorbed with the cell types indicated (center column) (2 times each except for the bottom line where the serum was absorbed once with ectoderm cells and once with endoderm cells). In the right column the bars represent the titer of agglutinated cells. N.A. = no agglutination at an initial serum dilution of 1:4.

cence (McClay and Marchase, 1979). Again, these paternal antigens were first detected during the early stages of gastrulation.]

Before continuing, it is necessary to comment on the utility of agglutination and immunofluorescence for studies of this sort. Agglutination assays can be used quantitatively to titer cell surface antigens if two criteria are met: The cells must not agglutinate in the absence of serum, and the agglutination must be shown to be a direct response to the presence of antiserum. Both of these criteria were met with the sea urchin cells: The cells did not agglutinate (nor did they aggregate) in calcium–magnesium-free seawater, and they were agglutinated only by specific antiserum (preimmune serum had no effect). If there is a fairly sharp end point to an agglutination titration, this approach can provide a fairly good quantitative impression of the antigens present on the surface of a population of cells (Bjerrum, 1977).

Immunofluorescence offers a second approach for detecting cell surface antigens. With this technique a fluorescent probe is chemically attached directly to the antiserum, or the probe is attached indirectly to a second antiserum which can then bind to the first. Immunofluorescence can be titered although it is generally harder to quantitate than agglutination assays (Bjerrum, 1977). The advantage of immunofluorescence is that in a mixed population of cells the technique can distinguish whether all or some of the population of cells contain an antigen. If, for example, in a population of cells there were a small number of cells containing a cell surface antigen that was recognized by an antiserum, that population would be detected by immunofluorescence but might not be detected by agglutination assays.

Going back to Table I, agglutination studies established that TL cells expressed L-specific antigens at gastrulation. Immunofluorescent studies then established that most, if not all, the TL cells expressed the L antigens.

The next study asked whether the antigens detected were age-specific and/or germ layer-specific. Tables II and III summarize the results of agglutination and indirect immunofluorescent studies testing anti-LL and anti-TT sera on cells of different ages and of different germ layer specificities. The tests coupled a quantitative preabsorption of the sera with cells followed by the determination of agglutination titer or fluorescence. These tests identified four classes of antigens on cell surfaces at the gastrula stage. First, maternal antigens, defined as those antigens present during early cleavage, were also present on gastrula cells, and at least some of those maternal antigens detected at gastrulation were produced on embryonic mRNAs. The evidence is as follows: LL gastrula stage cells were agglutinated in the presence of

TABLE II

Agglutination of LL Cells from Different Embryonic Ages with Antiserum against LL Gastrula Cell Membranes[a]

CELLS	SERUM	RECIPROCAL SERUM DILUTION (4, 8, 16, 32, 64, 128, 256, 512)
LL BLASTULA CELLS	PREIMMUNE SERUM	N.A.
	LL ANTISERUM UNABS.	bar to 128
	LL ANTISERUM ABSORBED WITH	
	TT GASTRULA CELLS	bar to 128
	LL BLASTULA CELLS	N.A.
	LL PLUTEUS CELLS	N.A.
LL GASTRULA CELLS	PREIMMUNE SERUM	N.A.
	LL ANTISERUM UNABS.	bar to 512
	LL ANTISERUM ABSORBED WITH	
	TT GASTRULA CELLS	bar to 512
	LL GASTRULA CELLS	N.A.
	TT GASTRULA MEMBRANES	bar to 256
	LL GASTRULA MEMBRANES	N.A.
LL PLUTEUS CELLS	PREIMMUNE SERUM	N.A.
	LL ANTISERUM ABSORBED WITH	
	TT GASTRULA CELLS	bar to 512
	LL BLASTULA CELLS	bar to 64
	LL PLUTEUS CELLS	N.A.
	LL ECTODERM CELLS	N.A. – ECTODERM CELLS; bar to 128 ENDODERM CELLS
	LL ENDODERM CELLS	bar to 256 ECTODERM CELLS; short bar ENDODERM CELLS

[a] The serum was preabsorbed with the cell types indicated (center column) and the agglutination titer determined (right column). N.A. = No agglutination at an initial serum dilution of 1:4.

antiserum directed against cell membranes of LL gastrulae. The agglutinating activity was completely absorbed by LL gastrula cells; it was partially absorbed by cleavage stage cells. Thus, the serum identified antigens present on cleavage stage cells that were also present on gastrula stage cells. Cleavage stage LL cells and mesenchyme blastula stage LL cells also were agglutinated by the LL gastrula antiserum; the agglutinating activity could be completely absorbed from

TABLE III

Agglutination of TT Cells of Different Ages with Antiserum Preabsorbed with Different Cell Types[a]

CELLS	SERUM	RECIPROCAL SERUM DILUTION 4 8 16 32 64 128 256 512
TT BLASTULA CELLS	PREIMMUNE SERUM	N.A.
	TT ANTISERUM UNABS.	▬▬ (to 16)
	TT ANTISERUM ABSORBED WITH	
	TT GASTRULA CELLS	N.A.
	TT BLASTULA CELLS	N.A.
	LL GASTRULA CELLS	▬▬ (to 16)
TT GASTRULA CELLS	PREIMMUNE SERUM	N.A.
	LL ANTISERUM UNABS. ABSORBED WITH	▬▬ (to 16)
	TT GASTRULA CELLS	N.A.
	LL GASTRULA CELLS	N.A.
	TT ANTISERUM UNABS. ABSORBED WITH	▬▬▬▬▬▬ (to 512)
	LL GASTRULA CELLS	▬▬▬▬▬▬ (to 256)
	TT GASTRULA CELLS	N.A.
	TT BLASTULA CELLS	▬▬ (to 16)
	TT ENDODERM CELLS	▬▬▬▬ ECTODERM CELLS N.A.− ENDODERM CELLS
	TT ECTODERM CELLS	N.A.− ECTODERM CELLS ▬▬ ENDODERM CELLS

[a] Agglutination titers are shown in the right column. N.A. = No agglutination at an initial serum dilution of 1:4.

the serum either with cleavage-stage cells or with gastrula-stage cells. Again, this supported the observation that antigens on cleavage-stage cells were also on gastrula-stage cells. As indicated previously, hybrid TL gastrula cells were agglutinated by the LL antiserum. The agglutinating activity was absorbed by LL gastrula cells, by TL gastrula cells, partially by LL cleavage stage cells, but not at all by either TL cleavage stage cells nor by TT gastrula cells. This evidence supports the notion that L-specific maternal antigens and embryonic antigens (discussed below) are both produced on L-specified mRNAs in the TL hybrid embryos. It further suggests that embryonic genes for these antigens are not activated until gastrulation. This latter suggestion should be interpreted with caution, however, since the methods cannot detect "first appearance" of an antigen; they apply only when antigen concentrations reach the limits of detection of an antiserum and/or the technique used. It is possible, therefore, that the embryonic genome for new cell surface antigens is activated earlier than gastrulation. The caution, however, has no bearing on the conclusion with TL cells that production of maternal cell surface antigens are produced on embryonic templates at gastrulation.

The second class of cell surface antigens are "embryonic antigens," defined as those not detected on cleavage-stage cells. Embryonic antigens were detected on LL gastrula cells and on TL gastrula cells. In this case the evidence is as follows: Cleavage-stage cells of TL and TT did not remove the LL gastrula-stage cell agglutinating activity from the LL gastrula serum. Cleavage-stage LL cells removed some but not all of the agglutination activity. Thus, new antigens must have been synthesized. These antigens may have been synthesized on embryonic mRNAs or by a delayed expression of a maternal mRNA, but this latter possibility is eliminated in the hybrid study: Expression of L embryonic antigens in the TL hybrid could have resulted only from the expression of embryonic genes.

The third and fourth classes of embryonic antigens detected were ectoderm-specific and endoderm-specific cell surface antigens (actually those probably are subclasses of the embryonic antigen class). Detection of these antigens was possible through development of a method for quantitative separation of ectoderm and endoderm cells (McClay and Marchase, 1979). Kane (1973) showed that glycine–EDTA removed the hyaline membrane coat surrounding the embryo. With the hyaline coat removed, it was possible to separate ectoderm cells from the more tightly adhering endoderm (mesoderm could be selected against but it was not possible to isolate mesoderm with these methods). The ability to separate two germ layers now permitted an immunochemical examination of hyaline, the germ layers, and the basement membrane lying under the ectoderm layer. The protein hyalin (Kane, 1973) was isolated from the hyaline membrane. Purified hyalin did not absorb immune activity from the anti-LL serum. After removing the hyaline membrane and releasing the ectoderm, the ectoderm-free embryo was surrounded by a basement membrane (Fig. 3). Immunofluorescent studies showed that the basement membrane contained ectoderm-specific antigens; it did not contain endoderm-specific antigens.

Endoderm cells were isolated by a second treatment with glycine–EDTA. Using a histochemical test (McClay and Marchase, 1979), it was shown that the ectoderm and endoderm cell preparations were each better than 90–95% pure. Ectoderm and endoderm cells were observed to contain qualitatively different antigens in agglutination studies and by immunofluorescence (Tables II and III).

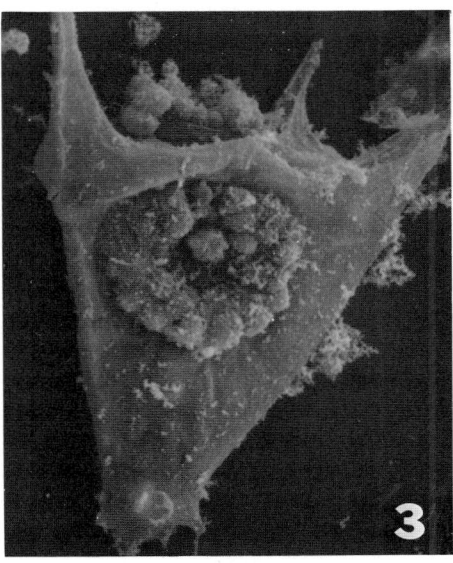

FIG. 3. Ectoderm-free embryo. A pluteus-stage embryo of *Arbacia punctulata* was treated with glycine–EDTA and the ectoderm was released. The ectoderm-free embryo was fixed and processed for scanning electron microscopy (McClay and Marchase, 1979). The basement membrane is seen covering the embryo. The view is from the aboral surface showing a ring of cells surrounding the anus. The stomodael opening can be seen at the top. In living embryos the basement membrane is transparent and, after removing the ectoderm, the endoderm remains intact. From McClay and Marchase (1979).

IV. Relationship between Cell Surface Antigens and Cell–Cell Interactions

These studies had demonstrated a parallel between the detection of new adhesive specificites at gastrulation and the detection of new cell surface antigens at gastrulation. The next step was to ask whether the two phenomena were related or coincidental. The approach used by Gerisch and co-workers (Beug et al., 1970, 1973a,b) was adapted for these tests. Gerisch (see chapter in this series, Vol. 14) found that adhesion of aggregation-competent *Dictyostelium* cells was inhibited selectively in the presence of Fab antibody fragments of serum directed against aggregation-competent cells. The monovalent Fab fragments inhibited aggregation presumably by blocking adhesive sites present on aggregation competent cells but not present on aggregation-incompetent growth phase cells. Aggregation-competent cells absorbed away the Fab-inhibiting activity; the activity was not absorbed by growth phase cells. It was concluded that aggregation-competent cells had specific contact sites that were recognized by the antiserum.

The protocol of Gerisch was adapted to test the activity of anti-LL serum Fab fragments in the sea urchin system (McClay *et al.*, 1977). Fab fragments were made from the same anti-LL serum that had been shown to identify the four classes of antigens in the gastrula. Figure 4 shows the results of aggregate collection experiments in which LL cells were pretreated with Fab fragments at several concentrations (Fig. 4A) or after absorption of the Fabs with LL or TT cells (Fig. 4B). The cells were washed and then added to flasks containing LL aggregates. Collection was inhibited and there was a dose dependence to the inhibition (Fig. 4A). Fab fragments of preimmune serum did not inhibit collection. The inhibiting activity was removed from the anti-LL Fab fragments by preabsorption with LL gastrula cells (Fig. 4B); activity was not removed by preabsorption with TT cells nor with LL cleavage stage cells. These results suggested, therefore, a relationship between the embryonic ectoderm–endoderm class of antigens and the new adhesive specificity detected at gastrulation (Fig. 5).

The results have several shortcomings, however. Gerisch showed that Fabs directed against aggregation-incompetent cells would bind to aggregation-competent cells and not affect adhesion. Fab preparations against aggregation-competent cells could bind to cells at one-thirtieth the concentration of bound aggregation-incompetent Fabs

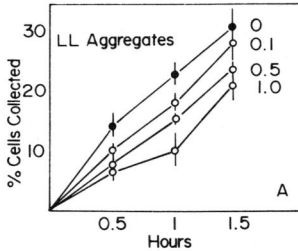
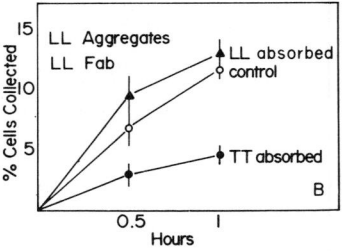

FIG. 4. Collection of LL gastrula-stage cells to LL gastrula-stage aggregates. Fab fragments of antiserum directed against LL gastrula membranes were tested at several concentrations (A) and after being absorbed with LL or TT cells (B). In Fig. 4A cell suspensions of LL cells were incubated in LL Fab (at concentrations 0, 0.1, 0.5, and 1 mg/ml); after 20 minutes the cells were added to flasks containing collecting aggregates. In Fig. 4B the Fabs were preabsorbed with LL or TT cels, then added to LL cell suspensions at a concentration of 1 mg/ml. After 20 minutes the cell suspensions were harvested and added to flasks of collecting aggregates. Collection measurements were as described in Fig. 1.

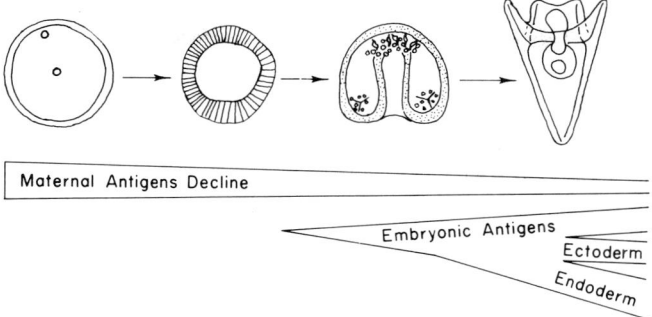

FIG. 5. Summary of the four classes of antigens detected. Maternal antigens were present on all stages tested. Some antigens of this class were synthesized by the embryonic genome beginning at the gastrula stage. Embryonic antigens were not detected until gastrulation and they were products of the embryonic genome. Ectoderm-specific and endoderm-specific antigens appeared at gastrulation and were defined by their germ layer location.

and completely block aggregation. In the sea urchin system this kind of control was not available so it was possible the anti-LL Fabs bound to antigens unrelated to adhesion and adhesion was inhibited coincidentally by accidentally covering adjacent adhesive sites. The same criticism holds for the *Dictyostelium* inhibition but is less likely there because of the lack of an effect with aggregation-incompetent Fabs.

A second shortcoming for Fab experiments is polyspecific antiserum. Even after absorption with TT cells and with cleavage-stage cells the serum may have remained polyspecific. Therefore, if there were only one important adhesive antigen, the serum would require further absorptions in order to identify that single antigen. Of course this assumes a one-antigen adhesive mechanism although there is no evidence to indicate how many molecules might participate in any adhesive interaction. Furthermore, there is no *a priori* reason to assume that recognition antigens must be related to the mechanism of adhesion. Indeed, some recognition antigens might signal a release of an adhesive interaction as evidenced by delaminations and cavitation movements.

V. Synthesis and Conclusions

Two experimental approaches for studying gastrulation have been reviewed. The adhesive behavior data demonstrated changes in specificity coincident with the onset of gastrulation. The immunochemical data demonstrated cell surface antigen distributions that were age-specific and tissue-specific, and these observations also correlated well with the onset of gastrulation. Data using Fab antibody fragments suggested a functional link between antigen appearance and adhesive behavior changes. The strength of the data is that they correlate well with a specific morphogenetic event, yet it is still premature to suggest a mechanism. Before morphogenesis can be described at the molecular level, several problems must be solved. One concern often expressed is that morphogenetic determinants might be present in such small numbers and/or with such a great diversity that an immune or biochemical approach would not detect them. This dilemma was faced by Sperry in his analysis of retina–optic tectum synaptogenesis (Sperry, 1963). In order to eliminate the genetic load requirement for a one-cell–one-specificity hypothesis, most models suggest gradients of only a few determinants for morphogenetic organization. If gradients are indeed present and/or if the number of determinants per cell is large enough, there is reason for optimism that an immune approach coupled with studies on cell interactions can explain the role of cell surface specificities in morphogenesis.

A second concern is the heterogeneity of a tissue. Ideally, in order to study membrane biochemistry it would be useful to have large numbers of cells of a homogeneous type. Unfortunately, the embryo works against this requirement both spatially and temporally. This is a major problem with analyses of the retina–optic tectum system and is a problem associated with most embryonic systems. This problem is reduced though not eliminated in early sea urchin development in that large numbers of embryos can be grown synchronously, and then separated into germ layer-specific populations.

The serology presents a third problem. Fortunately for this study, sea urchin antigens produced a brisk immune response when injected into rabbits. The antiserum, though polyspecific, was of value because the number and kinds of absorptions available permitted detection of a number of antigen classes and also permitted rigorous controls. When coupled with methods for isolating membrane antigens a polyspecific serum can be reduced to monospecificity (Thiery *et al.*, 1977). It can then be used to probe the function of individual antigens. The realization of this goal is complicated by the insolubility of most membrane antigens. Membranes can be solubilized with detergents and the antigens may retain immune activity, but it becomes technically very difficult to relate solubilized antigens back to possible morphogenetic function. Isolated antigens can be used to generate monospecific antiserum, perhaps easing this problem. This goal also requires that relevant antigens be in sufficient numbers to be analyzed immunologically.

Going back to the hypotheses presented at the outset of this chapter, the data presented here support the concepts that new recognition determinants appear on the cell surface at gastrulation. Further, the demonstrated changes in specific cell interactions at gastrulation support the notion that specific adhesive interactions help govern morphogenetic movements.

Additional tests of the specificity hypothesis are needed in two areas, however, before these assertions can be made with any strength. First, the functional relationship between cell surface antigens and the mechanism of cell adhesion must be established. Second, the cell surface data must be related to cell shape changes, movements, and to other phenomena thought to participate in morphogenesis. The approaches presented here offer both a model system and a technology for asking the critical though difficult questions.

VI. Concluding Remarks

A species-specific pattern of cellular recognition was identified experimentally between embryonic cells of *Lytechinus variegatus* and *Tripneustes esculentus*. Hybrid embryos, formed as a cross between these species, were studied to determine their pattern of cellular recognition. Prior to gastrulation, hybrid cells adhered to cells of the maternal genotype only. After gastrulation had begun in the hybrid embryos, hybrid cells adhered to cells both the maternal and paternal genotype.

A serological study identified a similar pattern of cell surface antigen expression: Paternal antigens were first observed on hybrid cell surfaces at the gastrula stage. The serology further identified four classes of antigens present on embryonic cell surfaces: (1) maternal antigens present on cleavage-stage cells, (2) embryonic antigens appearing at gastrulation and specified by embryonic templates, (3) ectoderm-specific antigens, and (4) endoderm-specific antigens. Fab fragments of the antisera were found to inhibit adhesion of cells selectively, suggesting a relationship between the antigens identified and the adhesive specificity demonstrated.

REFERENCES

Beug, H., Gerisch, G., Kempff, S., Riedel, V., and Cremer, G. (1970). *Exp. Cell Res.* **63**, 147.
Beug, H., Katz, F. E., and Gerisch, G. (1973a). *J. Cell Biol.* **56**, 647–659.

Beug, H., Katz, F. E., Stein, A., and Gerisch, G. (1973b). *Proc. Natl. Acad. Sci. U.S.A.* **70,** 3150–3154.
Bjerrum, O. J. (1977). *Biochim. Biophys. Acta* **472,** 135–196.
Burnside, B. (1971). *Dev. Biol.* **26,** 416–441.
Burnside, B. (1973). *Am. Zool.* **13,** 989–1006.
Gibbins, J. R., Tilney, L. G., and Porter, K. R. (1969). *J. Cell Biol.* **43,** 201–226.
Giudice, G. (1962). *Dev. Biol.* **5,** 402–411.
Giudice, G., and Mutolo, V. (1970). *Adv. Morphogen.* **10,** 115–158.
Giudice, G., Mutolo, V., Donatum, G., and Bosco, M. (1969) *Exp. Cell Res.* **54,** 279–281.
Gustafson, T. (1963). *Exp. Cell Res.* **32,** 570–589.
Gustafson, T., and Wolpert, L. (1962). *Exp. Cell Res.* **27,** 260–279.
Gustafson, T., and Wolpert, L. (1963). *Int. Rev. Cytol.* **15,** 139–214.
Gustafson, T., and Wolpert, L. (1967). *Biol. Rev.* **42,** 442–498.
Holtfreter, J. (1939). *Arch. Exp. Zellforsch.* **23,** 169–209.
Holtfreter, J. (1943a). *J. Exp. Zool.* **94,** 261–318.
Holtfreter, J. (1943b). *J. Exp. Zool.* **95,** 171–212.
Holtfreter, J. (1948). *Ann. N.Y. Acad. Sci.* **49,** 709–760.
Hood, L., Huang, H. V., and Dreyer, W. J. (1977). *J. Supramol. Struct.* **1,** 531–559.
Kane, R. E. (1973). *Exp. Cell Res.* **81,** 301–311.
Marchase, R. B. (1977). *J. Cell Biol.* **75,** 237–257.
Marchase, R. B., Barbera, A. J., and Roth, S. (1975). *Ciba Found. Symp.* **15,** 315–341.
Marchase, R. B., Vosbeck, K., and Roth, S. (1976). *Biochim. Biophys. Acta* **457,** 385–416.
McClay, D. R. (1971). *Biol. Bull.* **141,** 319–330.
McClay, D. R. (1976). *J. Cell Biol.* **70,** 130a.
McClay, D. R., and Baker, S. R. (1975). *Dev. Biol.* **43,** 109–122.
McClay, D. R., and Hausman, R. E. (1975). *Dev. Biol.* **47,** 454–460.
McClay, D. R., and Chambers, A. F. (1978). *Dev. Biol.* **63,** 179–186.
McClay, D. R., and Marchase, R. B. (1979). *Dev. Biol.* **71,** 289–296.
McClay, D. R., Chambers, A. E., and Warren, R. H. (1977). *Dev. Biol.* **56,** 343–355.
Moscona, A. A. (1968). *Dev. Biol.* **18,** 250–277.
Moscona, A. A. (1976). *In* "Neuronal Recognition." Plenum, New York.
Roth, S. (1968). *Dev. Biol.* **18,** 602–631.
Roth, S. (1973). *Q. Rev. Biol.* **48,** 451–563.
Roth, S. A., and Weston, J. A. (1967). *Proc. Natl. Acad. Sci. U.S.A.* **58,** 974–980.
Sperry, R. W. (1963). *Proc. Natl. Acad. Sci. U.S.A.* **50,** 703–710.
Spiegel, M., and Spiegel, E. S. (1975). *Am. Zool.* **15,** 583–606.
Steinberg, M. S. (1970). *J. Exp. Zool.* **173,** 395–434.
Thiery, J.-P., Brackenbury, R., Rutishauser, U., and Edelman, G. M. (1977). *J. Biol. Chem.* **252,** 6841–6845.
Townes, P. L., and Holtfreter, J. (1955). *J. Exp. Zool.* **128,** 53–120.
Trelstad, R. L., Hay, E. D., and Revel, J. P. (1967). *Dev. Biol.* **16,** 78–106.
Trinkaus, J. P. (1967). *In* "Major Problems in Developmental Biology" (M. Locke, ed.), pp. 125–176. Academic Press, New York.
Tyler, A. (1947). *Growth* **10,** 6–7.
Weiss, P. (1947). *Yale J. Biol. Med.* **19,** 235–278.

CHAPTER 10

CELL SURFACE ANTIGENS OF CULTURED NEURONAL CELLS

Richard Akeson

DEPARTMENTS OF PEDIATRICS AND BIOLOGICAL CHEMISTRY
CHILDREN'S HOSPITAL RESEARCH FOUNDATION
CINCINNATI, OHIO

I.	Introduction	215
II.	Neural-Specific Antigens of Cultured Neuronal Cells	216
III.	Do Mouse Neuroblastoma Cells Differentiate *in Vitro*?	224
IV.	Approaches to Neural Cell–Cell Interaction	226
	A. Adhesion Assays	226
	B. Effects of Antibodies on Neural Embryogenesis	229
V.	Perspectives on Future Research Directions	230
VI.	Concluding Remarks	232
	References	233

I. Introduction

The biological mechanisms underlying the development and organization of multiple cell types into the vertebrate brain have attracted a great deal of attention from many scientific disciplines. It has seemed a reasonable working hypothesis that the interactions among cells in the developing nervous system are mediated at least in part by components present on their exterior surfaces. Cell surface molecules played a key role in Ehrlich's early theories of the immune response and have also long been postulated to play a role in tissue development (Tyler, 1947; Weiss, 1947). It has further seemed reasonable to hypothesize that the cell surface molecules mediating the specific cell–cell interactions and organizations of the nervous system should be unique to the nervous system and probably unique to a class or subclass of cells within the nervous system. In the past 5 years immunological approaches to testing these hypotheses have made significant progress primarily due to the availability of clonal neural* cell lines of defined cell type. This chapter reviews available data on surface antigens† of

* The term "neural" will be used to refer to all nervous system cell types.

† As was emphasized to me several years ago by Charles Todd, what is actually described in the literature are patterns of antibody reactivity from which the presence of an "antigen" is inferred. Using the standard definition that an antigen is a molecule

cultured neuronal lines, discusses approaches to determining the role of surface components in cell–cell interactions, and indicates some perspectives on future work in this area. Other neural-specific antigens such as myelin basic protein, S-100, 14-3-2, glycolipid antigens, or synaptosomal antigens will not be discussed (for review, see Bock, 1978). Chapters in this volume by Drs. M. Schachner and K. L. Fields review additional aspects of neural cell surface antigens.

II. Neural-Specific Antigens of Cultured Neuronal Cells

Although some human neuroblastoma cell lines had been available earlier, the establishment in culture in 1969 of the mouse neuroblastoma C1300 by three laboratories (Schubert *et al.*, 1969; Augusti-Tocco and Sato, 1969; Klebe and Ruddle, 1969) stimulated a flurry of studies on cultured neuronal tumors. Clones from this tumor were shown to synthesize and respond to neurotransmitters, generate action potentials in response to electrical stimulation, and extend cellular processes up to 3000 μm long which morphologically resemble axons (Fig. 1; Seeds *et al.*, 1970). Despite this striking change in morphology to an apparently more differentiated state, measurements of biochemical parameters such as choline acetylase and tyrosine hydroxylase (Kates *et al.*, 1971) tended to show rather modest (2- to 3-fold) increases accompanying this morphologic change.

We chose to use immunologic methods to search for neuron-specific changes in cell surface composition which might accompany this morphologic differentiation (Akeson and Herschman, 1974a,b). Cells of clone N18 of the C1300 tumor were grown in medium lacking serum for 5 days (Fig. 1G) to induce morphologic differentiation. After treatment with $0.001\ M$ $ZnCl_2$ which preserved the processes in an extended state, the cells were scraped from the dish and injected into rabbits. Without this fixation, processes on scraped cells retract into the cell body.

---

which combines with an antibody, it is generally inferred that the observed pattern of cell and tissue reactivity is due to the presence of a unique single macromolecule (or small group of macromolecules) on the reactive tissues. However, this is not necessarily the case. All assays can be misleading particularly when values for positive cells are not greatly above "control" values. "Antigens" apparently unique to a tissue may be only quantitatively enriched on that tissue when studied with refined methods. Bearing these potential problems in mind, only two "antigens" defined by reactivity patterns have been adequately characterized so that a specific macromolecule can be identified as the actual antigen [NS2 (Yuan *et al.*, 1977) and band 1 (Akeson and Hsu, 1978)]. Having made this distinction, for economy of space it will be disregarded and patterns of reactivity will be called antigens. The reader, however, should keep this distinction in mind when evaluating the significance of the "antigens" described.

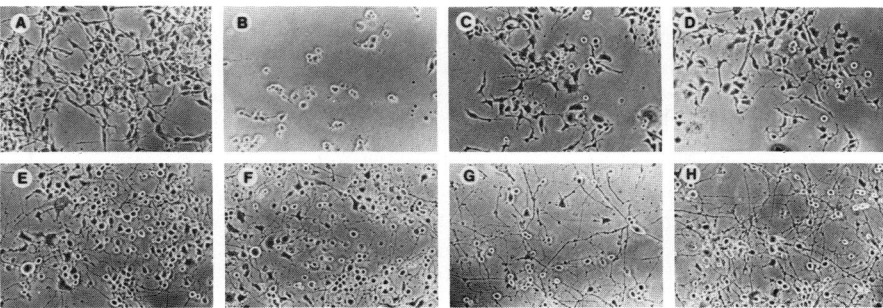

Fig. 1. Mouse neuroblastoma clone N18 cells in late logarithmic growth (A), and after replating in serum-free media for 2 hours (B), 1 (C), 2 (D), 3 (E), 4 (F), 5 (G), or 6 (H) days. Bar: 50 μm.

Prior to adsorption, the antiserum chosen for further characterization reacted more strongly with brain than with other murine organs and more strongly with neuroblastoma cells than with other cultured mouse cell lines. Furthermore, the antiserum reacted more strongly with morphologically differentiated neuroblastoma cells (induced by growth in serum-free media; hence N18-SF) in the complement-fixation assay. After adsorption with particulate material from mouse liver and suspension-culture neuroblastoma cells, no specific binding of immune ^{125}I-labeled IgG could be detected with adult mouse liver, spleen cells, or heart. Brain bound significant amounts of ^{125}I-labeled IgG and slight binding was observed with kidney. Morphologically differentiated N18-SF cells bound IgG, while both N18 cells grown in suspension culture and the 10D clone of neuroblastoma cells, which does not extend processes in serum-free media, bound merely background amounts of IgG. Further analysis of the IgG molecules that specifically bound to N18-SF cells by both adsorption and direct binding experiments with mouse organs indicated that about two-thirds of these molecules reacted with mouse brain but not liver, spleen, heart, or skeletal muscle. Again, kidney appeared to have some activity (~1/7 that of brain). Combined indirect immunofluorescence and adsorption experiments documented the cell surface reactivity and brain specificity of the IgGs which reacted with the N18-SF cells. Although both cell bodies and processes were fluorescent, there was no apparent antigen localization to the growth cone or other portions of the processes. Two classes of neural-specific antigens were postulated based on these experiments. The first is shared by cultured N18 cells and adult mouse brain but not other cultured mouse cell lines or the organs previously mentioned. The second antigen or antigens is also found on morphologically differentiated N18 cells and adult brain but not on undifferentiated N18 cells, other cultured cells, or other organs. By virtue of their location on the cell surface and apparent nervous system specificity, these antigens appeared to be interesting candidates for a role in cell–cell interactions in the nervous system.

Further studies extended the correlation between the ability to differentiate morphologically and an increased binding of IgG to additional neuroblastoma clones (Akeson and Herschman, 1975). In other direct binding experiments with adsorbed IgG (previously unreported), adult mouse superior cervical ganglia showed binding similar to that of brain. These cells are presumably a normal analog of the cultured neuroblastoma tumor lines. IgG binding to cerebrum, cerebellum, olfactory bulb, and brain stem did not vary more than threefold on a per milligram protein basis. Strong binding was observed with synaptosomal preparations.

More recently we have begun to characterize the nervous system-specific cell surface antigens found on the N18 murine neuroblastoma line. The solubilization of membrane-bound antigens is presently a rather empirical science (but see Umbreit and Strominger, 1973); hence, in our initial study we took advantage of the fact that cell surface polypeptide antigens can be obtained from culture media (or body fluids) in water-soluble form (Kapellar et al., 1973). Whether the mechanism of appearance of these antigens in the media is via classical secretion pathways, shedding, or exfoliation of individual surface components or small pieces of membrane, proteolytic or glycolytic cleavage of surface components, cell death and lysis, or other means must be determined for each individual component. However, the presence of antigenically active surface components in water-soluble form aids significantly in purification and isolation. Concentration and iodination of N18 proteins exfoliated into serum-free media followed by immune precipitation and electrophoresis on SDS-polyacrylamide gels revealed a major component of apparent molecular weight ~200,000 (Fig. 2; Akeson and Hsu, 1978). Adsorption analysis indicated that normal adult murine brain contained this polypeptide (named band 1) but not adrenal, heart, kidney, liver, lung, and spleen. The absence of band 1 from adrenal, a neural crest-derived tissue, and its presence on central nervous system tissue are of interest. Antisera produced by other workers to mouse neuroblastoma (see Table I) have also detected antigens common to the peripheral nervous system, neuroblastoma tumor, and central nervous system tissue. Four other cultured mouse cell lines including the G26 glioma (Sundarraj et al., 1975) were nega-

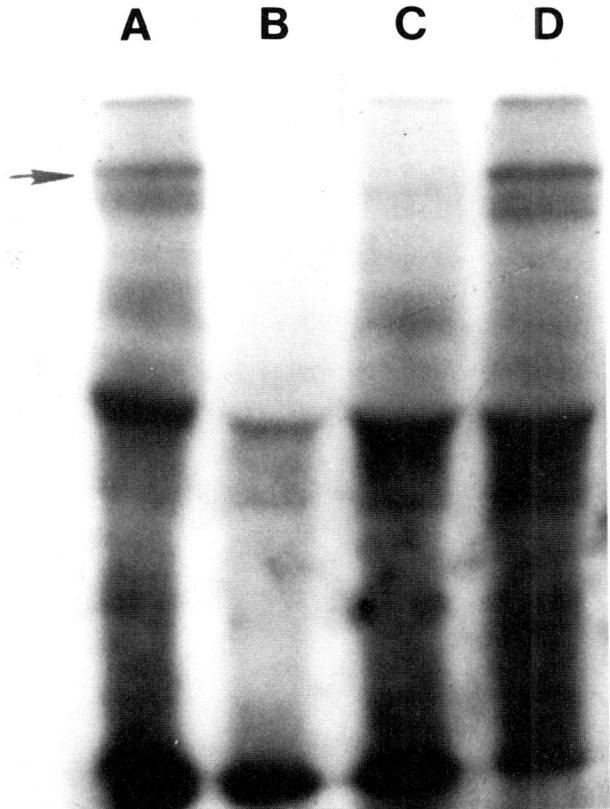

FIG. 2. SDS-polyacrylamide gel of immune precipitates with ^{125}I-labeled N18 shed proteins and antiserum to N18 (A), preimmune serum (B), antiserum to N18 first adsorbed with mouse brain (C), or liver (D). The arrow marks the position of band 1 polypeptide. The polypeptides of slightly lower apparent molecular weight than band 1 are present in some preparations of N18 shed proteins and are believed to be proteolytic fragments of band 1.

tive for band 1 by adsorption, but the independently derived NB2a clone of the mouse neuroblastoma was positive. Solubilization of enzymatically iodinated cells in detergent and immune precipitation yielded a component of somewhat higher molecular weight than band 1 which also appeared to be brain-specific. Our present working hypothesis is that the form of band 1 found in culture media is a proteolytic product of a larger membrane-bound molecule.

Band 1 is a polypeptide whose characteristics are identical to those of the first group of antigens previously described: It is present on the cell surface of cultured neuroblastoma cells and in adult murine brain, but not on other cultured murine cells or adult tissues. Band 1 is, however, clearly present on morphologically undifferentiated neuroblastoma cells and its level is increased less than 10-fold in morphologically differentiated cells (Akeson and Graham, unpublished data). Thus, it is not a member of the second group of antigens previously described.

Band 1 has not been detected on cultured human neuroblastoma cells by adsorption analysis. In general, our own experience with rabbit antisera to cell surface antigens produced by xenogeneic immunization is that they are relatively species-specific. Cross-reactivity between closely related species such as mouse and rat can sometimes be detected, but cross-reactivity between human and mouse is very weak or undetectable. However, both aggregation assays (Moscona, 1965) and studies of synapse formation in cell culture (Crain et al., 1970) indicate that neural cells from one species do recognize nerve or muscle cells from other species.

Two reports of polypeptides similar to band 1 have appeared. McGuire et al. (1978) have shown that nerve growth factor treatment of the cultured rat pheochromocytoma line PC12 results in increased incorporation of fucose and glucosamine into a polypeptide (NILE) of apparent molecular weight 230,000. NGF promotes process formation in monolayer PC12 cells although increased NILE glycosylation can also occur in suspension culture. Anti-band 1 serum immunoprecipitates from PC12 cells a polypeptide in this molecular weight range which can be biosynthetically labeled with leucine or glucosamine. The relationship between band 1 and NILE deserves further study. Gordis et al. (1978b) have studied the reactivity between an antiserum produced to mouse cerebellar cells and lactoperoxidase iodinated cells from primary cultures of cerebellum and cerebral hemispheres. Detergent-soluble nervous system specific components of MW = 145,000 and 200,000 and 145,000 and 250,000 were obtained from these two sources, respectively. The larger components have molecular weights in the same range as band 1 and a component of MW = 145,000 could be obtained by trypsinization of the 200,000 MW polypeptide. These authors suggest that similar antigenic sites are found on these apparently structurally different polypeptides.

The molecular weight of 145,000 obtained by Gordis and co-workers is (perhaps) coincidentally identical to that obtained by Edelman and co-workers for CAM. CAM is a very interesting chicken nervous system-specific cell surface polypeptide first obtained from chick retina culture media which has been shown to apparently play a role in neurite fasiculation in chick ganglia primary cultures (Rutishauser 1978a,b). Anti-CAM does not react with murine tissues. However, one cannot help but speculate that one of the murine polypeptides described in this section may have the same or similar functions in mammals. Continued studies of the similarity and biological function(s) of these murine polypeptides should bring interesting results.

The initial hypothesis of this chapter, that cell–cell recognition is mediated by specific cell surface components, could perhaps then be refined to state that at least some of these components are conserved phylogenetically. In a direct effort to detect nervous system-specific cell surface antigens which had been conserved during evolution from a precursor organism to both mouse and man, we immunized rabbits with human neuroblastoma cells and after a rest period boosted the animals with mouse neuroblastoma cells (Akeson and Seeger, 1977). This protocol should enhance the immune response to antigens common to the two species. An extensive analysis of the sera from eight rabbits immunized with this protocol indicated that boosting with cultured murine neuroblastoma cells specifically enhanced the response to human neuroblastoma cells. One of these sera was chosen for further characterization. After extensive adsorption with nonneural cells and tissues from both species, IgG binding assays detected antigen(s) common to adult human and murine brain, two of four cultured murine neuroblastomas, and three of four cultured human neuroblastomas. Murine kidney, lung, heart, liver, and spleen; human colon, liver, and kidney; two nonneural murine cell lines; and seven nonneural human cell lines were negative. These interspecies neural membrane antigen(s) (or INMA) were subsequently detected by direct binding of IgG to cat, dog, goat, rat, and rabbit brains (livers were negative in all cases). Significant anti-INMA activity was found only after immunization with cells from both species, as has been observed by other workers (Caspar *et al.*, 1977). INMA could be detected on chick brain by adsorption, but not on brains from frog or snake or the major ganglion from a blueshell crab by either assay. An additional experiment with an anti-INMA serum is presented below.

Since the initial report on anti-N18 sera, several other groups have

TABLE I
ANTIGENS DETECTED BY IMMUNIZATION WITH NEURONAL CELL LINES AND TUMORS

Antigen	Immunizing cell line	Adult brain	Fetal brain	Other species	Other organs	Glial tumor cell lines	Reference
N18-differentiated	N18	+	+	NR[a]	Kidney?	0/1	Akeson and Herschman (1974a,b)
MBA1	C1300	+	NR	NR	Kidney	NR	Martin (1974)
NS3	C1300 tumor	+	−	NR	Retina	5/5	Schachner and Wortham (1975)
MBA-2	Autoantibody	+	NR	Many	Kidney	0/5	Martin and Martin (1975a,b)
N1	B103	NR	NR	NR	NR	0/12	Stallcup and Cohen (1976)
N2	B103	NR	NR	NR	NR	0/12	Stallcup (1977a,b)
N3	B35	−	+	NR	NR	0/12	Stallcup (1977a,b)
PC12-brain	PC12	+	NR	NR	−	NR	Lee et al. (1977)
PC12-adrenal	PC12	−	NR	NR	Adrenal	NR	Lee et al. (1977)
INMA	Human, murine neuroblastoma	+	+	Many	−	NR	Akeson and Seeger (1977)
Band 1	N18	+	+	rat	−	0/3	Akeson and Hsu (1978)

[a] NR = not reported.

detected antigens common to brain and cultured neuronal tumor lines after immunization with these lines (Table I). None of these has been isolated or implicated in a neuronal function, thus current analysis of their potential significance must be speculative. One of the most interesting of these is the N1 antigen (Stallcup and Cohen, 1976; Stallcup, 1977a,b) which was found on all six rat central nervous system neuronal tumors tested but not on 18 other rodent lines including 12 glial tumors. The N3 antigen of the same study is specific to fetal brain. Our implicit assumption has been that neural antigens important in development are expressed on neural tumors and adult brain. However, it is also possible that developmentally important antigens may be reexpressed on tumor cells but absent from normal adult brain. The PC12 antigens are also of particular interest as this cell line forms neuromuscular synapses in culture (Schubert *et al.*, 1977). Since the other antigens listed in Table I have not been isolated, it is somewhat dangerous to make comparisons with band 1. However, the cell and tissue distribution of band 1 make it appear to be distinct from N1 and N2 (band 1 absent from B103), N3 (band 1 present on adult rat brain), NS3 (band 1 absent from 3T3 cells) (Doyle *et al.*, 1977), PC12 adrenal (band 1 absent from adrenal), and MBA-2 (band 1 present on N18). According to the reported data, band 1 cannot presently be distinguished from PC12-brain or MBA-1. Band 1 isolated from culture media does not react with an anti-INMA serum. Present experiments in our laboratory are testing the possibility of a more subtle relationship between INMA and band 1.

The apparent presence on kidney of some of these otherwise nervous system-specific antigens is difficult to interpret. One interpretation of this observation is that these antigens are not truly organ-specific and hence not worth further study. Alternatively, one could argue that a single antigen could function in mediating cellular interaction and organization in more than one organ system without confusion due to anatomic isolation. A similar "problem" with some antigens is their presence on tumor cells of both apparently neuronal and glial phenotype or even nonneural tumors but not the corresponding normal tissues (Zeltzer and Seeger, 1978). These apparently anomalous distributions could be due to reexpression of hypothetical antigens present on a precursor cell leading to both neurons and glia, expression of "inappropriate" antigens by tumor cells (M. Raff, personal communication), or other factors. Both these issues are discussed in greater detail in the articles by Drs. M. Schachner and K. L. Fields in this volume.

III. Do Mouse Neuroblastoma Cells Differentiate *in Vitro*?

As previously mentioned, the morphologic differentiation of cultured murine neuroblastoma cells is accompanied by changes in biochemical parameters which are relatively modest. We have found the nervous system-specific band 1 protein to increase less than 10-fold in morphologically differentiated cells. However, several membrane differences between differentiated and undifferentiated cells have been observed. Changes in the rate of turnover of two membrane glycoproteins (Mathews *et al.*, 1976), changes in the characteristics of glycopeptides released from intact cells by trypsin (Brown 1971; Glick *et al.*, 1973), changes in the relative synthesis of membrane proteins isolated on sodium dodecyl sulfate containing acrylamide gels (Schubert *et al.*, 1971a), and in the distribution of concanavalin A receptors (Graham *et al.*, 1974) have been observed on "differentiated" neuroblastoma cells. The polypeptide profiles of morphologically undifferentiated (suspension culture) and differentiated (dibutyryl cAMP-treated) clone N2aE cells have been examined on SDS-polyacrylamide gels (Truding *et al.*, 1975). A component on differentiated cells of MW = 105,000 was preferentially labeled using [^{3}H]glucosamine, while enzymatic iodination preferentially labeled a component of MW = 78,000 on these cells. Morphologically differentiated cells also have a different secreted protein profile (Truding and Morell, 1977). On the other hand, expression of several surface antigens (Schachner, 1973) and also the glycolipid composition of monolayer and suspension culture cells (Yogeeswaran *et al.*, 1973) are similar. Unfortunately, none of the components in which changes were observed are known to be specific to neurons. If we adopt the practical definition of differentiation as a 10-fold increase in the expression of a gene product related to the unique functions of an organ or cell type, there is no strong biochemical evidence available to support the view that the morphologic differentiation of neuroblastoma cells is also a biochemical differentiation. Two proteins involved in acetylcholine function—true acetylcholinesterase (Blume *et al.*, 1970; Kates *et al.*, 1971) and the acetylcholine receptor (Simantov and Sachs, 1973)—do show more than 10-fold or nearly 10-fold increases in differentiated cells, respectively. However, the changes in acetylcholinesterase have been interpreted by some workers to be a nonspecific effect of suboptimal culture conditions (Schubert *et al.*, 1971b).

Electrophysiologic studies have shown more dramatic changes accompanying morphologic differentiation. Differentiated cells are sensitive to acetylcholine only on the cell body and process tips (Harris and Dennis, 1970). Mature neurons also have restricted chemosensitivity.

Selection of large nondividing morphologically mature cells with aminopterin (Peacock et al., 1972) or dimethyl sulfoxide (DMSO) (Kimhi et al., 1976) yields a population with a greater resting potential and ability to generate an action potential. In the study of Peacock et al. (1972), the changes induced by aminopterin selection in a process-forming clone were more pronounced than those induced in a clone unable to form processes. Objections to the idea of correlative morphologic and electrophysiologic differentiation have been raised by Schubert et al. (1973). These workers grew neuroblastoma cells in suspension culture for 20 generations and then monitored them for electrical activity immediately after plating. Such cells could generate an action potential. Therefore, the ability to generate an action potential was not acquired during *in vitro* morphologic differentiation. These authors also criticized earlier electrophysiologic studies on technical grounds. More recently, the aminopterin selection method has been used to demonstrate a shift from a Ca^{2+}-mediated action potential characteristic of immature neurons to the Na^+-mediated action potential of mature neurons (Miyake, 1978). This is the first instance of a qualitative rather than quantitative change in action potential generating capability or transmitter sensitivity (Chalazonitis and Greene, 1974). It is conceivable that the observed quantitative changes could be mediated by a reorganization of preexisting cell surface components. Unfortunately, these electrophysiologic studies may suffer from a selection bias in that large, easily penetrated cells with stable resting potentials are usually studied in such experiments. These cells are similar to the larger cells in Fig. 1G and H while many biochemical assays of "differentiated" cells have been done on cells like those in Fig. 1A or C. It is not clear that the entire population of a culture dish achieves the electrophysiologic capabilities described. With the exception of the Na^+ action potential, all the observed changes seem to be enhancements of preexisting capabilities rather than the development of new cell type-specific capabilities.

Thus, although some changes in surface components measured electrophysiologically clearly occur, the preponderance of biochemical evidence does not support the hypothesis that morphological changes in cultured neuroblastoma cells represent complete differentiation to a neuronal cell type. This is somewhat surprising in view of the facts that morphologically differentiated cells have more polyA-containing RNA (presumably messenger RNA; Prasad et al., 1975), exhibit more protein synthesis in cell-free systems (Zucco et al., 1975), and have different patterns of nonhistone chromosomal proteins (Zornetzer and Stein, 1975). It is certainly possible that additional evidence for correlative morphologic and biochemical differentiation in cultured mouse neuroblastoma cells will accumulate, but current data are not fully persuasive. Some authors have suggested that expression of differentiated functions and components of the plasma membrane are uncoupled from those of the cytoplasm in these tumor cells. Whereas this hypothesis may explain the restriction of "differentiation" of these cells to surface membrane changes, it would also seem to restrict their utility greatly as a model system for neuronal differentiation.

In addition, it should be noted that all the other available neuron like cell lines also do not show marked signs of differentiation *in vitro*. Several groups have induced tumors with ethylnitrosourea (Schubert *et al.*, 1974; Fields *et al.*, 1975; Stahn *et al.*, 1975) which have neuron-like characteristics, but no studies of differentiation with these lines have been reported. The interesting PC12 pheochromocytoma-derived cell line characterized by Greene and co-workers (Greene and Tischler, 1976) expresses several neuron-like properties including a marked process formation in response to nerve growth factor, but does not show substantial changes in adrenergic neurotransmitter levels or metabolism or total polypeptide composition under the same conditions (McGuire and Greene, 1977). Perhaps the best available culture model of the differentiation of cells committed to the neuronal pathway is the primary culture of newborn rat superior cervical ganglion cells (Patterson, 1978; Varon and Bunge, 1978). The cell lines derived from fetal brain by Bulloch *et al.* (1977, 1978) may prove to be better clonal cell models. However, the fundamental problem is that a cell line must have neuron-like characteristics (primarily electrical excitability as measured by ion flux or electrophysiologic methods) in order to be recognized as being neuronal, but cells having these properties seem to be differentiated already. The characteristics of an uncommitted or undifferentiated neuronal precursor cell are presently unclear.

IV. Approaches to Neural Cell–Cell Interaction

A. ADHESION ASSAYS

Studies of cell adhesion or aggregate formation have long played a useful role in analyzing cell–cell recognition. Moscona and co-workers used such an assay in their now classic work on recognition between different species of sponges. Extending their studies into the nervous system, this group has demonstrated and isolated factors which specifically enhance the size of aggregates of dissociated cells from various brain regions (Garber and Moscona, 1972a,b; Hausman and Moscona, 1975; see also McDonough and Lilien, 1975). As previously mentioned, the assay of self-adhesion by chick embryo retina cells has been combined with immunologic approaches to isolate a family of polypeptide cell surface components involved in this self-adhesion (Rutishauser et al., 1976; Brackenbury et al., 1977; Thiery et al., 1977). This combination has yielded some stimulating hypotheses on the role of complementary surface components and proteolysis in cell–cell recognition and it will be interesting to see it applied to other systems. The major polypeptide currently characterized, CAM, has been postulated to mediate aggregation by all neural cells. Its role in cellular interactions beside fasiculation and aggregation remains to be determined.

Stallcup (1977a) has attempted to simplify cell adhesion studies by using cultured tumor cell lines of known phenotype. Assaying the collection of biosynthetically labeled probe cells by monolayers of other cell types, he has concluded that unmodified neural cells will adhere to monolayers of most other neural cells. However, modification of the probe cells can have differential effects on adherence to various monolayers. Subclasses of neural cells can be demonstrated which appear to have adhesive components which are sensitive to trypsin, temperature (inactive at $0°C$), or antiserum to neural cell lines. The data are summarized in a general hypothesis of multiple pairs of complementary components mediating adhesion. One or both of each pair may be expressed on a given cell line.

It may not be possible, however, to extrapolate from adhesion assays to other types of cell–cell interaction. For example, there is no marked preferential binding of nerve cells to clonal muscle cells in Stallcup's study. Such a preference might not have been observed because the neuronal cells used are not known to be capable of synapsing or otherwise interacting with muscle. Additionally, all nonmuscle cells used for assay were of central nervous system origin and may reasonably be suspected of having biologically meaningful specific interactions with each other. As a preliminary test of the relationship between adhesion and the capability for synaptic interaction, we examined the rate of adhesion of PC12 neural cells to muscle and nonmuscle monolayers. PC12 is a rat pheochromocytoma which forms cholinergic synapses with the L6 muscle line (Schubert et al., 1977). Schmidt-Ruppin sarcoma RR1022 (American Type Culture Collection) was used as a control monolayer. No difference in the rate of adherence of PC12 cells to the muscle and nonmuscle line could be detected (Fig. 3). In a related, more extensive study, nerve growth factor which increases the cholinergic character of PC12 (Dichter et al., 1977) was found to in-

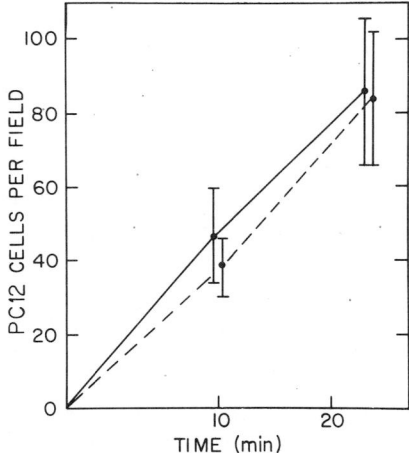

FIG. 3. Adhesion of PC12 cells to monolayers of fused L6 myotubes (solid line) or sarcoma RR1022 (dashed line). PC12 cells were labeled with fluorescein diacetate (Bodmer et al., 1967) and adhesion was carried out at 37°C in 35-mm plates without shaking. At 20 minutes, 50–60% of the input cells adhere. Vertical bars indicate standard errors of the means.

crease the rate of adhesion of PC12 cells to smooth muscle and heart muscle cells and also chondrocytes and fibroblasts, but to decrease the rate of adhesion to L6 cells (Schubert and Whitlock, 1977). Thus, there does not appear to be a straightforward relationship between adhesion rate and the potential for synaptic interaction with this pair of clonal cell lines. The preferential adherence of neural cells to each other has, however, been reproduced using cultured rat neuronal tumor lines. In the monolayer binding assay B103 cells (Schubert et al., 1974) adhere preferentially to chick retina, tectum, or telencephalon and rat brain and cerebral cortex, when compared with chick or rat liver, chick embryo fibroblasts, or Chinese hamster ovary cells (Glaser et al., 1976; Santala et al., 1977). The preferential binding of B103 cells can be reproduced using B103 plasma membranes prepared by density gradient fractionation. Analysis of the sensitivity of binding to trypsin and aldehyde fixation led these authors to independently hypothesize complementary pairs of surface components mediating adhesion. In a more recent study the adhesiveness of C6 glioma cells was shown to decline at high cell density (Santala and Glaser, 1977). Homologous cell–cell contact was postulated to modulate adhesion at high cell density, a growth condition previously shown to modify cell surface antigenicity (Pfeiffer et al., 1971).

In summary, assays of cell adhesion show substantial promise as a means of determining general mechanisms by which cells bind to each other and may also lead to greater understanding of some forms of specific neural cell–cell interaction. However, extrapolations from *in vitro* adhesion experiments to other forms of cell–cell interaction must be made with great care.

B. Effects of Antibodies on Neural Embryogenesis

Although cultured cells and other simple model systems have been very useful in approaching the role of cell surfaces in the development of the nervous system, ultimately hypotheses developed in these systems must be tested *in vivo*. Specific intervention in normal embryogenesis with exogenous antibodies has been attempted by several workers (reviewed in Brent, 1971). The openness of the developing neural tube to external agents makes intervention in the development of the nervous system with added antibodies seem particularly feasible. As a preliminary experiment to test this approach, we windowed fertile chicken eggs and added IgG (from normal rabbit serum or anti-INMA serum from rabbit C5, bleed MB1 of Akeson and Seeger, 1977) to the blastodisk, resealed the eggs, and incubated them for 7 days. Preliminary analysis of externally visible defects in head (presumably brain) and body indicated that the anti-INMA serum induced more defects in development, particularly in the head region. Two general types of head defects seemed to be present in the INMA-treated group. The first of these was an apparent enlargement of the head rostral to the telencephalon with failure of the telencephalon to form individual lobes. Defects with the same external appearance were also observed in a small percentage of animals treated with other IgGs. The second type of defect that was seen almost exclusively in the anti-INMA-treated group was an undersized brain with possible exposure of brain tissue to the external surface. Histologic analysis of a selected animal from each group confirmed the presence of two types of defects: The animal with the anterior head enlargement shows both an abnormal cortical folding pattern and also a poorly formed cellular layering pattern (Fig. 4C,D). The cell-rich germinal zone adjacent to the ventricle is not observed in this animal to the degree seen in the sham-operated animal (Fig. 4A,B). The cells are rather disorganized and loosely packed and the cell-poor layer adjacent to the germinal zone has not formed. In contrast, the animal with the undersized head had a relatively normal germinal zone and cell-layering pattern (Fig. 4F). However, the gross organization of cortical folds is disrupted and there is apparent exposure of germinal zones to the exterior surface in both telencephalon

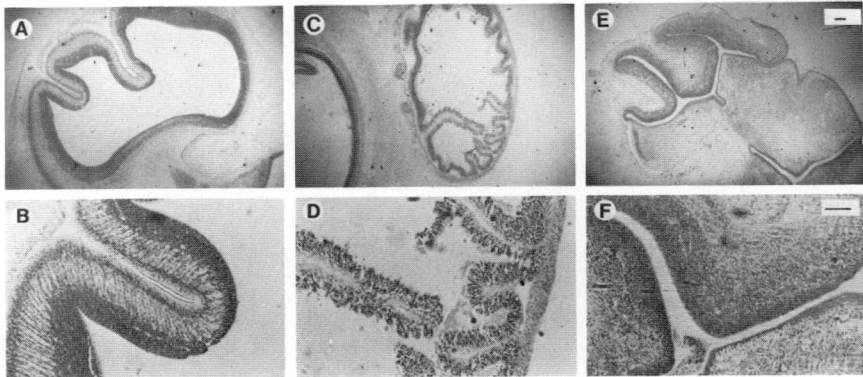

FIG. 4. Hematoxylin–eosin staining of representative 7-μm sections of 7-day embryonic chicks after sham-operated (A,B) or treated with 225 μg anti-INMA serum (C–F). Bars: 100 μm for A, C, and E; and B, D, and F, respectively.

(Fig. 4E) and more caudal structures. Much more work must be done to confirm these observations. This type of approach does seem to hold promise for elucidating the role of particular cell surface components or cell types in neural embryogenesis.

V. Perspectives on Future Research Directions

The monoclonal antibody methodology developed by Milstein and co-workers (Milstein, this series, Vol. 14) will greatly facilitate the study of neural antigens. In much of the previous work, including our own, a large percentage of the total effort is devoted to adsorption of polyspecific antisera until they are "specific" under a defined set of conditions for the cell or tissue of interest. Typically, the sera are rather low-titer after this adsorption and reactivity with preimmune serum is a significant percentage of that with immune serum. The availability in quantity of monoclonal antibody will substantially reduce these problems and allow more rapid progress not only in the demonstration of "neural-specific antigens" but also in the firm identification of these antigens as specific macromolecules. The first steps in this procedure have recently been reported with embryonic chick brain (Gottlieb and Greve, 1978).

A problem that will become more acute as monoclonal methods come into use is the analysis of the relationship of newly described antigens to those previously described. Of the antigens previously described only band 1, NS2 (Yuan *et al.*, 1977), and the NS4 polypeptide have been adequately characterized to make comparison of physical properties possible. Initial publications on new cell surface antigens tend to describe reactivity patterns of antisera rather than physical properties of antigens. Therefore, in order to make useful comparisons with previously described antigens, it is appropriate to suggest that new antisera thought to delineate new cell surface antigens be tested against a standard battery of readily available cell lines. For neuronal cells these would include the human neuroblastoma IMR-32 and SK-N-SH, mouse neuroblastoma NB2a, rat neuronal lines B103 and PC12, rat myoblast line L6, and the rat gliomas C6 and 33B. A second criterion for comparison might be analysis of newborn and adult kidney, liver, spleen, or thymus and sperm from the appropriate species. While such experiments may add a few weeks to the time necessary to characterize new antisera, this time would be well spent in terms of clarity of preliminary antigen identification and ultimate progress of analysis of neural antigens.

One might reasonably ask whether the progress to date justifies the continued effort and use of new methods such as the monoclonal antibody techniques. The available data can be summarized as indicating that nervous system-specific antigens do exist. Furthermore, there seem to be many such antigens rather than few. Some interesting hints about the action of such antigens may be derived from studies of their tissue distribution (for example, see Gordis et al., 1978a; Kemler et al., 1977). However, with the exception of the studies of CAM, specific data on the role of any neural-specific antigen in a biological process are currently lacking. The functions of neural-specific antigens should clearly be one focus of future activity. Some potential uses for nervous system-specific antigens can be filled with existing markers. For example, tetanus toxin binding appears to be a neuron-specific marker (Mirsky et al., 1978) and the presence of the nerve growth factor receptor can be used to detect cells responsive to this neurotrophic agent (Zimmerman et al., 1978). The rationale for continuing studies of neural-specific antigens must be the same as that for initiating them: We know very little about mechanisms of nervous system development and it seems reasonable that nervous system-specific cell surface components should play a role in these processes. This approach is a bit like the proverbial drunk looking for his wallet under the lamp post. It is not certain that nervous system-specific antigens will shed light on the mechanisms of development of the nervous system but they seem to be a reasonable place to start the search. An alternative approach has been to analyze fully the composition of, for example, synaptosomes or neuronal growth cones. This "needle in the haystack" methodology also has merit. Both approaches yield information about the nervous system beyond their contributions to theories of development.

What are the potential uses of this information on neural surface antigen function? A first hope is that insight into neurologic defects including mental retardation may be gained. It seems intuitively obvious that failures of proper cellular organization must be one cause of neurologic deficit. The first steps in this direction in murine model systems have been taken (Mallet et al., 1976; Wallenfels and Schnacher, 1978). The availability in culture of cell lines which synapse (Schubert et al., 1977; Christian et al., 1977) and of systems showing apparent modulation of synapse formation (Ruffolo et al., 1978) should allow cell surface antigenic correlates of this process to be analyzed. Additional hopes are that a thorough understanding of cell–cell interaction in the nervous system would allow repair or restoration of neural function after trauma and also prevention of loss of function with advanced age.

In order to make progress toward these goals, more effort must go into analysis of factors (including specific cell surface antigens) which influence the expression of the neural phenotype. Cell–cell interaction is clearly such a factor, but analysis must continue beyond a simple "holding-hands" model of cell–cell interaction into detailed analysis of how cells interact and the metabolic consequences of their interaction. With the notable exception of synapse formation, neural tumor cells do not appear to interact appropriately. No tumor cell models of myelination, chemotropism, or neuronal biochemical differentiation, for example, are available. Thus, while tumor cells have been and will probably remain an important tool in the demonstration of neural-specific antigens, the analysis of the biological role of these antigens will primarily be made with *in vivo* and *in vitro* models of normal neural development. The blend of current immunologic methods with modern and classical embryological techniques should be a productive interaction.

VI. Concluding Remarks

In the past 5 years several laboratories have initiated efforts to identify nervous system-specific cell surface antigens on cultured neuronal cells. The working hypothesis has been that such antigens should play a key role in cell–cell interaction in both the developing and mature nervous system. Work utilizing neuronal tumor cell lines suggests that several such nervous system-specific cell surface antigens exist. The best characterized is the band 1 polypeptide initially identified on N18 neuroblastoma murine cells. Band 1 is a glycoprotein of apparent molecular weight greater than 200,000 found in culture media. A very similar polypeptide of somewhat higher molecular weight can be immunoprecipitated from detergent-solubilized biosynthetically labeled cells. Enzymatic iodination experiments indicate band 1 is a prominent component on the external surface of cultured neuroblastoma cells. Other individual antigens have been described which are expressed in culture only by neurons, are present only on fetal brain, or are present on cells capable of forming synapses in culture.

Neuronal tumor cells in culture, such as the murine neuroblastoma, express a variety of organ-specific characteristics including nervous system-specific antigens. However, despite a variety of evidence which indicates that the morphologic differentiation of cultured murine neuroblastoma cells is accompanied by changes in the organization and functional capacities of plasma membrane components, there is little conclusive evidence that this morphologic change represents biochemical differentiation. Neural tumor cells in culture do show preferential adhesion to other neural cells mimicking the behavior of normal neural cells. Such assays show promise for determining mechanisms of specific neural cell–cell adhesion. The relationship of these assays to other forms of cell–cell interaction such as synapse formation needs further study.

Continued careful characterization and comparison of nervous system-specific cell surface antigens will make available a battery of antisera to defined cell surface components. The recently developed monoclonal antibody technology will substantially aid this effort. These antisera should be very useful in probing the role of these components in both *in vitro* assays of neural cell–cell interaction and also normal nervous system development *in vivo*.

ACKNOWLEDGMENTS

I would like to thank Drs. Edelman, Fields, Glaser, Gottlieb, Morell, Nirenberg, Patterson, Seeger, Stallcup, and Zimmerman, and particularly Dr. Raff for copies of published and unpublished work. Drs. E. Zimmerman, J. Lessard, Mr. J. Sawyer, and Ms. A. Akeson critically reviewed the manuscript. In addition to assistance cited in previous individual publications, excellent assistance and collaboration was provided by Ms. K. Graham, Mr. T. Mellman, and Dr. J. Lessard. Ms. J. Youtsey gave excellent secretarial help. Work in the author's laboratory is supported by grants from the NINCDS and NICHD.

REFERENCES

Akeson, R., and Herschman, H. R. (1974a). *Proc. Natl. Acad. Sci. U.S.A.* **71**, 187–191.
Akeson, R., and Herschman, H. R. (1974b). *Nature (London)* **249**, 620–623.
Akeson, R., and Herschman, H. R. (1975). *Exp. Cell Res.* **93**, 492–495.

Akeson, R., and Hsu, W.-C. (1978). *Exp. Cell Res.* **115**, 367–377.
Akeson, R., and Seeger, R. C. (1977). *J. Immunol.* **118**, 1995–2003.
Augusti-Tocco, G., and Sato, G. (1969). *Proc. Natl. Acad. Sci. U.S.A.* **64**, 311–315.
Blume, A., Gilbert, F., Wilson, S., Farber, J., Rosenberg, H., and Nirenberg, M. (1970). *Proc. Natl. Acad. Sci. U.S.A.* **67**, 786–792.
Bock, E. (1978). *J. Neurochem.* **30**, 7–14.
Bodmer, W. F., Tripp, M., and Bodmer, J. (1967). *In* "Histocompatability Testing 1967" pp. 341–350. Munksgard, Copenhagen.
Brackenbury, R., Thiery, J.-P., Rutishauser, U., and Edelman, G. M. (1977). *J. Biol. Chem.* **252**, 6835–6840.
Brent, R. L. (1971). *Adv. Biosci.* **6**, 421–455.
Brown, J. C. (1971). *Exp. Cell Res.* **69**, 440–442.
Bulloch, K., Stallcup, W. B., and Cohn, M. (1977). *Brain Res.* **135**, 25–36.
Bulloch, K., Stallcup, W. B., and Cohn, M. (1978). *Life Sci.* **22**, 495–504.
Caspar, J. T., Borella, L., and Sen, L. (1977). *Cancer Res.* **37**, 1750–1756.
Chalozonitis, A., and Greene, L. (1974). *Brain Res.* **72**, 340–345.
Christian, C. N., Nelson, P. G., Peacock, J., and Nirenberg, M. (1977). *Science* **196**, 995–998.
Crain, S. M., Alfei, L., and Peterson, E. F. (1970). *J. Neurobiol.* **1**, 471–489.
Dichter, M. A., Tischler, A. S., and Greene, L. A. (1977). *Nature (London)* **268**, 501–504.
Doyle, J. M., Schachner, M., and Davidson, R. L. (1977). *J. Cell. Physiol.* **93**, 197–204.
Fields, K. L., Gosling, C., Megson, M., and Stern, P. L. (1975). *Proc. Natl. Acad. Sci. U.S.A.* **72**, 1296–1300.
Garber, B. B., and Moscona, A. A. (1972a). *Dev. Biol.* **27**, 217–234.
Garber, B. B., and Moscona, A. A. (1972b). *Dev. Biol.* **27**, 235–243.
Glaser, L., Merrell, R., Gottlieb, D. I., Littman, D., Pulliam, M. W., and Bradshaw, R. A. (1976). *In* "Surface Membrane Receptors" (R. A. Bradshaw, W. A. Frasier, R. C. Merrell, D. I. Gottlieb, and R. A. Hogue-Angeletti, eds.), pp. 109–131. Plenum, New York.
Glick, M. C., Kimhi, Y., and Littauer, U. Z. (1973). *Proc. Natl. Acad. Sci. U.S.A.* **70**, 1682–1687.
Gordis, C., Artzt, K., Wortham, K. A., and Schachner, M. (1978a). *Dev. Biol.* **65**, 238–243.
Gordis, C., Joher, M. A., Hirsch, M., and Schachner, M. (1978b). *J. Neurochem.* **31**, 531–539.
Gottlieb, D., and Greve, J. (1978). *Curr. Top. Microbiol. Immunol.* **81**, 40–44.
Graham, D. I., Gonatas, N. K., and Charalampous, F. C. (1974). *Am. J. Pathol.* **76**, 285–312.
Greene, L., and Tischler, A. S. (1976). *Proc. Natl. Acad. Sci. U.S.A.* **73**, 2424–2428.
Harris, A. J., and Dennis, M. J. (1970). *Science* **167**, 1253–1254.
Hausmann, E. F., and Moscona, A. A. (1975). *Proc. Natl. Acad. Sci. U.S.A.* **72**, 916–920.
Kapellar, M. R., Gal-oz, P., Grover, N. B., and Doljanski, F. (1973). *Exp. Cell Res.* **79**, 152–158.
Kates, J. R., Winterton, R., and Schlessinger, K. (1971). *Nature (London)* **229**, 345–347.
Kemler, R., Babinet, C., Eisen, H., and Jacob, F. (1977). *Proc. Natl. Acad. Sci. U.S.A.* **74**, 4449–4452.
Kimhi, Y., Palfrey, C., Spector, I., Barak, Y., and Littauer, U. Z. (1976). *Proc. Natl. Acad. Sci. U.S.A.* **73**, 462–466.
Klebe, R. J., and Ruddle, F. H. (1969). *J. Cell Biol.* **43**, 69a.
Lee, V., Shelanski, M. L., and Greene, L. A. (1977). *Proc. Natl. Acad. Sci. U.S.A.* **74**, 5021–5025.

Mallet, J., Huchet, M., Pongeois, R., and Changeux, J.-P. (1976). *Brain Res.* **103,** 291–312.
Martin, S. E. (1974). *Nature (London)* **249,** 71–73.
Martin, S. E., and Martin, W. J. (1975a). *Proc. Natl. Acad. Sci. U.S.A.* **72,** 1036–1040.
Martin, S. E., and Martin, W. J. (1975b). *Cancer Res.* **35,** 2609–2612.
Mathews, R. A., Johnston, T. C., and Hudson, J. E. (1976). *Biochem. J.* **154,** 57–64.
McDonough, J., and Lilien, J. (1975). *Nature (London)* **256,** 416–417.
McGuire, J., and Greene, L. A. (1977). *Neurosci. Meet. Abstr., Anaheim, California,* cited in Lee *et al.,* 1977.
McGuire, J. C., Greene, L. A., and Furano, A. V. (1978). *Cell* **15,** 357–365.
Mirsky, R., Wendon, L. M. B., Black, P., Stocklin, C., and Bray, D. (1978). *Brain Res.* **148,** 251–259.
Miyake, M. (1978). *Brain Res.* **143,** 349–354.
Moscona, A. A. (1965). "Cells and Tissues in Culture" (E. N. Willmer, ed.), Vol. 1, pp. 487–529. Academic Press, New York.
Patterson, P. H. (1978). *Annu. Rev. Neurosci.* **1,** 1–18.
Peacock, J. H., Minna, J., Nelson, P. G., and Nirenberg, M. W. (1972). *Exp. Cell Res.* **73,** 367–377.
Pfeiffer, S. E., Herschman, H. R., Lightbody, J. E., Sato, G., and Levin, L. (1971). *J. Cell. Physiol.* **78,** 145–152.
Prasad, K. N., Bondy, S. C., and Purdy, J. L. (1975). *Exp. Cell Res.* **94,** 88–94.
Ruffolo, R. R., Jr., Eisenbarth, G. S., Thompson, J. M., and Nirenberg, M. (1978). *Proc. Natl. Acad. Sci. U.S.A.* **75,** 2281–2285.
Rutishauser, U., Thiery, J.-P., Brackenbury, R., Sela, B.-A., and Edelman, G. M. (1976). *Proc. Natl. Acad. Sci. U.S.A.* **73,** 577–581.
Rutishauser, U., Gall, W. E., and Edelman, G. M. (1978a). *J. Cell Biol.* **79,** 382–393.
Rutishauser, U., Thiery, J.-P., Brackenbury, R., and Edelman, G. M. (1978b). *J. Cell Biol.* **79,** 371–381.
Santala, R., and Glaser, L. (1977). *Biochem. Biophys. Res. Commun.* **79,** 285–291.
Santala, R., Gottlieb, D. I., Littman, D., and Glaser, L. (1977). *J. Biol. Chem.* **252,** 7625–7634.
Schachner, M. (1973). *Nature (London), New Biol.* **243,** 117–119.
Schachner, M., and Wortham, K. A. (1975). *Brain Res.* **99,** 201–208.
Schubert, D., and Whitlock, C. (1977). *Proc. Natl. Acad. Sci. U.S.A.* **74,** 4055–4058.
Schubert, D., Humphreys, S., Baroni, C., and Cohn, M. (1969). *Proc. Natl. Acad. Sci. U.S.A.* **64,** 316–323.
Schubert, D., Humphreys, S., DeVitry, F., and Jacob, F. (1971a). *Dev. Biol.* **25,** 514–516.
Schubert, D., Tarikas, H., Harris, A. J., and Heinemann, S. (1971b). *Nature (London) New Biol.* **233,** 79–80.
Schubert, D., Harris, A. J., Heinemann, S., Kodokoro, Y., Patrick J., and Steinbach, J. H. (1973). *In* "Tissue Culture of the Nervous System" (G. Sato, Ed.), pp. 55–86, Plenum, New York.
Schubert, D., Heinemann, S., Carlisle, W., Tarikas, H., Kimes, B., Patrick, J., Steinbach, J. H., Culp, W., and Brandt, B. L. (1974). *Nature (London)* **249,** 224–227.
Schubert, D., Heinemann, S., and Kodokoro, Y. (1977). *Proc. Natl. Acad. Sci. U.S.A.* **74,** 2579–2583.
Seeds, N. W., Gilman, A. G., Amano, T., and Nirenberg, M. W. (1970). *Proc. Natl. Acad. Sci. U.S.A.* **66,** 160–167.
Simantov, R., and Sachs, L. (1973). *Proc. Natl. Acad. Sci. U.S.A.* **70,** 2903–2905.
Stahn, R., Rose, S., Sanborn, S., West, G., and Herschman, H. (1975). *Brain Res.* **96,** 287–298.

Stallcup, W. B. (1977a). *Brain Res.* **126,** 475–486.
Stallcup, W. B. (1977b). *In* "Progress in Clinical and Biological Research" (Z. Hall, R. Kelly, and C. F. Fox, eds.), (Vol. 15), pp. 165–178, Liss, New York.
Stallcup, W. B., and Cohn, M. (1976). *Exp. Cell Res.* **98,** 285–297.
Sundarraj, N., Schachner, M., and Pfeiffer, S. E. (1975). *Proc. Natl. Acad. Sci. U.S.A.* **72,** 1925–1931.
Thiery, J.-P., Brackenbury, R., Rutishauser, U., and Edelman, G. M. (1977). *J. Biol. Chem.* **252,** 6841–6845.
Truding, R., and Morell, P. (1977). *J. Biol. Chem.* **252,** 4850–4854.
Truding, R., Shelanski, M. L., Daniels, M. P., and Morell, P. (1975). *J. Biol. Chem.* **249,** 3973–3982.
Tyler, A. (1947). *Growth* **10**(Suppl.), 7–19.
Umbreit, J. M., and Strominger, J. L. (1973). *Proc. Natl. Acad. Sci. U.S.A.* **70,** 2997–3001.
Varon, S. S., and Bunge, R. P. (1978). *Annu. Rev. Neurosci.* **1,** 327–362.
Wallenfels, B., and Schachner, M. (1978). *Brain Res.* **148,** 269–277.
Weiss, P. (1947). *Yale J. Med. Biol.* **19,** 235–278.
Yogeeswaran, G., Murray, R. K., Pearson, M. L., Sanwal, B. D., McMorris, F. A., and Ruddle, F. H. (1973). *J. Biol. Chem.* **248,** 1231–1239.
Yuan, D., Vitetta, E. S., and Schachner, M. (1977). *J. Immunol.* **118,** 551–557.
Zeltzer, P., and Seeger, R. C. (1978). *Proc. Am. Assoc. Cancer Res.* **19,** 112.
Zimmerman, A., Sutter, A., Samuelson, J., and Shooter, E. M. (1978). *J. Supramol. Struct.* **9,** 351–362.
Zornetzer, M. S., and Stein, G. S. (1975). *Proc. Natl. Acad. Sci. U.S.A.* **72,** 3119–3123.
Zucco, F., Persico, M., Felsani, A., Metafora, S., and Augusti-Tocco, G. (1975). *Proc. Natl. Acad. Sci. U.S.A.* **72,** 2289–2293.

CHAPTER 11

CELL TYPE-SPECIFIC ANTIGENS OF CELLS OF THE CENTRAL AND PERIPHERAL NERVOUS SYSTEM

Kay L. Fields

DEPARTMENT OF NEUROLOGY
ALBERT EINSTEIN COLLEGE OF MEDICINE
BRONX, NEW YORK

I. Introduction	237
II. Specific Antigens for Different Neural Cell Types	238
A. Schwann Cells and Ran-1 Antigen	238
B. Neuronal Markers	241
C. Astrocyte Antigens	241
D. Oligodendroglia, Schwann Cells, and Galactocerebroside	243
III. Thy-1: A Developmental Antigen on Neuronal Cells	245
A. Thy-1 on Lymphocytes and Brain	245
B. Thy-1 Development on Brain and Brain Fractions	246
C. Thy-1 on Neuronal Cells in Culture	247
D. Expression of Thy-1 in Cerebellum Cultures	248
IV. Present and Potential Applications of Antibodies	252
A. Radioimmunoassays and Immunofluorescence	252
B. Cell Type-Specific Antigens as Markers	252
C. Use of Antibodies as Selective Agents	252
D. Positive Selection Using Antibody	253
E. Complex Cell Interactions	253
V. Concluding Remarks	254
References	255

I. Introduction

The most basic level of differentiation of the nervous system is the determination of the distinctive neuronal, astrocytic, and oligodendroglial cell types. We can now use immunological techniques to distinguish these cell types in tissue culture, where many of the clues of gross location and cellular pattern formation of *in vivo* material are completely missing. We review here the markers which are specific for Schwann cells, neurons, astrocytes, or oligodendroglia in cultures from the central and peripheral nervous system. In addition, the Thy-1 antigen is considered in some detail, and we present evidence that it is a differentiation antigen of neurons. Finally, practical use of cell surface antigens in preparing pure populations of Schwann cells is described.

Of course, the use of brain-specific antigens as markers is only one application of immunological techniques to the study of the nervous system. The location of transmitters such as substance P or the encephalins and of transmitter-related enzymes such as dopamine β-hydroxylase or choline acetyltransferase is possible now with *in vivo* material and may become important with *in vitro* systems. The specificity of immunological reactions, the ability to work at a single cell level for many antigens, and the possibilities for quantitation of very low levels of antigen are powerful reasons for the rapid spread and diversity of the use of antibodies. The use of antibodies *in vivo* is limited by poor penetration of immunoglobulin to sites within the brain. In cell cultures antibodies can diffuse to all the cells, and the increased knowledge of cell surface antigens and interest in cell interactions should encourage the use of antibodies with live cells, rather than simply as tools of histological or quantitative analysis. One hopes to use antisera to analyze the cell interactions leading to synapse formation and to myelination, and perhaps to look at the earlier phases of cell migration and sorting that create ordered patterns of neurons. At present, such uses would appear to require a knowledge of functional cell surface molecules which we are only beginning to gather.

II. Specific Antigens for Different Neural Cell Types

A. Schwann Cells and Ran-1 Antigen

Schwann cells are the glia of the peripheral nervous system (PNS). In ganglia they are associated with neuronal cell bodies, and in nerves they surround all neuronal axons and form myelin, a specialized, multilayered cylinder of cell membranes around some of the axons. Schwann cells have been studied in explant tissue culture for many years and have a characteristic elongated shape and migratory behavior (Murray and Stout, 1942).

We dissociated cells from newborn rat sciatic nerve and found that, although the majority of cells had the bipolar elongated shape and fairly distinctive nuclear morphology which are characteristic of Schwann cells, approximately 20% of the cells were flatter and fibroblastic in appearance. However, under some culture conditions, cell morphology became too ambiguous for adequate classification of all cells. Fortunately, under all tested culture conditions, two cell surface antigens provided an unambiguous cell classification. The more fibroblastic cells, like fibroblasts from other tissues, had the surface glycoprotein antigen Thy-1 which we detected with mouse or rabbit antisera and indirect immunofluorescence. The elongated, putative Schwann cells had no Thy-1 antigen but were positively identified by the presence of an antigen we called Ran-1 (for rat neural antigen-1) (Fig. 1) (Brockes et al., 1977).

Ran-1 was detected using mouse antisera raised against 33B, a rat cell line derived from a tumor of the spinal cord and roots. By cytotoxic tests, anti-33B serum was specific for a surface antigen which was on all rat neural tumors, on brain and spinal cord, and on peripheral nerves, but which was not present on most other tissues or rat fibroblasts (Fields et al., 1975). This antigen, by virtue of its destruction by proteolytic digestion or heat, is likely to be a cell surface protein (Fields, 1977). By several criteria, the Ran-1 antigen on Schwann cells, as detected by indirect immunofluorescence, and the "common antigen," defined by cytotoxicity, are the same.

In sciatic nerve cultures, all the cells were either Ran-1-positive or Thy-1-positive, but no cells had both antigens, as indicated by immunofluorescent tests using two fluorescent derivatives. In cultures of other tissues such as muscle or adrenal, a few Ran-1-positive cells were present, but other cell types such as lymphocytes, myoblasts, or adrenal chromaffin cells were Ran-1-negative (K. L. Fields, unpublished observations). In central nervous system cultures, very few Ran-1-positive cells were detected, and most were due to adherent meningeal tissue (Raff et al., 1979).

In dissociated cultures of the rat dorsal root ganglion, there were many Ran-1-positive cells, and they were associated with the cell bodies and processes of large dorsal root ganglia (DRG) neurons. The neurons, the Thy-1-positive fibroblasts, and some other flat (endothelial?) cells were all Ran-1-negative (Fields et al., 1978). In a well-characterized system of DRG cultures, Schwann cells free of other cell types have been obtained and characterized by light and electron microscopy and by their ability to myelinate neuronal axons (Wood 1976; Bunge and Bunge, 1978). The Schwann cells of these cultures are also Ran-1-positive (K. L. Fields and R. P. Bunge, unpublished observation). These same cultures have been used by Bell and Seetharam (1977) to demonstrate the Schwann cell specificity of another antigen, using a rabbit anti-human tumor serum which had been seen to stain nerves in tissue sections.

Therefore, there are two antigens specific for Schwann cells. Ran-1 is specific for rat tissues, but the Bell and Seetharam (1977) antigen is present on rat and human nerves. A third, less well-characterized anti-Schwann cell antibody has been found in human sera from patients with Chagas' disease (Khoury et al., 1979; E. Khoury, personal communication).

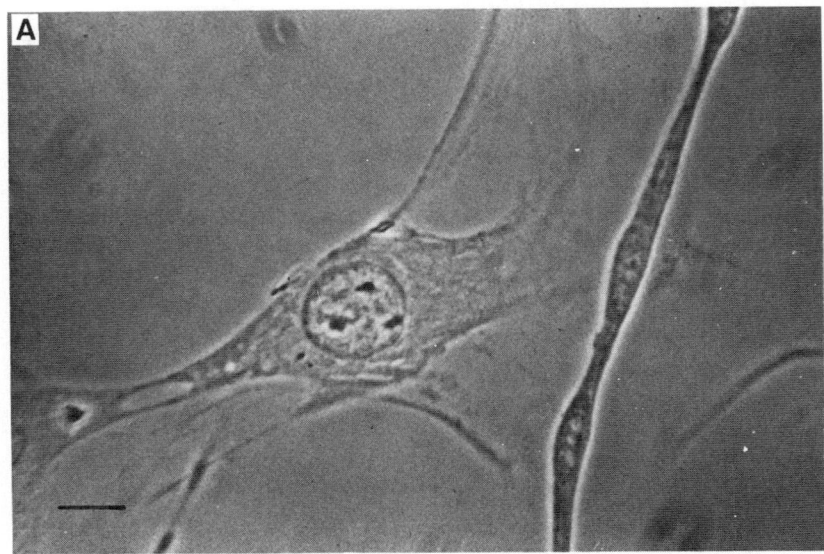

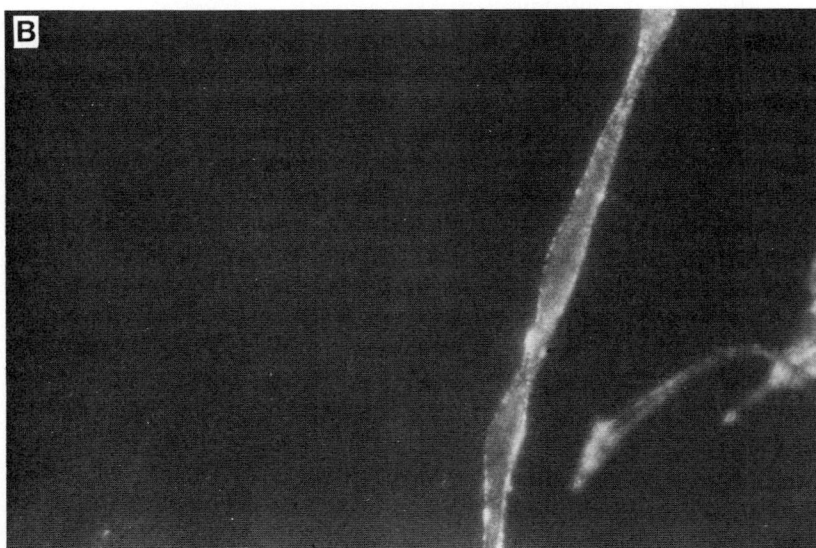

FIG. 1. Ran-1 antigen on Schwann cells from rat sciatic nerve. Surface antigen on live cells were detected by indirect immunofluorescence using mouse anti-Ran-1 serum and goat anti-mouse immunoglobulin conjugated to rhodamine. Cells were then fixed, and viewed with phase contrast optics (A), or UV-epiillumination and filters for rhodamine fluorescence (B). The flat fibroblastic cell does not stain with anti-Ran-1, but the elongated Schwann cells are positive. Scale bar, 10 μm. For further details, see Brockes *et al.* (1977).

B. Neuronal Markers

Tetanus toxin is a potent neurotoxin that normally enters the nervous system by uptake at nerve endings and transport back up the axons to cell bodies in the spinal cord, where it affects synaptic transmission. Scattered neurons in primary brain cultures and mouse neuroblastoma cells were shown by Dimfel et al. (1977) to bind iodinated tetanus toxin all along the cell body and processes. This finding has been extended using indirect immunofluorescence and cell culture to show that chick and rat neurons of several different regions all bind tetanus toxin, whereas nonneuronal cells such as fibroblasts, astrocytes, or Schwann cells do not (Mirsky et al., 1978). As far as we know, all neuronal cells in culture bind tetanus toxin, with the possible exception of rat chromaffin cells. The only nonneuronal cells known to bind this toxin are thyroid cells, where the toxin appears to bind to the thyrotropin receptor site. Like the brain tetanus toxin receptor, thyrotropin receptor is believed to be G_{D1b} or G_{T1} ganglioside or a glycoprotein analog (Ledley et al., 1977).

We have used tetanus toxin binding as a criterion of neuronal identity in cultures of the cerebellum (Currie et al., 1977). In addition to tetanus as a marker for all neurons, there are many culture situations in which an antibody marker for neuronal subclasses is desirable, especially where the neurons are not morphologically obvious. Thy-1 is one such marker and is discussed in Section III. Other possible markers which may be neuronal-specific are 14-3-2 (Cicero et al., 1970), C1300 neuroblastoma antigens (Akeson and Herschman, 1974; Y. Netter, personal communication), a cerebellar antigen (Seeds, 1975), and rat neuronal cell line antigens (Stallcup and Cohn, 1976). In several cases, there is information about the developmental appearance of these antigens in brain. Other intracellular markers which will probably be widely applied in tissue culture are antibodies for neurotransmitter-related enzymes. Several enzymes have been studied in this way in brain sections using light and electron microscopy.

C. Astrocyte Antigens

The best astrocyte marker is the glial fibrillary acidic protein (GFAP), an acidic protein purified from human astrocytic plaques and used to raise rabbit antisera (Uyeda et al., 1972). Such sera stain fibrous astrocytes specifically in sections (Bignami et al., 1972; Ludwin et al., 1976). At the ultrastructural level, GFAP is associated with glial filament bundles and with membranes (Schachner et al., 1977a). Anti-GFAP sera have been shown to bind to filaments in flat astrocytic cells in human brain cultures (Antanitus et al., 1975) and to flat cells in mouse brain cultures (Kozak et al., 1978).

Cultures of optic nerve, but not sciatic nerve, have cells with cytoplasmic filaments which bind the anti-GFAP serum of Dahl and Bignami (1976) (Raff et al., 1978a). Cultures of cerebellum, cerebral cortex, and corpus callosum all have a majority of flat or process-bearing cells which have prominent filamentous GFAP antigen (Fig. 2), but neither fibroblasts nor neurons have filaments which react significantly using immunofluorescence methods (Raff et al., 1979).

The LETS or fibroblast surface antigen, which is on fibroblasts and other cells (Vaheri et al., 1978), is another putative astrocyte marker, because it was found on human astrocyte cell lines and was made by cell lines derived from human gliomas (Vaheri et al., 1976). However, in our rat cultures LETS protein was not associated with astrocytes (GFAP-positive cells) or with neurons (tetanus toxin binding cells). LETS protein was on fibroblasts (flat, Thy-1-positive cells) and on flat, meningeal cells (Thy-1 and GFAP-negative) that were abundant in

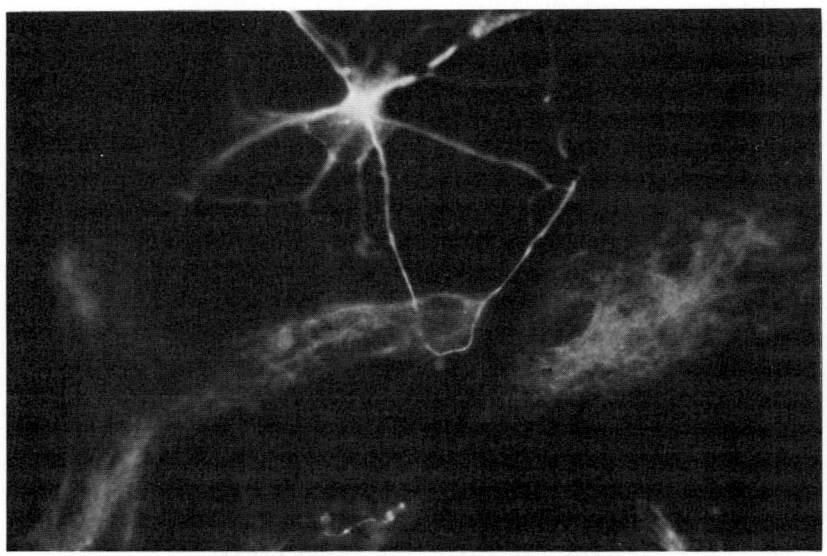

FIG. 2. Glial fibrillary acidic protein (GFAP)-positive cells (putative astrocytes) in cultures of rat corpus callosum. Cells were fixed with acid alcohol, and stained with rabbit anti-human GFAP serum (provided by Drs. D. Dahl and A. Bignami) and goat anti-rabbit immunoglobulin-rhodamine. GFAP-positive cells stain in a fibrous pattern and have both flat and process-bearing forms. Scale bar, 10 μm. For further details, see Raff et al. (1978a).

optic nerve and meninges cultures (Raff et al., 1979). LETS-positive cells were rare in cerebellum cultures when the meninges had been completely removed, whereas astrocytes (GFAP-positive cells) were plentiful. From this, we have concluded that astrocytes, neurons, and Schwann cells are all characterized by a LETS-free cell surface.

D. Oligodendroglia, Schwann Cells, and Galactocerebroside

Oligodendroglia are the cells that make myelin in the central nervous system. Purified myelin is rich in galactocerebroside (GC), a brain lipid also found in peripheral nerve (Norton and Autilio, 1966). Oligodendroglial cell fractions from bovine or calf brain also contain GC (Fewster and Mead, 1968) and they bind anti-GC sera (Poduslo, 1978). Cells with multiple processes in explant cultures have also been shown to bind anti-GC sera (Johnson and Bornstein, 1978). Antisera with high anti-GC titers showed a specific antibody binding to cells in dissociated cell cultures of optic nerve (see Fig. 3). By double-label fluorescence, the GC-positive cells were shown to have no Thy-1 or GFAP antigen, and they were absent from cultures of sciatic nerve or the meninges (Raff et al., 1978a). The cells were likely to be oligodendroglia since they were found in the central but not the peripheral nervous system in a region without neuronal cell bodies, and they were not astrocytes since they had no GFAP. Anti-GC sera bound to similar cells in cultures of rat cerebellum, cerebral cortex, and corpus callosum, all of which should contain oligodendroglia. Direct evidence that these cells are oligodendroglia will require examination by electron microscopy.

Sciatic nerve or dorsal root ganglia cultures had no positive cells. Since peripheral myelin is rich in GC, we had expected Schwann cells to contain this lipid. Indeed, Schwann cells were found to have surface GC if assayed immediately after dissociation, or after a short time in culture, but the cells stopped making or retaining antigen. By 3 days in culture, all the Schwann cells appeared negative by immunofluorescent assays (Raff et al., 1979). Apparently both oligodendroglia and Schwann cells contained surface GC, but only oligodendroglia remained positive in culture. We do not know the physiological significance, nor the details of the Schwann cell loss of this antigen; nor do we know if it is related to the *in vivo* Schwann cell response to nerve injury. The observation of this loss may be related to a report that oligodendroglioma tumor cell lines had high GC levels, but that Schwannoma tumor cell lines did not (Dawson et al., 1977).

Anti-oligodendroglioma sera have been reported which bind to the isolated cell bodies of oligodendroglia but not to neuronal perikarya

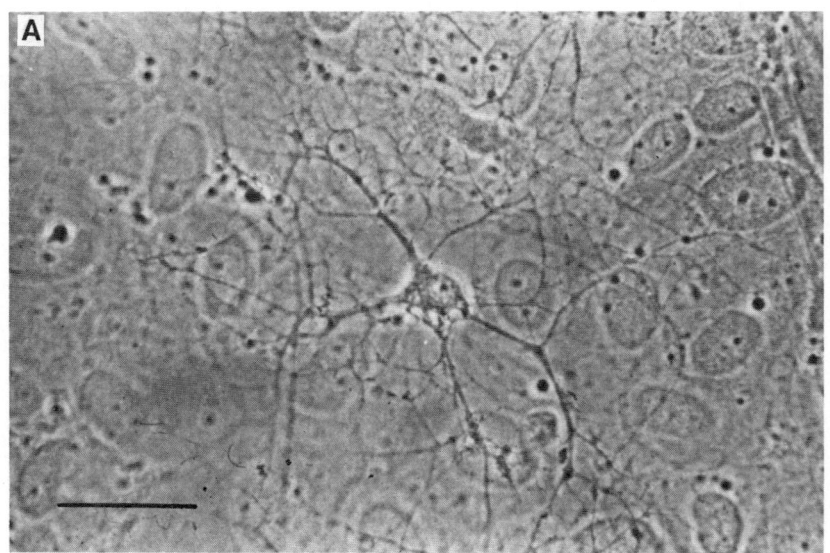

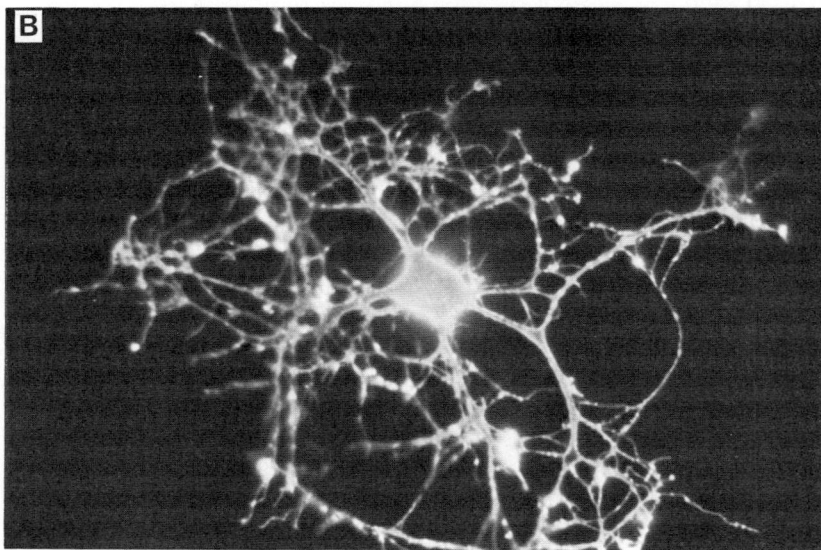

Fig. 3. Galactocerebroside-positive cell (putative oligodendrocyte) on top of a monolayer of antigen-negative fibroblast-like cells in a culture of rat optic nerve. Live cells were stained with rabbit anti-galactocerebroside serum followed by goat anti-rabbit immunoglobulin-rhodamine, fixed in acid alcohol, and viewed with phase contrast (A) or fluorescence (B) optics. Scale bar, 10 μm. From Raff et al. (1978a), with permission of *Nature (London)*.

(Poduslo et al., 1977). Similar sera bind to adult human or bovine oligodendroglia in sections (Abramsky et al., 1978; Traugott et al., 1978). Anti-corpus callosum serum (Schachner et al., 1977b), anti-myelin basic protein (Sternberger et al., 1978), and anti-Wolfgram protein (Roussel et al., 1978) have all been shown to bind to oligodendroglia or to white matter in sections of brain tissue and may be useful as markers of differentiation or myelin formation *in vitro*.

Although some of these cell type-specific antibodies, especially anti-GFAP, have been known for some years, others have just been defined or only recently shown on normal cells in tissue culture. Although better markers will surely be found, there are now markers for Schwann cells, neurons, astrocytes, and oligodendroglia which are specific enough to single out these cells in culture from accompanying cell types. These may not be the most appropriate markers for developmental studies, for they have been used in newborn rat cultures where differentiation is not complete and in dissociated monolayer cultures where cell interactions may well be minimal. However, in the following section the markers will be shown to be useful for localizing the Thy-1 antigen, long known to be a developmental brain antigen, to several identifiable types of neurons in culture.

III. Thy-1: A Developmental Antigen on Neuronal Cells

A. Thy-1 on Lymphocytes and Brain

Thy-1 is a cell surface antigen which is probably the best studied brain antigen known. It occurs on thymocytes and thymus-derived lymphocytes of mice, and two antigenic specificities (Thy-1.1 and Thy-1.2) are governed by allelic genes (Reif and Allen, 1964). Thy-1 is a differentiation antigen of lymphocytes, absent from cells in mouse bone marrow, induced to a high density on lymphocytes in the thymus, and present in reduced amounts on thymus-derived lymphocytes in the peripheral circulation or lymphoid organs (Reif and Allen; 1964, Raff, 1970). Although its function on the cell surface is unknown, Thy-1 is important because it corresponds to functional differences between classes of lymphocytes, and it is used to kill off the thymus-derived lymphocytes of heterogeneous lymphocyte populations.

Thy-1 is not limited to lymphocytes. It is on skin, on epidermal cells, and on fibroblasts. Neither is it exclusive to mouse cells: Thy-1.1 antigen, but not Thy-1.2, is on rat thymocytes. There are differences between the distribution of Thy-1 on mouse and rat lymphocytes, but Thy-1.1 has been shown on rat fibroblasts and myoblasts (for a recent review and references, see Barclay et al., 1976).

In both mice and rats, brain is nearly as rich in Thy-1 antigen as thymus. The antigen was first purified from rat brain then from rat thymus and mouse brain. The antigen from all sources was found to be a glycoprotein of molecular weight 19,000–28,000 (reviewed in Barclay et al., 1976). A similar 25,000 MW protein (T-25) is the Thy-1 antigen on mouse thymus and lymphoma cells (Trowbridge et al., 1975). The protein purification has been done using strong rabbit antisera which detect different antigenic sites on the glycoprotein that also carries the alloantigens Thy-1.1 and Thy-1.2 defined by mouse sera. Sera raised against the purified rat glycoprotein (Barclay et al., 1975) also precipitate mouse T-25 antigen (Trowbridge and Mazauskas, 1976). This work has established that thymus and brain share a cell surface glycoprotein which is defined by very well-tested antisera.

B. Thy-1 Development on Brain and Brain Fractions

Reif and Allen (1964) showed that Thy-1 is a developmental antigen in mouse brain, increasing 50- to 100-fold from negligible levels at birth to adult levels in the first 3–4 weeks of life. The increase is similar in rat brain (Douglas, 1972). Reif and Allen (1964) also measured Thy-1.1 in different regions of AKR mouse brain. Antigen-specific activity varied 5-fold, but in general Thy-1 was in all regions including peripheral nerve. More striking differences have been reported for regions of the rat nervous system; in particular, optic, sciatic, and other nerves had no detectable Thy-1 (Acton and Pfeiffer, 1978).

Other adsorption studies have failed to indicate clearly whether Thy-1 was on only one, or equally on all cell types. Neurological mutants with very little myelin have undiminished Thy-1 content (Schachner, 1973). Myelin is therefore unlikely to be the only site of Thy-1. Mouse brain synaptosomes contain substantial amounts of Thy-1 (Stohl and Gonatas, 1977; Acton et al., 1978), but the two studies differ as to whether myelin contains some or no Thy-1 and whether synaptosomes are depleted or enriched in antigen compared with the starting homogenate. The results are consistent with the presence of Thy-1 on synaptic plasma membranes but do not rule out its presence on neuronal cell membranes or membranes of other cell types such as astrocytes.

Barclay and Hydén (1978) used strong rabbit anti-Thy-1 sera to localize Thy-1 in rat brain sections by immunofluorescence. Strong fluorescence was found in gray matter, in the neuropil, in synaptic glomeruli, and around large neurons. However, these studies at the light microscopic level did not establish that Thy-1 antigen was only on neuronal membranes, for glial and neuronal membranes are closely associated at these sites. However, the antigen distribution was suggestive of neuronal localization and radically different from that of myelin proteins.

Tumor cells have also been assayed for Thy-1. If neurons were the only brain cell type to express Thy-1, we might expect that only neuroblastomas and no gliomas would have surface Thy-1. Among rat brain tumor cells, there is no clear-cut correlation of Thy-1 antigen and markers of neuronal differentiation. One out of 6 neuronal lines, 6 out of 13 nonneuronal (probably glial) cell lines, and 1 transformed muscle cell line were positive (Lesley and Lennon, 1978). A cell line PC12, which was derived from an adrenal medullary tumor, shows several highly differentiated neuronal properties (Greene and Tischler, 1976; Schubert et al., 1977) and has high levels of Thy-1.1 as shown by immunofluorescence (K. L. Fields, unpublished observations) or by adsorption (Morris et al., 1978). However, RN-1, a rat cell line derived from a Schwann cell tumor (neurinoma), has also been found to be Thy-1.1-positive (Acton and Pfeiffer, 1978).

These methods have not given a strong indication of cell type specificity. Thy-1 was found on neuronal membranes and did not behave like a myelin-specific protein, but direct evidence for or against its presence on astrocytes was lacking. If the tumor evidence is taken literally, we would expect Thy-1 to be made by some (or all) neurons, glia, and Schwann cells, as well as by muscle cells. The predictive value of these results is questionable, but direct examination of normal cells in tissue culture has given clearer results.

C. Thy-1 on Neuronal Cells in Culture

Mirsky and Thompson (1975) used dissociated cultures of fetal mouse cerebral hemispheres in an immunofluorescence study of Thy-1 expression. Relying on morphological criteria for the identification of different cell types, they found that anti-Thy-1.2 bound specifically to the cell bodies and processes of about 30% of the neuronal-type cells and that it required 4–11 days in culture for the antigen to develop. Some of the flat, fibroblastic cells also had antigen, but process-bearing glial cells did not.

Fields et al. (1978) tested the hypothesis that neurons were Thy-1-positive, using dissociated cultures of newborn rat DRG. Such cultures contain large, morphologically distinct sensory neurons which have been extensively studied by electrophysiological and morphological techniques. All the large neurons bound anti-Thy-1 serum, which was detected by immunofluorescence, and the same cells had surface receptors for tetanus toxin. By antigenic markers three other cell types were present: Ran-1-positive Schwann cells; Thy-1-positive flat fibroblastic cells; and some flat cells which had no Thy-1, no Ran-1, and no toxin receptors. By autoradiography combined with immunofluorescence, the tetanus toxin binding Thy-1-positive cells rarely showed incorporation of [^{3}H]thymidine, a result expected for DRG neurons. The large neuronal cells were Thy-1-positive even if tested in cell suspension before culturing. Since DRG neurons have no synapses on their cell bodies *in vivo* or *in vitro* (Bunge, 1976), Thy-1 may have no special relation to synapses, and the high concentration of Thy-1 in synaptosomes may simply reflect an enrichment for neuronal plasma membranes. This is also more consistent with the observation of Thy-1 on cell bodies and processes of neurons in culture.

D. Expression of Thy-1 in Cerebellum Cultures

We have made a detailed study of Thy-1 expression on rat cerebellar cells in culture (Currie *et al.*, 1977). The cerebellum is an attractive region for analysis because it contains relatively few neuronal cell types (Purkinje, Golgi, basket, stellate, and granule cells) in the cortex, whose synaptic connections are known in detail. Synapse formation, neuronal maturation, and the generation and migration of the small granule cell neurons are entirely postnatal events (Altman, 1970, 1972). Finally, dissociated monolayer (Lasher and Zagon, 1972; Lasher, 1974; Messer, 1977; Burry and Lasher, 1978) and reaggregate cultures (Seeds, 1973; Trenkner and Sidman, 1977) have been thoroughly described, and progress has been made toward identifying some of the cell types in these cultures by light and electron microscopy and by biochemical criteria.

We cultured cerebellar cells from rats ranging in age from embryonic day 18 to postnatal day 5, using a standard dissociation method, polylysine-coated coverslips, medium with high KCl content (Lasher and Zagon, 1972), and added 2'-deoxy-5-fluorouridine (FUdR). Cells bound to the polylysine-coated surface and put out thin processes in 12 hours. Some cells flattened out and formed a monolayer upon which round neuronal cells developed longer processes. For neuronal survival it was important to inhibit cell division of the flat cells with FUdR. Cultures were continued with surviving neurons for 2–4 weeks, but most experiments were performed between 4 and 14 days in culture.

At first, using postnatal cerebellum, the only Thy-1-positive cells we found were flat cells of fibroblastic appearance. The great majority of flat cells had no Thy-1 antigen in the first 2 weeks. More recently,

Raff et al. (1979) found that most of the Thy-1-negative flat cells were positive for the astrocyte-specific GFAP. If the meninges were completely removed from 1-day cerebella, very few flat Thy-1-positive cells were seen, but the GFAP-positive cells persisted. The evidence is very good that the Dahl and Bignami (1976) anti-GFAP serum we used is specific for astrocytes (see Section II,A). Therefore, our conclusion was that astrocytes in these cultures did not have Thy-1 antigen (Currie et al., 1977; Raff et al., 1979).

R. Pruss (personal communication) has examined astrocytes more carefully in various central nervous system cultures; she found that GFAP-positive astrocytes do develop surface Thy-1.1, but only after the first week in culture. During the second and third weeks, more and more GFAP-positive cells become Thy-1-positive, until a majority have both antigens. Under the conditions used to study cerebellar neurons, I have found a maximum of 17% of GFAP-positive cells to be clearly Thy-1.1-positive after 35 days in culture, confirming Pruss' observation.

The vast majority of neuronal cells of the cerebellum are granule cells. Although most granule cells are formed by division of neuroblasts in the external granular layer during the second and third postnatal week, even in the embryonic cerebellum cultures there were many small, phase-bright, process-bearing cells. Mirsky et al. (1978), using similar cultures, showed that the small cell bodies and their processes bound tetanus toxin. However, the small neuronal cells did not express Thy-1.1 during the first week in culture. After more prolonged culture, 2-4 weeks, many of the surviving small cells and the bundles of processes did finally become Thy-1-positive. Various controls indicated that the binding was specific for Thy-1.1 (Currie et al., 1977).

The identification of these cells as granule cells is supported by their size and similarity to cells in mouse (Messer, 1977) and rat (Lasher, 1974) cerebellum cultures, which have been examined by light and electron microscopy and by uptake of [^{3}H]γ-aminobutyric acid (GABA). Thus, we concluded that a substantial proportion of granule cells which survive for 2-4 weeks in culture expressed Thy-1 antigen.

However, we were also interested in other neuronal cell classes. Purkinje cells, in particular, would be interesting in such cultures, since granule cells normally form synapses onto dendritic spines of these cells. Purkinje cells undergo their last cell division much earlier than granule cells, migrate, and start to form dendrites by about the third day in the rat (Altman, 1972). When cultures of *embryonic* cerebellum were examined after 1 week in culture, a minority of large, process-bearing cells were found to be Thy-1-positive (Fig. 4). Newborn rat cerebellar cultures sometimes had a few such cells, but they were rare or absent in cultures from 2-, 3-, or 5-day-old rats. The Thy-1-positive cells had a large cell diameter and a prominent cell nucleus, but it was not possible to identify them by morphology alone. Such cells were generally multipolar with long processes and a cell body of roughly 12 µm diameter. Fibroblasts in the culture occasionally had processes but these were thicker, shorter, and easily distinguished from processes of the multipolar cell type. Thy-1-positive cells of this type were not observed before 4 days or after 10 days in culture. Cells nearby and the great majority of neuronal cell processes were not Thy-1-positive. This class of positive cells was GFAP-negative and bound tetanus toxin. By these criteria, we concluded they were neuronal cells (Currie et al., 1977).

When cultures containing these cells were exposed to [^{3}H]GABA and examined by autoradiography, a cell with similar morphology was found to take up GABA. Counts of parallel cultures showed fewer Thy-1-positive cells than similar-shaped GABAergic cells. [^{3}H]GABA uptake by these cells was eliminated by added aminocyclohexanecarboxylic acid but not by β-alanine (inhibitors of neuronal or glial GABA uptake, respectively) (Schon and Kelly, 1975; Bowery et al., 1976).

Could these large Thy-1-positive neurons be Purkinje cells? All the cerebellar cortex neurons except granule cells are thought to be GABAergic. Lasher (1974) studied GABAergic cells of 2-day-old rat cerebellum cultures and concluded that most of the very positive cells were stellate or basket cell neurons. Our Thy-1 neurons had larger cell bodies and could be cultured only from rats younger than 2 days of age, which may suggest that they are Purkinje or Golgi cells or neurons from the deep cerebellar nuclei, for these are the only neurons which have had their last cell division before birth (Altman, 1970). It is probably necessary to try to characterize the cells by combined immuno- and electron microscopy to see whether the Thy-1-positive cells have structural properties of a distinct class of neurons.

The main conclusions from the study of Thy-1 in the cerebellum were that both granule cells and some larger neurons became Thy-1-positive. Astrocytes also developed this antigen slowly in culture. However, there is probably an enormous discrepancy between the small amount of Thy-1 seen in the cultures and the large amount in brain. It seems unlikely that a minority of neurons could account for all the *in vivo* antigen. Perhaps all neurons *in vivo* mature and have Thy-1, whereas few do so *in vitro*. Astrocytes, since they develop Thy-1 in culture, may also do so *in vivo* as they become more differentiated.

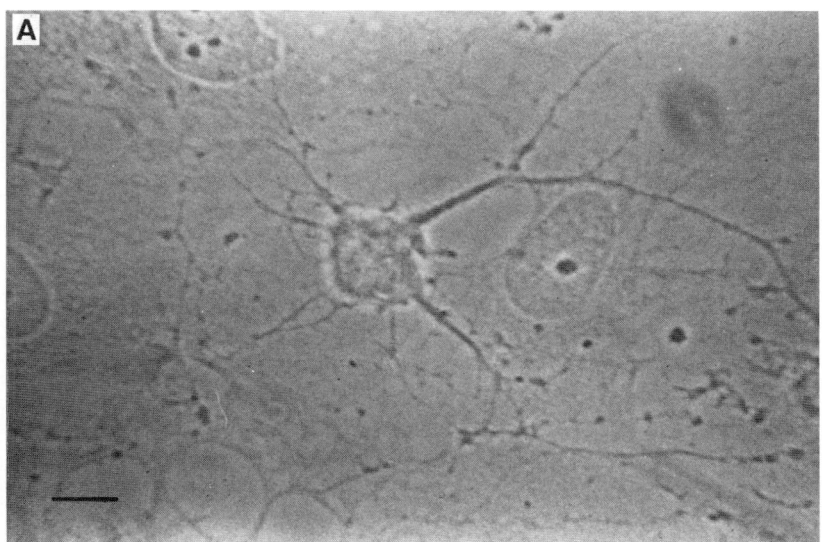

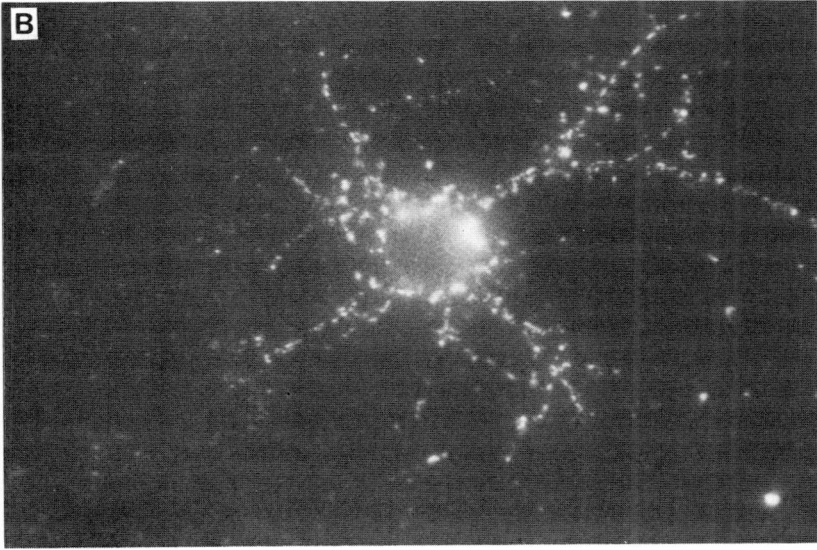

Fig. 4. Thy-1 antigen on a process-bearing cell (putative neuron) from rat cerebellum. Cells from newborn rats were dissociated and maintained in culture for 7 days. Live cells were stained with mouse anti-Thy-1.1 serum (provided by Dr. P. Lake) and goat anti-mouse immunoglobulin-rhodamine. Cells were then fixed and viewed with phase contrast (A) or fluorescence optics (B). Scale bar, 10 μm. Photographs courtesy of K. Fields, N. Currie, and G. Dutton.

IV. Present and Potential Applications of Antibodies

A. Radioimmunoassays and Immunofluorescence

The enormous increase in use of radioimmunoassay systems has made us aware of the potential specificity and sensitivity of immunological assays. Where pure protein enzymes or other potentially antigenic molecules are available, immunological assays may increase the sensitivity enough to make developmental studies feasible. Immunofluorescent techniques sacrifice quantitation for information at the level of single cells and can provide information about the local distribution of surface or cytoplasmic antigens. Both quantitative radioimmunoassays and visualization by indirect immunofluorescence are likely to have many direct and obvious uses in laboratories which are not otherwise concerned with immunology.

B. Cell Type Specific Antigens as Markers

Our effort has gone into developing some valid cell type markers which have been described here. Each new culture system has expanded our confidence, or modified our claims for the specificity of the markers. We have one or more for Schwann cells, astrocytes, oligodendroglia, and neurons; and these can be used to establish new antigenic markers, analyze new *in vitro* cultures, or test the cell type of separated cells. There are many antigens which have been characterized with tumor cells (see Goridis and Schachner, 1978) which may have some cell type specificity.

An advantage of antigenic markers for heterogeneous cultures is their easy combination with autoradiography. It is convenient to score cells for Ran-1 or Thy-1 antigen and for incorporation of [^{3}H]thymidine in sciatic nerve or DRG cultures. (Brockes *et al.*, 1979; Fields *et al.*, 1978). In pure cultures of a single type of cell, the identification by antibody binding would be less advantageous.

C. Use of Antibodies as Selective Agents

We have concentrated on surface antigens with the expectation that we could use specific antibodies in combination with complement to select against cell types. Brockes *et al.* (1979) used this method in combination with cytosine arabinoside treatment to produce pure Schwann cells. The mixture of Schwann cells and fibroblasts from sciatic nerve was first treated with cytosine arabinoside for 2 days, then grown in the presence of a pituitary extract (which stimulated the slowly growing Schwann cells to divide more rapidly). After a week, they were treated with anti-Thy-1 serum plus complement to kill off remaining fibroblasts. The surviving cells, judged by the presence of surface Ran-1 antigen, were >99% Schwann cells. These pure Schwann cells have been used for an analysis of growth factors: Raff *et al.* (1978b,c) have shown that the Schwann cells, unlike fibroblasts, are stimulated to divide by a crude pituitary extract, dibutyryl cyclic AMP, or treatment with cholera toxin.

Antibody selection was unnecessary for other studies of pure Schwann cells by Wood (1976) and Wood and Bunge (1975). Their Schwann cells were obtained as an outgrowth from DRG explants after 2 pulses of cytosine arabinoside. Pure neuronal cultures also were obtained from explants, and a stimulation of Schwann cell division by neuronal axons was shown by autoradiography and light microscopy (Wood and Bunge, 1975).

Other selective procedures have also been developed using nutritional selection (Mains and Patterson, 1973) or selective adhesion (Varon and Raiborn, 1972) to obtain neurons free of nonneuronal cells.

D. Positive Selection Using Antibody

A fluorescence-activated cell sorter has been used with specific antisera to separate T-cells and B-cells, or T-cell subpopulations either analytically or preparatively (Cantor *et al.,* 1975). Campbell *et al.* (1977) were the first to use this method with cells from young mouse brain. They sorted cells which bound anti-corpus callosum serum, and showed an enrichment for a glial antigen and a glial enzyme.

It is possible that antisera against cell surface lipids will be more useful than antisera to cell surface proteins for cell selection since most cell preparations of the nervous system involve proteolytic digestion. Therefore, the specificity of anti-GC for oligodendroglia may lead to an immunoselection of a pure population of these cells. Tetanus toxin binding may also work with fresh neurons if the toxin receptor is a glycolipid. Cholera toxin may be useful, but physiological side effects of either tetanus or cholera toxin may make their use as markers for positive selection less desirable.

E. Complex Cell Interactions

Aggregation of developing cells has been shown to exhibit cell type and organ specificities, and antisera also show similar distinctions (Goldschneider and Moscona, 1972). We hope that antisera will help to define the cell surface units important for aggregation and adhesion. Stallcup (1977) has shown that antisera to clonal cell lines can block adhesion in a manner which can be used to classify the number of adhesion components. The use of antisera to study the role of the cell surface in early embryonic development is fully discussed elsewhere in this volume.

One of the most important of all interactions in the development of the nervous system is specific synapse formation. Antibody to acetylcholine receptor has been used to study synapse formation *in vitro* on muscle (Anderson and Cohen, 1977). Other *in vitro* systems show electrophysiological synapse formation within 2 hours (Ruffolo *et al.*, 1978), and it is possible that antisera which block synapse formation will be found, and help determine the molecular basis of this event.

However, antisera specific for cells or events in the nervous system are themselves at an early stage of development and exploitation. It is very likely that immunological techniques will contribute to our understanding of the cell biology of the special cell types of the nervous system.

V. Concluding Remarks

Antigens specific for different types of cells of the nervous system have been described. Schwann cells can be recognized by the presence of a surface antigen Ran-1 or by several other antisera. Fibroblasts, which contaminate cultures of Schwann cells, can be recognized by surface LETS protein, or the glycoprotein Thy-1. Antisera cytotoxic for cells with Thy-1 antigen can be used to obtain pure Schwann cell cultures. Other glia also have distinctive surface or cytoplasmic antigens. Oligodendroglia in culture bind antibodies to GC but do not have surface Ran-1, Thy-1, or other markers. Astrocytes can be distinguished from other glia by plentiful cytoplasmic filaments binding antibodies to glial fibrillary astrocytic protein. Astrocytes in culture at first have no Thy-1 antigen, but slowly more and more of them acquire this glycoprotein.

Neurons have also been characterized with antisera. Dorsal root ganglion neurons bind tetanus toxin and also express Thy-1 *in vivo* and in culture on all visible plasma membranes. Other neurons from the central nervous system also develop Thy-1 antigen in culture. Large cerebellar neurons develop Thy-1 during their first week in culture, but small, granule cell neurons take longer to become Thy-1-positive.

The development of Thy-1 on brain tissues *in vivo* and brain cells *in vitro* is one example of the use of immunological and tissue culture techniques to study neuronal maturation. Other uses of antibodies are as selective agents for obtaining pure cell populations, and for the determination of the arrangements of molecules such as glycoproteins and glycolipid or neurotransmitter receptors on surface membranes of cells. As more brain proteins are defined and purified, antibodies may be useful in studies of their functions.

ACKNOWLEDGMENTS

This work was supported in Great Britain by a Medical Research Council grant to the Neuroimmunology Project, University College London, and by The Open University. In New York, K.L.F. is supported by United States Public Health Service grants NS 02476 and NS 03356.

REFERENCES

Abramsky, O., Lisak, R. P., Pleasure, D., Gilden, D. H., and Silberberg, D. H. (1978). *Neurosci. Lett.* **8,** 311–316.
Acton, R. T., and Pfeiffer, S. (1978). *Dev. Neurosci.* **1.**
Acton, R. T., Addis, J., Carl, G. F., McClain, L. D., and Bridgers, W. F. (1978). *Proc. Natl. Acad. Sci. U.S.A.* **75,** 3283–3287.
Akeson, R., and Herschman, H. R. (1974). *Nature (London)* **249,** 620–623.
Altman, J. (1970). *In* "Developmental Neurobiology" (W. A. Himwich, ed.), pp. 197–237. Thomas, Springfield, Illinois.
Altman, J. (1972). *J. Comp. Neurol.* **145,** 399–464.
Anderson, M. J., and Cohen, M. W. (1977). *J. Physiol. (London)* **268,** 757–773.
Antanitus, D. S., Choi, B. H., and Lapham, L. W. (1975). *Brain Res.* **89,** 363–367.
Barclay, A. N., and Hydén, H. (1978). *J. Neurochem.* **31,** 1375–1391.
Barclay, A. N., Letarte-Muirhead, M., and Williams, A. F. (1975). *Biochem. J.* **151,** 699–706.
Barclay, A. N., Letarte-Muirhead, M., Williams, A. F., and Faulkes, R. A. (1976). *Nature (London)* **263,** 563–567.
Bell, C. E., and Seetharam, S. (1977). *J. Immunol.* **118,** 826–831.
Bignami, A., Eng, L. F., Dahl, D., and Uyeda, C. T. (1972). *Brain Res.* **43,** 429–435.
Bowery, N. G., Jones, G. P., and Neal, M. J. (1976). *Nature (London)* **264,** 281–284.
Brockes, J. P., Fields, K. L., and Raff, M. C. (1977). *Nature (London)* **266,** 364–366.
Brockes, J. P., Fields, K. L., and Raff, M. C. (1979). *Brain Res.* **165,** 105–118.
Bunge, R. P. (1976). *In* "Neuronal Recognition" (S. H. Barondes, ed.), pp. 109–128. Plenum, New York.
Bunge, R. P., and Bunge, M. B. (1978). *J. Cell Biol.* **78,** 943–950.
Burry, R. W., and Lasher, R. S. (1978). *Brain Res.* **151,** 1–17.
Campbell, G. LeM., Schachner, M., and Sharrow, S. O. (1977). *Brain Res.* **127,** 69–86.
Cantor, H., Simpson, E., Sato, V., Fathman, G., and Herzenberg, L. A. (1975). *Cell. Immunol.* **15,** 180–196.
Cicero, T. J., Cowan, W. M., Moore, B. W., and Suntzeff, V. (1970). *Brain Res.* **18,** 25–34.
Currie, D. N., Fields, K. L., and Dutton, G. R. (1977). *Proc. Int. Soc. Neurochem.* **6,** 635.
Dahl, D., and Bignami, A. (1976). *Brain Res.* **116,** 150–157.
Dawson, G., Sundarraj, N., and Pfeiffer, S. E. (1977). *J. Biol. Chem.* **252,** 2777–2779.
Dimfel, W., Huang, R. T. C., and Habermann, E. (1977). *J. Neurochem.* **29,** 329–334.
Douglas, T. C. (1972). *J. Exp. Med.* **136,** 1054–1062.
Fewster, M. E., and Mead, J. F. (1968). *J. Neurochem.* **15,** 1041–1052.
Fields, K. L. (1977). *Prog. Clin. Biol. Res.* **15,** 179–190.
Fields, K. L., Gosling, C., Megson, M., and Stern, P. L. (1975). *Proc. Natl. Acad. Sci. U.S.A.* **72,** 1296–1300.
Fields, K. L., Brockes, J. P., Mirsky, R., and Wendon, L. M. B. (1978). *Cell* **14,** 43–51.
Goldschneider, I., and Moscona, A. A. (1972). *J. Cell Biol.* **53,** 435–449.
Goridis, C., and Schachner, M. (1978). *In* "Biology of Brain Tumors" (O. D. Laerum, D. D. Bigner, and M. F. Rajewsky, eds.), pp. 158–171. International Union Against Cancer, Geneva.
Greene, L. A., and Tischler, A. S. (1976). *Proc. Natl. Acad. Sci. U.S.A.* **73,** 2424–2428.

Johnson, A. B., and Bornstein, M. B. (1978). *Brain Res.* **159,** 173–182.
Khoury, E., Ritacco, V., Cossio, P. M., Laguens, R. P., Szarfman, A., Diez, C., and Arana, R. M. (1979). *Clin. Exp. Immunol.* **36,** 8–15.
Kozak, L. P., Dahl, D., and Bignami, A. (1978). *Brain Res.* **150,** 631–637.
Lasher, R. S. (1974). *Brain Res.* **69,** 235–254.
Lasher, R. S., and Zagon, I. S. (1972). *Brain Res.* **41,** 482–488.
Ledley, F. D., Lee, G., Kohn, L. D., Habig, W. H., and Hardegree, M. C. (1977). *J. Biol. Chem.* **252,** 4049–4055.
Lesley, J. F., and Lennon, V. A. (1978). *Brain Res.* **153,** 109–120.
Ludwin, S. K., Kosek, J. C., and Eng, L. F. (1976). *J. Comp. Neurol.* **165,** 197–208.
Mains, R. E., and Patterson, P. H. (1973). *J. Cell Biol.* **59,** 329–345.
Messer, A. (1977). *Brain Res.* **130,** 1–12.
Mirsky, R., and Thompson, E. J., (1975). *Cell* **4,** 95–101.
Mirsky, R., Wendon, L. M. B., Black, P., Stolkin, C., and Bray, D. (1978). *Brain Res.* **148,** 251–259.
Morris, R., Colby, W., Betschart, B., and Pfeiffer, S. (1978). *Soc. Neurosci. Abstr.* **4,** 390.
Murray, R. M., and Stout, A. P. (1942). *Anat. Rec.* **84,** 275–294.
Norton, W. T., and Autilio, L. A. (1966). *J. Neurochem.* **13,** 213–222.
Poduslo, S. E. (1978). *In* "Myelination and Demyelination" (J. Palo, ed.), pp. 71–94. Plenum, New York.
Poduslo, S. E., McFarland, H., and McKhann, G. (1977). *Science* **197,** 270–272.
Raff, M. C. (1970). *Immunology* **19,** 637–650.
Raff, M. C., Mirsky, R., Fields, K. L., Lisak, R. P., Dorfman, S. H., Silberberg, D. H., Gregson, N. A., Liebowitz, S., and Kennedy, M. C. (1978a). *Nature (London)* **274,** 813–816.
Raff, M. C., Hornby-Smith, A., and Brockes, J. P. (1978b). *Nature (London)* **273,** 672–673.
Raff, M. C., Abney, E., Brockes, J. P., and Hornby-Smith (1978c). *Cell* **15,** 813–822.
Raff, M. C., Fields, K. L., Hakomori, S., Mirsky, R., Pruss, R. M., and Winter, J. (1979). *Brain Res.* (in press).
Reif, A. E., and Allen, J. M. V. (1964). *J. Exp. Med.* **120,** 413–433.
Reif, A. E., and Allen, J. M. V. (1966). *Nature (London)* **209,** 523.
Roussel, G., Delaunoy, J. P., Mandel, P., and Nussbaum, J.-L. (1978). *J. Neurocytol.* **7,** 155–163.
Ruffolo, R. R., Eisenbarth, G. S., Thompson, J. M., and Nirenberg, M. (1978). *Proc. Natl. Acad. Sci. U.S.A.* **75,** 2281–2285.
Schachner, M. (1973). *Brain Res.* **56,** 382–386.
Schachner, M., Hedley-Whyte, E. T., Hsu, D. W., Schoonmaker, G., and Bignami, A. (1977a). *J. Cell Biol.* **75,** 67–73.
Schachner, M., Wortham, K. A., Ruberg, M. Z., Dorfman, S., and Campbell, G. LeM. (1977b). *Brain Res.* **127,** 87–97.
Schubert, D., Kidokoro, Y., and Heinemann, S. (1977). *Proc. Natl. Acad. Sci. U.S.A.* **74,** 2579–2583.
Schon, F., and Kelly, J. S. (1975). *Brain Res.* **86,** 243–257.
Seeds, N. W. (1973). *In* "Tissue Culture of the Nervous System" (G. Sato, ed.), pp. 35–54. Plenum, New York.
Seeds, N. W. (1975). *Proc. Natl. Acad. Sci. U.S.A.* **72,** 4110–4114.
Stallcup, W. B. (1977). *Brain Res.* **126,** 475–486.
Stallcup, W. B., and Cohn, M. (1976). *Exp. Cell Res.* **98,** 285–297.
Sternberger, N. H., Itoyama, Y., Kies, M. W., and Webster, H. deF. (1978). *J. Neurocytol.* **7,** 251–263.

Stohl, W., and Gonatas, N. K. (1977). *J. Immunol.* **119,** 422–427.
Traugott, U., Snyder, S., Norton, W. T., and Raine, C. S. (1978). *Ann. Neurol.* **4,** 431–439.
Trenkner, E., and Sidman, R. L. (1977). *J. Cell Biol.* **75,** 915–940.
Trowbridge, I. S., and Mazauskas, C. (1976). *Eur. J. Immunol.* **6,** 557–562.
Trowbridge, I. S., Weissman, I. L., and Bevan, M. J. (1975). *Nature (London)* **256,** 652–654.
Uyeda, C. T., Eng, L. F., and Bignami, A. (1972). *Brain Res.* **37,** 81–89.
Vaheri, A., Ruoslahti, E., Westermark, B., and Ponten, J. (1976). *J. Exp. Med.* **143,** 64–72.
Vaheri, A., Ruoslahti, E., and Mosher, D. F., eds. (1978). "Fibroblast Surface Protein" *Ann. N.Y. Acad. Sci.* **312,** 1–456.
Varon, S., and Raiborn, C. W., Jr. (1972). *J. Neurocytol.* **1,** 211–221.
Wood, P. M. (1976). *Brain Res.* **115,** 361–375.
Wood, P., and Bunge, R. P. (1975). *Nature (London)* **256,** 662–664.

CHAPTER 12

CELL SURFACE ANTIGENS OF THE NERVOUS SYSTEM

Melitta Schachner

DEPARTMENT OF NEUROBIOLOGY
HEIDELBERG UNIVERSITY
HEIDELBERG, FEDERAL REPUBLIC OF GERMANY

I.	Introduction	259
II.	Cell Type-Specific Cell Surface Antigens	261
III.	Developmental Expression of Antigens	268
IV.	Subcellular Distribution of Antigens	273
V.	Concluding Remarks	276
	References	277

I. Introduction

Cell interactions in the nervous system have been developed to a high degree. No other organ displays a similar complexity of cellular connections diverging and converging from one cell to another. The coordination of these interactions represents important events during the formation of multicellular organisms and may be due to genetic and epigenetic mechanisms. The molecular strategies underlying these phenomena, including cell proliferation, migration, differentiation, and cell death, have remained largely obscure. It is assumed, however, that morphogenetic signals may arise from diffusible gradients (Wolpert, 1969; Lawrence *et al.*, 1972; Gierer and Meinhardt, 1972) or selective cell surface contacts between neighboring cells (see, e.g., Lash and Burger, 1977; Lerner and Bergsma, 1978).

If one chooses to study the molecular and cellular biology of cell surface components in the nervous system, the search for cell surface molecules might preferentially be geared toward the detection of surface constituents unique to nervous tissue, to a particular cell class, or to a particular developmental event. To achieve these aims, we have used a combination of immunological and molecular biological techniques. This approach was prompted by the success of immunogeneticists in characterizing differentiation antigens of the mouse lymphoid system in its normal and malignant states (see, e.g., Boyse and Old, 1969). A combined immunological and biochemical approach is potentially a very versatile one, and its possibilities can be summarized as follows.

1. Identification and immunological characterization of new and sometimes minor cell surface constituents allow one to distinguish developmentally regulated and cell type-specific surface antigens. A methodological refinement in the study of neural cell surface antigens is available with the hybridoma technique of monoclonal antibody production (Köhler and Milstein, 1975; Milstein, this series, Vol. 14).

2. Antibodies may serve as analytical reagents to assay the molecular nature of the corresponding antigens characterized initially by purely immunological methods (see 1). These antigens can then be isolated and purified by affinity chromatography using specific immunoglobulins coupled to solid state carriers.

3. Cell type-specific antibodies may serve to isolate antigen-positive and -negative cell classes from their mixtures, provided that cells can be obtained as single cell suspensions. The unique advantage of surface antigens is that antibody can bind to live cells without destroying their integrity and viability. In contrast, intracellular antigens become accessible to antibody only after disruption of the plasma membrane, for instance, by aldehyde fixation or lipid extraction. After such treatments cells are naturally of reduced value for structural and functional characterization.

4. When conjugated to visual labels (e.g., fluorescein, ferritin, peroxidase, colloidal gold) antibodies can be used to localize the cellular and subcellular distribution of antigens in histological sections. In view of the fact that nerve cells are organized in special domains of functional and structural properties (e.g., pre- and postsynaptic sites, axon, dendrite, cell body), visually labeled antibodies are of particular value in probing for the topographical distribution of antigens.

5. Antibodies are used as modifiers of functional properties of the molecules to which they are attached and may therefore be viewed as homemade antibiotics or toxins which influence various physiological functions of cells *in vitro* and even *in vivo*. Of particular interest with regard to the study of cell surface interactions is the influence of antibodies on the adhesion and reaggregation behavior of dissociated neural cells.

While encouraging first steps in developmental neuroimmunology have been taken toward some of these objectives, much still remains to be achieved. It should be emphasized that further progress in this field will depend on successfully emulgating various methodologies encompassing also biochemistry, cell biology, anatomy, and electrophysiology. It is by multidisciplinary research that insights into developmental mechanisms can be hoped for.

II. Cell Type-Specific Cell Surface Antigens

Crucial to the investigation of cell–cell interactions in the nervous system is the availability of preparative amounts of isolated cell populations of a particular type consisting of a high degree of homogeneity and functional and morphological integrity. Methods have been described for the separation and enrichment of neural cell classes by sedimentation and buoyant density centrifugation techniques (Barkley *et al.*, 1973; Sellinger *et al.*, 1974; Rose 1967; Hamberger *et al.*, 1975; Norton and Poduslo, 1970). However, cell populations which meet the criteria of morphological, biochemical, and pharmacological homogeneity are difficult to achieve by these methods. Immunological techniques might therefore offer an alternative and complementary approach to the isolation of neural cell classes (Campbell *et al.*, 1977).

Several cell surface antigens have now been attributed to the three major neural cell classes: the neuronal, astroglial, and oligodendroglial cells (Tables I and II). Various immunological methods were used to investigate the cell type specificity of surface antigens. Localization of a particular antigen on the cell surface is unequivocally identifiable by the complement-dependent cytotoxicity test or immunofluorescence using live single cells as antigen-carrying targets (Schachner *et al.*, 1975). Visually labeled antibodies have served as assay reagents to recognize cell surface antigens at the light and electron microscopic levels on freshly dispersed single cells, on cells cultured *in vitro* under monolayer conditions, or in tissue sections, where a correlation between antigen localization and cell type is more easily achieved. It is also possible to analyze the cell type specificity of antibodies by physically isolating cells first and then probing the separated cell populations for morphological and physiological properties and cell type-specific biochemical markers, such as S-100 protein, glial fibrillary acidic protein, neurofilament protein, or enzymes known to be restricted to or predominantly expressed in certain neural cell types.

For the detection of nervous system-specific cell surface antigens (NS antigens, see Table III), neural tumors have served an important role as sources of immunogenic material. They were used because they may retain some properties characteristic of normal cells, because they grow in relatively homogeneous form, and because the cells are available in preparative amounts. It was hoped that the use of these tumors would circumvent a major difficulty in delineating antigenic surface components that distinguish among classes of brain cells since normal

TABLE I

REGIONAL DISTRIBUTION OF ANTIGENS IN SECTIONS OF 8-DAY-OLD AND ADULT MOUSE CEREBELLUM

Region	NS-1	Galacto-cerebroside	Corpus callosum antigen	NS-4	Cholera toxin receptor (G_{M1} ganglioside)	Tetanus toxin receptor (G_{D1b} and G_{T1} gangliosides)	LETS protein
Molecular layer	−	−	+	+	+	+	−
Granular layer	−	−	+	+	+	+	−
White matter	+	+	+	−	−	−	−
Meninges	−	−	−	−	−	−	+
Choroid plexus	−	−	−	−	−	−	+

TABLE II

CELLULAR DISTRIBUTION OF ANTIGENS IN MONOLAYER CULTURES OF 7-DAY-OLD MOUSE CEREBELLUM

Cell type	NS-1	Galacto-cerebroside	Corpus callosum antigen	NS-4	Cholera toxin receptor (G_{M1} ganglioside)	Tetanus toxin receptor (G_{D1b} and G_{T1} gangliosides)	LETS protein
Neuron	−	−	−	+++[a]	+++	+++	−
Astrocyte	−	−	++[b]	+[c]	++[d]	+[d]	−
Oligodendrocyte	+++	+++	+++	+	+++	−	−
Fibroblast	−	−	−	−	−	−	+++

[a] +++ = strongly positive immunofluorescence.
[b] ++ = moderately positive immunofluorescence.
[c] + = weakly positive immunofluorescence.
[d] Not all astrocytes are positive.

TABLE III
TISSUE DISTRIBUTION OF NERVOUS SYSTEM SSPECIFIC CELL SURFACE ANTIGENS

Tissue	NS-1 mouse anti-glioma G26 (Schachner, 1974)	NS-2 rabbit anti-glioblastoma (Schachner and Carnow, 1975; Yuan et al., 1977)	NS-3 rabbit anti-neuroblastoma (Schachner and Wortham, 1975; Doyle, Schachner and Davidson, 1977)	NS-4 rabbit anti-cerebellum (Schachner et al., 1974; Schachner, 1976; Ruberg and Carnow, 1976; Solter and Schachner, 1977; Goridis et al., 1978a,b)	NS-5 mouse anti-cerebellum (Zimmermann and Schachner, 1976; Zimmermann, Schachner and Press, 1977; Goridis et al., 1978a)	NS-6 mouse anti-MUNTAD[a] (Chaffee and Schachner, 1978a)	NS-7 rabbit anti-MUNTAD[a] (Chaffee and Schachner, 1978b)
Brain							
Adult	+	+	+	+	+	+	+
Fetal	−	−	−	+	+	+	+
Retina	−	−	+	+	+	+	+
Kidney	−	−	−	−(+)[b]	+	+	+
Sperm	−	−	−	+	+	+	+
Other tissues	−	−	−	−	−	−	−

[a] MUNTAD is mouse undifferentiated tumor adenovirus induced.
[b] Sera from some, but not all, rabbits react with kidney.

neural cell populations cannot be obtained in homogeneous and pure forms. Indeed, one is caught in a vicious circle since very pure cell populations are needed as antigens in order to raise specific antisera, while at the same time the immunological approach itself is expected to provide the desired cell separations. The first antigen described in my laboratory, NS-1, is expressed on malignant as well as normal oligodendroglia but is not present on nonneural cells, astroglia, and neurons. Other cell surface antigens, for which neural tumors have served as sources of antigens, for instance NS-3, NS-6, and NS-7 (see Table III), have not been found to be differentially expressed among neural cell types.

The cell surface of oligodendrocytes in the adult and early postnatal mouse brain is not only characterized by expression of NS-1 antigen (Schachner, 1974; Schachner and Willinger, 1979a), but also of galactocerebroside (Schachner and Willinger, 1979a) and "corpus callosum" antigen (Schachner et al., 1977b). In addition to oligodendrocytes, astrocytes also express "corpus callosum" antigen. This antigen therefore stands as marker for all macroglia (Schachner et al., 1977b; Campbell et al., 1977). All classes of neuronal cells are characterized by receptors for cholera toxin (Willinger and Schachner, 1978) and tetanus toxin (Dimpfel et al.; 1977; Mirsky et al., 1978; Schnitzer and Schachner, 1979) which represent, respectively, G_{M1} ganglioside (Cuatrecasas, 1973) and gangliosides G_{D1b} and G_{T1} (Van Heyningen, 1963). Antibodies to these toxins have been used as convenient assay reagents. Nervous system antigen-4 (NS-4) also serves as neuronal marker since, similar to the cholera toxin and tetanus toxin binding sites, it seems to be absent from glial cells when assayed for in tissue sections (Schachner et al., 1976; Willinger and Schachner, 1979) and is expressed predominantly on neuronal, but also on some glial cells in tissue culture (Willinger and Schachner, 1979; Schnitzer and Schachner, 1979). A distinction between neuronal cell types, such as granule, Purkinje, Golgi type II, stellate, and basket cells in the cerebellum, has so far not been possible using solely immunological methods.

Accessory cells in the nervous system such as choroid epithelial cells, endothelial cells, leptomeningeal cells, and fibroblasts are characterized in tissue sections by the presence of large external transformation-sensitive (LETS) protein on their surfaces (Schachner et al., 1978a). In vitro, LETS protein is confined to cells with epitheloid and fibroblastic morphological appearance. It is therefore possible that in vitro endothelial and meningeal cells convert to a more fibroblast-like character.

It is noteworthy that several surface components which in histological sections are present uniquely on neurons become apparent on some, but not all, glial cells when cultured *in vitro*. A comparison of Tables I and II shows that particularly the neuronal markers (the cholera and tetanus toxin binding sites and NS-4 antigen) show this discrepancy. Although it is difficult at present to assess the basis for these findings and how they relate to previous observations (Holmgren *et al.*, 1975; Manuelidis and Manuelidis, 1976; De Baecque *et al.*, 1976), the neuronal markers have served, despite the lack of strict biochemical specificity, a useful role in the physical isolation of neural cell types (Willinger and Schachner, 1979).

Several methods are available to obtain pure cell populations using antibodies which distinguish among classes of neural cells. A very simple procedure is the complement-dependent cytotoxic kill of antigen-positive cell types which leads to immunolysis of antigen-positive cells but leaves antigen-negative cells untouched. This method works very well with cells accessible to both complement and complement-fixing antibody such as single cell suspensions and sparsely seeded cells in monolayer culture. Complement-mediated cytotoxicity has not been tried successfully with intact tissue where free diffusion of high-molecular-weight compounds is more restricted. Immunolysis of particular classes of neural cells in the intact brain by chronic *in situ* application of immunolytic agents would serve as a powerful tool in the study of cell–cell interactions.

Several cytotoxic antisera have been used in our laboratory to enrich for certain classes of cerebellar cells. Antisera to LETS protein, corpus callosum antigen, tetanus toxin, and cholera toxin have yielded at least a 10-fold depletion of the antigen-carrying cells after the first immunolytic treatment of single cells in suspension or in monolayer cultures. A further 10-fold depletion of antigen-positive cells is effected in a subsequent second round of cytolysis performed 24–48 hours after the first immunocytolysis. Two consecutive steps of immunoselection therefore lead to a depletion of at least 98% of the antigen-positive cells. Early postnatal cerebellar cells survive these procedures generally well and continue to survive in tissue culture. While cerebellar cultures depleted of LETS protein and cholera and tetanus toxin binding sites carrying cells show no diminution in healthiness and number of surviving cells, cultures which have been depleted of fibroblast-like cells and astro- and oligodendroglia show signs of premature degeneration of residual neuronal cells. The significance of this finding is not clear at present but could represent a dependence of neurons on glia

12. NEURAL CELL SURFACE ANTIGENS 267

which has been observed in the peripheral nervous system (Varon and Bunge, 1978).

A technically more demanding way to isolate neural cells consists in antibody-mediated affinity chromatography. Antibodies can be attached to solid state carriers, such as nylon fibers, Sepharose beads, or plastic surfaces. There they can act as immobilized affinity reagents which attract antigen-positive cells while leaving antigen-negative cells free in suspension. The advantage of this method is that both antigen-positive and -negative cell populations can be obtained while immunocytolysis leads to loss of the antigen-positive cell population. Antigen-positive cells can be retrieved by elution with excess of antigen or antibody, which can be carried out under physiological conditions and permit recovery of viable cells. Inasmuch as isolation speed is critical when live cells are involved, this approach is worthwhile since it can be carried out in relatively short time (approximately 1–2 hours) and therefore compares favorably in this respect with another immunoselective method, fluorescence-activated cell sorting. However, affinity chromatography is not without pitfalls when applied to neural cells because these have a natural tendency to attach to solid surfaces and extend processes in a relatively short time. This process therefore must be prevented from interfering with specific attachment to antibody.

Fluorescence-activated cell sorting is the most refined, although more time-consuming, procedure to separate antigen-positive from -negative cell classes. A costly machine, the fluorescence-activated cell sorter (FACS) has been developed by immunologists to isolate cell classes of the immune system carrying a specific surface marker by use of fluorescein-labeled antibody. The sorter is constructed such that cells are separated from each other on a single cell basis according to the presence or absence of the fluorescence label (Hulett *et al.*, 1965). The advantage of the sorting method over all other immunoselective measures is not only the capability to sort into two major categories— namely antigen-positive and -negative cell classes—but the possibility of making finer distinctions, such as antigen-positive cells with high levels of fluorescence as opposed to cells carrying intermediate or lower amounts of fluorescence. In addition to the fluorescence signal, other parameters can concomitantly be monitored by the machine, e.g., light scatter, Coulter volume, fluorescence anisotropy, and a second laser wavelength for the simultaneous detection of two different fluorochromes (Jovin *et al.*, 1976). As a first step toward the utilization of the sorter for neurobiological purposes, early postnatal cerebellar cells were isolated using corpus callosum antigen as a marker (Campbell *et* al., 1977). To assess cell type identity and degree of enrichment in astro- and oligodendrocytes, independent glial markers were used such as S-100 (Moore, 1965; Moore and Perez, 1968) and glial fibrillary acidic (Eng et al., 1971; Bignami et al., 1972; Schachner et al., 1977) proteins for astrocytes and 2′,3′-CNPase (Mandel et al., 1972) for oligodendrocytes. All three markers are enriched in the corpus callosum positive cell fraction. The sorting process is efficient in that more than 99% of fluorescent antibody-labeled cells are contained in the antigen-positive cell fraction. Similarly, the antigen-negative cell fraction contains less than 1% fluorescent cells. The machine's mechanical setup does not distort the proportion of fluorescent to nonfluorescent cells during sorting, which means that the sorting process does not preferentially destroy one cell population relative to another. After sorting, cells are recovered in viable form by the criteria of trypan blue dye exclusion and by a normal plating efficiency of approximately 50%. Provided that sterility can be maintained during sorting, cells can be cultured for a period of 2 weeks (Schachner, unpublished results).

A prerequisite for successful isolation and continued viability of isolated cells is the possibility of preparing single cells from brain with a high degree of structural and functional integrity. It is difficult to dissociate nervous tissue into viable single cells and in good yields if the tissue has progressed beyond a certain stage of development. While we found that postnatal day 10 is the upper limit for the dissociation of mouse cerebellum, other brain parts, e.g., cerebral cortex and medulla, are amenable to dissociation only at embryonic ages. It is expected that adult brain tissue will also be isolated by immunological means, once mild dissociation procedures are available which do not destroy the cells' integrity and surface markers.

III. Developmental Expression of Antigens

Cues to the importance of cell surface molecules in developmental processes such as cell migration and differentiation may be perceived through studies investigating the temporal and spatial occurrence of these cell surface constituents. While a developmentally regulated expression of cell surface antigens is in itself an interesting phenomenon, the functional significance of such surface molecules remains elusive unless *in vivo* or *in vitro* correlates of cell–cell interactions are shown to be modified by the corresponding antisera. *In vitro* cell adhesion and reaggregation assays have been used as useful parameters in the study of surface-mediated cell contacts (Rutishauser et al., 1976; Stallcup, 1977). It is through the application of such assays that we are presently attempting to characterize the involvement of several nervous system antigens in cerebellar cell interactions.

Cell surface antigens expressed at crucial developmental stages seem more plausible candidates for such investigations. Whereas some nervous system antigens are not present at early developmental stages but appear in parallel with the brain's general functional and structural maturation [e.g., NS-1, NS-2, NS-3 (see Table III for references to NS antigens), and Thy-1 (Fields, this volume)], other antigens (NS-4, NS-5, NS-6, and NS-7) are found not only in the adult brain but also— and at elevated levels—in early postnatal and embryonic brain. A striking feature of NS-4 to NS-7 antigens is their presence not only on adult and fetal brain but also on kidney and sperm. All other tissues that have been tested are found to be antigen-negative. An undifferentiated neural mouse tumor (MUNTAD) which is probably of subventricular cell origin (Chaffee and Schachner, 1978a,b) expresses high levels of NS-4, NS-5, NS-6, and NS-7 antigens while other more differentiated neural tumors of mouse and rat do not express these embryonic antigens. In our first characterization of NS-4 antigen, kidney was found to be antigen-negative, and this clearly distinguished it from NS-5, NS-6, and NS-7. However, subsequent immunizations in 12 rabbits have shown that kidney is recognized by the majority of the antisera and that NS-4 and NS-5 antigens may therefore be more similar to each other than initially anticipated. The significance of neural cell surface antigen expression on kidney which has also been observed by other laboratories (Akeson and Herschman, 1974a,b; Seeds, 1975; Stern *et al.*, 1975) is not understood. Whether NS-4, NS-5, NS-6, and NS-7 antigens are also structurally related to each other remains to be established by detailed biochemical investigations on the molecular nature of these antigens. By purely immunological criteria it is not yet possible to discern structural similarities or differences between these antigens.

An interesting aspect of NS-4 and NS-5 antisera is their reactivity not only with developing neuroectoderm, which could be traced back to embryonic day 9 (the earliest stage tested), but the antigens' presence on all cells of the early mouse embryo (Solter and Schachner, 1976; Zimmermann *et al.*, 1976). Both antigens are distributed uniformly on all cells of cleavage stage embryos and cells of the trophoblast and inner cell mass of the mouse blastocyst. NS-4 antiserum reacts strongly with cells of trophoblast and inner cell mass and, in order of decreasing reactivity, with 4–8 cell-stage embryos, zygotes, unfertilized eggs, and 2-cell-stage embryos. An embryonal carcinoma cell line, F9, also expresses NS-4 and NS-5 antigens (Goridis *et al.*, 1978a).

Another antigen, CB1, described by Seeds (1975), is very similar to NS-4 by serological criteria and could also be shown to be present on spermatozoa and preimplantation mouse embryos. (Solter and Schachner, 1976). Still another antiserum raised in syngeneic mice against pluripotent but largely undifferentiated teratocarcinoma cells reacts with surface antigens shared by early mouse embryos, brain, kidney, and sperm (Stern et al., 1975).

The fact that NS-4 and NS-5 antigens are expressed on spermatozoa and, in the case of NS-4, also on unfertilized eggs, might give rise to speculations about the biological basis and significance of continued surface antigen expression on germ line cells, on all cells of the early embryo, and—more restricted later in development—on neuroectoderm and kidney. The relationship between sperm and the nervous system may be more than coincidental, since several neurological mutants in the mouse also deviate from wild type with respect to spermiogenesis or composition of the sperm surface. In the quaking mouse, with its central myelin deficiency, Bennett et al. (1971) describe aspermia, probably due to loss of cell interactions between spermatogonia and Leydig cells. In the pcd mutant mouse, outstanding features are a rapid degeneration of Purkinje cells in the cerebellum and male sterility (Mullen et al., 1976), as well as slower degeneration of photoreceptors (Mullen and La Vail, 1975). It is interesting, in this context, that NS-4 and also NS-5 are most prominently expressed in cerebellum and retina. Hunt and Johnson (1971) and Ramamurthy and Fawcett (1975) describe the abnormal spermiogenesis of pink-eyed sterile mice whose locomotor disorder was noted by Hollander et al. (1960). Another set of genes which affect the nervous system, where they interrupt early neuroectodermal development, and spermatozoa are the T mutations associated with the *T-t* (Brachyury locus) (see Bennett, 1980).

No single developmental disorder is known which selectively affects brain, kidney, and spermatozoa in conjunction. However, kidney function may be less dependent on small structural modifications than the more complex construction of brain or spermatozoa. Kidney defects may therefore be more difficult to recognize. Several disorders of the nervous system (see e.g., Butterfield et al., 1977) and the neuromuscular junction (Rowland, 1976) are more generally detectable by more refined methods also in nonneural tissues. It is known that brain and kidney share some specialized cell surface constituents; for example, both are rich in complex glycolipids (Esselmann et al., 1973), which may indicate common features of membrane structure and organization. Whether these are also shared by spermatozoa is not known.

The partial biochemical characterization of NS-4 antigen has been carried out using lactoperoxidase-catalyzed radioiodination and immunoprecipitation with anti-NS-4 serum (Goridis et al., 1978b). The immunoprecipitation from solubilized surface-iodinated cerebellar cells contain two prominent proteins with apparent molecular weights of 200,000 and 145,000. These proteins can also be biosynthetically labeled with [^{3}H]leucine. The 145,000 molecular weight constituent is present in embryonic brain but the 200,000 peak is replaced by a broader peak with an apparent molecular weight of 250,000. These cell surface molecules are not present on thymocytes or spleen cells. They are sensitive to mild trypsin treatment of the intact cell and convert to a molecule of approximately 70,000 molecular weight. Trypsin treatment of viable cells confirms the surface location of the antigens. Experiments are in progress to determine the molecular nature of NS-4 antigens on sperm, teratocarcinoma cell lines, and embryonic brain.

The intriguing question is whether NS-4 is a surface molecule which contains a core region with developmentally constant antigenic specificity which is recognized by NS-4 antiserum. This nonvariable core would be common to all NS-4 positive cells, sperm, unfertilized eggs, cleavage stage embryos, and embryonic and adult brain. Variable, and possibly even hypervariable, regions on the NS-4 molecule which are not recognized by NS-4 antiserum might lead to a structural and concomitant functional modification of NS-4 antigen. An indication that such modification may indeed take place on one of the NS-4 molecules stems from the observation that in embryonic brain the 200,000 molecular weight peak is replaced by a peak of 250,000.

That NS-4 antigen is subject to developmental regulation in one particular neuronal cell type, the cerebellar granule cell, can be shown by immunohistological methods. When sections of 8-day-old cerebellum are treated with anti-NS-4 serum by indirect immunofluorescence methods, nondifferentiated pre- and postmitotic granule cells in the external granular layer (Miale and Sidman, 1961) appear to express relatively lower levels of NS-4 antigen than more differentiated postmitotic granule cells in the internal granular layer (Schnitzer et al., 1979).

Several lines of evidence support the concept that G_{M1}, another granule cell surface component, also increases with differentiation (Willinger and Schachner, 1978; Willinger and Schachner, 1979). Examination of histological sections incubated with choleragenoid, the G_{M1} binding subunit of cholera toxin, followed by indirect immunofluorescence, reveals that the external side of the external granule cell layer of 8-day-old cerebellum is relatively unstained compared with the internal half, consisting of postmitotic cells ready to migrate to the internal granule cell layer. Postmitotic granule cells can be shown to acquire G_{M1} ganglioside also by a different method. Mice were injected with [^{3}H]thymidine and sacrificed 3 hours or 3 days later. In the former case, only the external granule cells incorporate radioactive thymidine into DNA, which can be visualized by autoradiography on cerebellar sections. In the latter, a few cells in the external layer, but most in the internal layer, incorporate radioactivity, with the internal granule cells being most prominently labeled. Cerebellar cells are cultured from mice injected in this manner and scored simultaneously for the presence of [^{3}H]thymidine incorporated into DNA by autoradiography and ganglioside by immunofluorescence. None of the external cells of the external granular layer (cells labeled for 3 hours) possess cholera toxin binding sites. On the other hand, 70% of the toxin-positive cells of the animal injected 3 days earlier contain [^{3}H]thymidine. Therefore, the differential labeling of granule cells observed in dissociated cell preparations and at early times *in vitro* probably reflects *in vivo* development.

A recurring observation in our homologous (intraspecies) immunizations (e.g., injection of mouse brain tissue into mice) pertains to the problem of self-recognition of brain cell surface constituents. Although, in the case of NS-5, antiserum production could not be evoked in a syngeneic immunization protocol (cerebellar tissue of C57BL/6J mouse strain injected into C57BL/6J mice), NS-1 and NS-6 are recognized in a syngeneic or histocompatible situation precluding detection of allogeneic differences and allocation of gene loci by Mendelian genetics. This phenomenon may best be exemplified in the case of NS-6.

NS-6 is apparently not an alloantigen (for discussion, see Lake and Mitchison, 1976) because it is fully expressed in all mouse strains tested so far, including the mice producing antisera. While NS-1 self-antigenicity could be explained by assuming an efficient protection from the general circulation by the blood–brain barrier, the situation with NS-6 seems more complex since it is present also on kidney and spermatozoa. In testis NS-6 also would be relatively unavailable to the immune system, but in kidney it should be fully exposed and expected to elicit tolerance. Not only do mice develop hyperimmune titers of anti-NS-6, but preliminary investigation discloses neither behavioral nor histological evidence of tissue damage in brain or in kidney in these mice.

Is NS-6 really a normal self-product of the tissues that express it? Obvious alternative explanations are not compelling. NS-6 is not, for instance, a tumor antigen per se, even though the antiserum used to detect it was made against the MUNTAD tumor cell; nor is any relationship apparent between NS-6 and the adenovirus used to induce MUNTAD [although the example of the viral glycoprotein gp 69/70 which Tung *et al.* (1975) find present in normal tissues shows that the possibility of a viral origin is not simple to exclude]. It is also unlikely that NS-6 is distributed via the circulation to other tissues, since brain and testis are relatively isolated from other organs behind blood–brain and blood–testis barriers. However, kidney is not isolated from the blood circulation, and it is known to acquire exogenous antibody–antigen complexes. These appear in glomeruli, while NS-6 is localized on the kidney cell surface. Also, NS-6 is expressed at adult levels at birth and does not subsequently increase, whereas glomerular deposits accumulate with age, first seen in mice at 4–5 months of age and reaching a maximum at about 1.5 years (Linder *et al.,* 1972).

IV. Subcellular Distribution of Antigens

Since it is likely that nerve and glial cells are organized in functionally and structurally specialized domains, a differential distribution of surface antigens among subcellular compartments is of special interest, not only because inferences may be possible as to the functional role of cell surface antigens, but also with respect to developmental processes, where a rearrangement of subcellular domains might be instrumental in contact guidance and topographical specificity of synapse formation.

Some antigens such as NS-1 and corpus callosum antigen show a uniform distribution over cell body and processes on oligodendrocytes cultured *in vitro* (Schachner and Willinger, 1979a,b). Tetanus and cholera toxin binding sites are also more or less uniformly distributed over the entire cell, when assayed on cultured cells. In histological sections of adult mouse cerebellum, however, choleragenoid binding sites are not detectable on axonal processes of granule cells. While the molecular layer of the adult cerebellum appears to be void of binding sites, they are present in the nascent molecular layer of the early postnatal developing cerebellum (Willinger and Schachner, 1979). Such developmentally regulated expression has also been found for concanavalin A receptors (Hatten *et al.* 1979).

NS-4 is expressed on all parts of the neuronal plasma membrane of young cerebellar cells, but in histological sections of adult cerebellum (Schachner *et al.* 1976) molecular layer and synaptic glomeruli in the granular layer express particularly high amounts of antigen. It is not known, however, which cellular and subcellular constituent of the synaptic complex carries the antigen.

Extreme examples of differential distribution of cell surface anti-

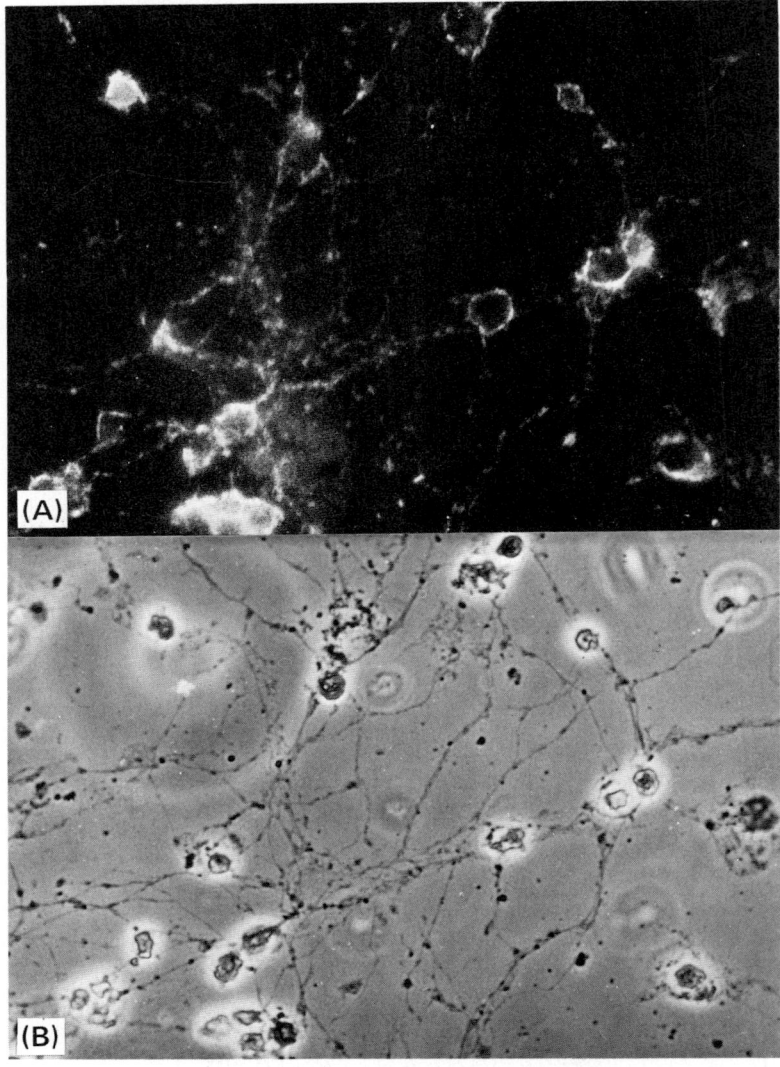

Fig. 1. Expression of NS-7 antigen on cerebellar cell surfaces of 7-day-old mice. ×600. (A) Antigen-bearing cells are visualized by indirect immunofluorescence performed on live cells maintained *in vitro* under monolayer conditions for 2 days. (B) Phase contrast image of Fig. 1a. ×510.

gens in topographical compartments are molecules recognized by an antiserum against adult mouse retina (Wortham and Schachner, in preparation) and anti-NS-7 serum. While NS-7 antigen is expressed more strongly on cell body than on processes of cerebellar granule cells

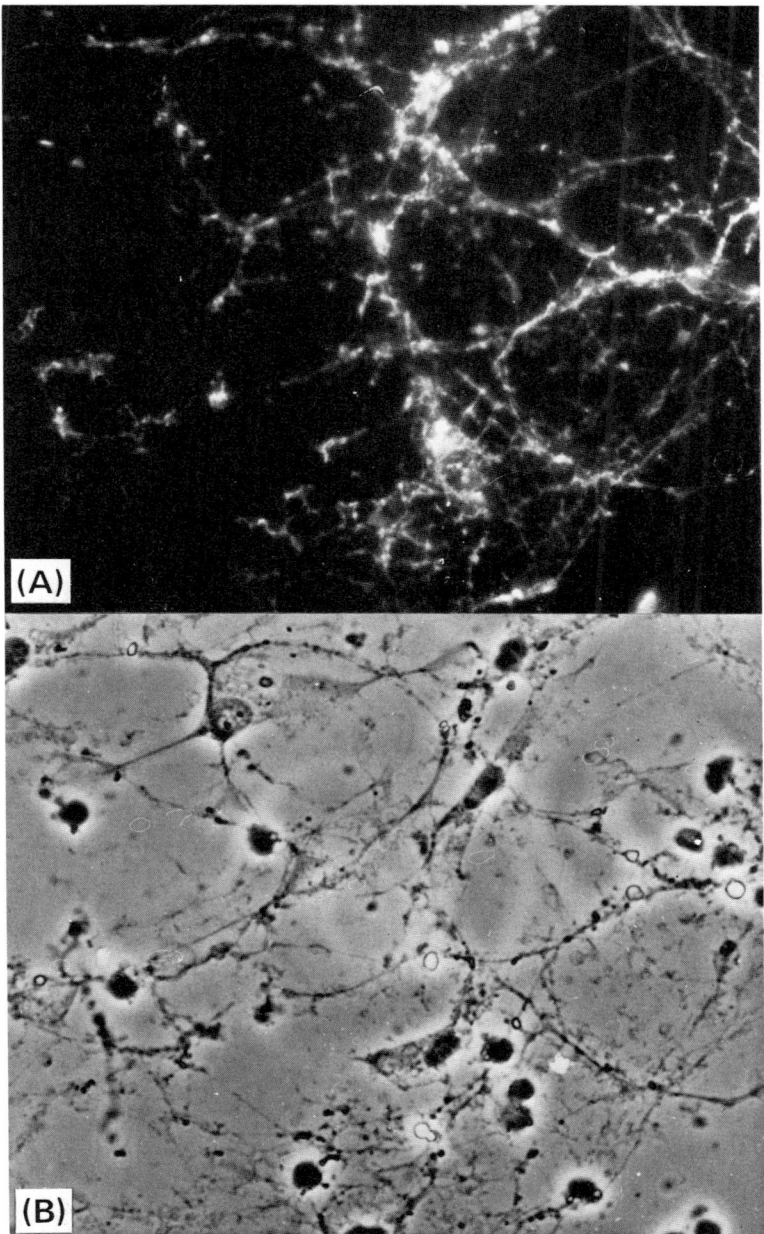

FIG. 2. Expression of retina antigen(s) on cerebellar cell surfaces of 7-day-old mice. ×600. (A) Antigen-carrying cells are visualized by indirect immunofluorescence in parallel and under identical conditions as in the experiment represented in Fig. 1. (B) Phase contrast image of Fig. 2a. ×600.

in vitro (Fig. 1), surface components recognized by retina antiserum are completely absent from cell bodies and expressed only on cellular extensions (Fig. 2). The latter antiserum may be useful in immunosurgical treatments similar to the one described by Solter and Knowles (1975) for manipulations of the trophoblast. If the cell body is able to survive after its processes are shorn off, the regenerated cell would then be capable of extending new cellular processes. It is anticipated that immunosurgery will play an important role in the study of the regenerative events *in vitro* and possibly *in vivo* also.

V. Concluding Remarks

An immunological approach to the study of cell surfaces in the nervous system has, until recently, served several aims while leaving others unattained.

The distinction of neuronal, oligo-, and astroglial cell types has been achieved by immunologically detectable cell surface markers. While antigenic differences between neuronal cell types have not been observed by these methods, it is hoped that with further refinements in immunological methodology such differences might become apparent.

Immunoselective methods have led to the physical separation and isolation of the major neural cell populations, neuronal and glial cells, using fluorescence-activated cell sorting and complement-dependent cytotoxicity. The isolated cell populations are viable and can be cultured *in vitro*. For cell biological studies which depend on the integrity of the whole cell, the improvement of sorting speed will be helpful.

Developmentally regulated cell surface antigens have been detected and characterized. Some antigens (NS-1, NS-2, and NS-3) reach maximal levels of expression in the adult nervous system, and their appearance correlates with the general functional and structural maturation of the brain. Other antigens are maximally expressed at embryonic and early postnatal ages (NS-4 and NS-5). Biochemical evidence points to the possibility that NS-4 antigen undergoes structural modification during development. The structural characterization of embryonic antigens shared by brain, kidney, sperm, and all cells of cleavage-stage embryos and the elucidation of their functional role in possible *in vitro* correlates of cell–cell interactions, such as cell adhesion and reaggregation, are of primary importance.

Some cell surface constituents are expressed in specialized domains on a particular cell. Ganglioside G_{M1} is expressed at high levels on immature axons of cerebellar granule cells, but disappears from axonal processes at later developmental stages to become restricted predominantly to the cell body. An antiserum to adult mouse retina reacts specifically with the extensions of cerebellar cells but not with the cell body. Such antiserum may allow excision of these processes by immunosurgery and observation of regeneration under controlled conditions *in vitro*. It is also possible that immunosurgical manipulations may be carried out *in vivo* with circumvention of the blood–brain barrier and permit precise deletions of cells or cellular extensions.

ACKNOWLEDGMENTS

The author is indebted to her collaborators who contributed to the research presented in this chapter. Support from the following sources is gratefully acknowledged: National Institutes of Health, National Science Foundation, National Foundation–March of Dimes, Deutsche Forschungsgemeinschaft, Land Baden-Württemberg, Stiftung Volkswagenwerk, and Gemeinnützige Hertie Stiftung.

REFERENCES

Akeson, R., and Herschman, H. R. (1974a). *Proc. Natl. Acad. Sci. U.S.A.* **71**, 187–191.
Akeson, R., and Herschman, H. R. (1974b). *Nature (London)* **249**, 620–623.
Barkley, D. S., Rakic, L. L., Chaffee, J. K., and Wong, D. L. (1973). *J. Cell Physiol.* **28**, 271–280.
Bennett, W. I., Gall, A. M., Southard, J. L., and Sidman, R. L. (1971). *Biol. Reprod.* **5**, 30–58.
Bennett, D. (1980). In "Current Topics in Developmental Biology" (A. A. Moscona and A. Monroy, eds.), Vol. 14. Academic Press, New York (in press).
Bignami, A., Eng, L. F., Dahl, D., and Uyeda, C. T. (1972). *Brain Res.* **43**, 429–435.
Boyse, E. A., and Old, L. J. (1969). *Annu. Rev. Genet.* **3**, 269–291.
Butterfield, D. A., Oeswein, J. Q., and Markesbery, W. R. (1977). *Nature (London)* **267**, 453–455.
Campbell, G. LeM., Schachner, M., and Sharrow, S. O. (1977). *Brain Res.* **127**, 69–86.
Chaffee, J. K., and Schachner, M. (1978a). *Dev. Biol.* **62**, 173–184.
Chaffee, J. K., and Schachner, M. (1978b). *Dev. Biol.* **62**, 185–192.
Cuatrecasas, P. (1973). *Biochemistry* **12**, 3558–3566.
De Baecque, C., Johnson, A. B., Naiki, M., Schwarting, G., and Marcus, D. M. (1976). *Brain Res.* **114**, 117–122.
Dimpfel, W., Huang, R. T. C., and Habermann, E. (1977). *J. Neurochem.* **29**, 329–334.
Eng, L. F., Vanderhaeghen, J. J., Bignami, A., and Gerstl, B. (1971). *Brain Res.* **28**, 351–354.
Esselmann, W. J., Ackermann, J. R., and Sweeley, C. C. (1973). *J. Biol. Chem.* **248**, 7310–7317.
Fields, K. L. (1979). In "Current Topics in Developmental Biology" (A. A. Moscona and A. Monroy, eds.), Vol. 13, pp. 237–257. Academic Press, New York.
Gierer, A., and Meinhardt, H. (1972). *Kybernetik* **12**, 30–39.
Goridis, C., Artzt, K., Wortham, K. A., and Schachner, M. (1978a). *Dev. Biol.* **65**, 238–243.
Goridis, C., Joher, J. A., Hirsch, M., and Schachner, M. (1978b). *J. Neurochem.* **31**, 531–539.
Hamberger, A., Hansson, H. A., and Sellström, A. (1975). *Exp. Cell Res.* **92**, 1–10.
Hatten, M. E., Schachner, M., and Sidman, R. L. (1979). *Neuroscience* (in press).
Hollander, W. F., Bryan, J. H. D., and Gowen, J. W. (1960). *Genetics* **45**, 413–418.

Holmgren, J., Lönnroth, I., Mansson, J. E., and Svennerholm, L. (1975). *Proc. Natl. Acad. Sci. U.S.A.* **72,** 2520–2524.
Hulett, H. R., Bonner, W. A., Barrett, J., and Herzenberg, L. A. (1965). *Science* **166,** 747–749.
Hunt, D. M., and Johnson, D. K. (1971). *J. Embryol. Exp. Morphol.* **26,** 111–121.
Jovin, T. M., Morris, S. J., Striker, M., Schultens, H. A., Digweed, M., and Arndt-Jovin, D. J. (1976). *J. Histochem. Cytochem.* **24,** 269–283.
Köhler, G., and Milstein, C. (1975). *Nature (London)* **256,** 495–497.
Lake, P., and Mitchison, N. A. (1976). *Cold Spring Harbor Symp. Quant. Biol.* **41,** 589–595.
Lash, J. W., and Burger, M. M., eds. (1977). "Cell and Tissue Interactions." Raven, New York.
Lawrence, P. A., Crick, F. H. C., and Munro, M. (1972). *J. Cell Sci.* **11,** 815–853.
Lerner, R. A., and Bergsma, D., eds. (1978). "The Molecular Basis of Cell-Cell Interaction." Liss, New York.
Linder, E., Pasternack, A., and Edginton, T. S. (1972). *Clin. Immunol. Immunopathol.* **1,** 104–121.
Mandel, P., Nussbaum, J. L., Neskovic, N. M., Sarlieve, L. L., and Kurihara, T. (1972). *In* "Advances in Enzyme Regulation" (G. Weber, ed.), Vol. 10, pp. 101–118. Pergamon, New York.
Manuelidis, L., and Manuelidis, E. E. (1976). *J. Neurocytol.* **5,** 575–589.
Miale, I. L., and Sidman, R. L. (1961). *Exp. Neurol.* **4,** 277–296.
Milstein, C. (1980). *In* "Current Topics in Developmental Biology" (A. A. Moscona and A. Monroy, eds.), Vol. 14. Academic Press, New York (in press).
Mirsky, R., Wendon, L., Black, P., Stolkin, C., and Bray, D. (1978). *Brain Res.* **148,** 251–259.
Moore, B. W. (1965). *Biochem. Biophys. Res. Commun.* **19,** 739–744.
Moore, B. W., and Perez, V. J. (1968). *In* "Physiological and Biochemical Aspects of Nervous Integration" (F. D. Carlson, ed.), pp. 343–359. Prentice-Hall, New York.
Mullen, R. J., and La Vail, M. M. (1975). *Nature (London)* **258,** 528–530.
Mullen, R. J., Eicher, E. M., and Sidman, R. L. (1976). *Proc. Natl. Acad. Sci. U.S.A.* **73,** 208–212.
Norton, W. T., and Poduslo, S. E. (1970). *Science* **167,** 1144–1146.
Ramamurthy, G. V., and Fawcett, D. W. (1975). *Anat. Rec.* **181,** 457 (Abstr.).
Rose, S. P. R. (1967). *J. Biochem.* **102,** 33–43.
Rowland, L. P. (1976). *Arch. Neurol.* **33,** 315–321.
Rutishauser, U., Thiery, J.-P., Brackenbury, R., Sela, B.-A., and Edelman, G. M. (1976). *Proc. Natl. Acad. Sci. U.S.A.* **73,** 577–581.
Schachner, M. (1974). *Proc. Natl. Acad. Sci. U.S.A.* **71,** 1795–1799.
Schachner, M., and Willinger, M. (1979a). *In* "Menarini International Symposium of Immunopathology" (P. A. Miescher, ed.). (in press).
Schachner, M., and Willinger, M. (1979b). *Prog. Brain Res.* (in press).
Schachner, M., Wortham, K. A., Carter, L. D., and Chaffee, J. K. (1975). *Dev. Biol.* **44,** 313–325.
Schachner, M., Ruberg, M. Z., and Carnow, T. B. (1976). *Brain Res. Bull.* **1,** 367–377.
Schachner, M., Hedley-Whyte, E. T., Hsu, D. W., Schoonmaker, G., and Bignami, A. (1977a). *J. Cell Biol.* **75,** 67–73.
Schachner, M., Wortham, K. A., Ruberg, M. Z., Dorfman, S., and Campbell, LeM. G. (1977b). *Brain Res.* **127,** 87–97.
Schachner, M., Schoonmaker, G., and Hynes, R. O. (1978a). *Brain Res.* **158,** 149–158.

Schachner, M., Smith, C., and Schoonmaker, G. (1978b). *Dev. Neurosci.* **1,** 1–14.
Schnitzer, J., and Schachner, M. (1979). Submitted for publication.
Schnitzer, J. Sommer, I., and Schachner, M. (1979). Submitted for publication.
Seeds, N. W. (1975). *Proc. Natl. Acad. Sci. U.S.A.* **72,** 4110–4114.
Sellinger, O. Z., Legrand, J., Clos, J., and Ohlsson, J. W. (1974). *J. Neurochem.* **23,** 1137–1146.
Solter, D., and Knowles, B. B. (1975). *Proc. Natl. Acad. Sci. U.S.A.* **72,** 5099–5102.
Solter, D., and Schachner, M. (1976). *Dev. Biol.* **52,** 98–104.
Stallcup, W. B. (1977). *Brain Res.* **126,** 475–486.
Stern, P. L., Martin, G. R., and Evans, M. J. (1975). *Cell* **6,** 455–465.
Tung, J.-S. Vitetta, E. S., Fleissner, E., and Boyse, E. A. (1975). *J. Exp. Med.* **141,** 198–205.
Van Heyningen, W. E. (1963). *J. Gen. Microbiol.* **31,** 375–387.
Varon, S. S., and Bunge, R. P. (1978). *Annu. Rev. Neurosci.* **1,** 327–361.
Willinger, M., and Schachner, M. (1978). *J. Supramol. Struct. (Suppl.)* **2,** 128.
Willinger, M., and Schachner, M. (1979). *Dev. Biol.* (in press).
Wolpert, L. (1969). *J. Theoret. Biol.* **25,** 1–47.
Wortham, K. A., and Schachner, M. (1979). (in preparation).
Zimmermann, A., Schachner, M., and Press, J. L. (1976). *J. Supramol. Struct.* **5,** 417–429.

CHAPTER 13

ANTIBODY EFFECTS ON MEMBRANE ANTIGEN EXPRESSION*

Edward P. Cohen[†] and Weitze Liang

LA RABIDA-UNIVERSITY OF CHICAGO INSTITUTE AND
DEPARTMENT OF MICROBIOLOGY
UNIVERSITY OF CHICAGO
CHICAGO, ILLINOIS

I. Introduction 281
II. Antigenic Modulation as a Model System for Studying Ligand-Induced Effects on Receptor Expression 283
III. The Metabolism of TL Antigens in the Presence (and Absence) of TL Antiserum 285
   A. Synthesis and Degradation of TL Antigens 285
   B. Tumor-Associated Antigen of ASL-1 Cells................... 288
IV. Membrane Antigens of Somatic Hybrids of TL(+) and TL(−) Cells 290
V. Failure of TL Antigens of Hybrid Cells to Undergo Modulation ... 293
VI. Metabolism of Membrane Antigens of Hybrid Cells 294
VII. The Quantities of TL Antigens of Parental and Hybrid Cells 295
VIII. Discussion 297
IX. Similar Features of Down Regulation and Antigenic Modulation . 300
X. Concluding Remarks 301
   References 303

I. Introduction

Specialized cells of complex organisms respond to a variety of external signals through a diversity of membrane-associated receptor molecules. Hormones and foreign antigens are examples of ligands whose effects on responsive cells depend upon interactions with appropriate, preexistent receptors. Extensively described alterations in cellular physiology occur after hormone stimulation including changes in cell metabolism, stimulation of cell division, and the synthesis and secretion of specialized products. Antibody formation is initiated after antigenic materials interact with preexistent immunoglobulin-like receptors on suitably differentiated lymphoid cells. In this instance, the

---

* Supported by Grant CA-19265-03 from the University of Chicago Cancer Research Center, Cancer Biology Program.

[†] Present address: Department of Microbiology and Immunology, University of Illinois at the Medical Center, Chicago, Illinois 60612.

extraordinarily large number and diversity of foreign substances capable of eliciting a specific immune response serves to emphasize the wide variety of membrane-bound receptor molecules formed by cells of the organism. Formation by the cells of specific receptors may be viewed as one essential aspect of functional maturity, the outgrowth of cellular differentiation and development.

In many instances, the expression of certain receptors by differentiated cells is altered after their association with specific ligands. The quantity of receptors present on surface membranes may diminish as a result, reducing the cells' subsequent sensitivity to stimulation. The hormonal effects of insulin, for example, are reduced following insulin stimulation, correlating with a quantitative reduction in the number of insulin receptors present, a phenomenon termed "down regulation" by Roth and his colleagues (Kahn, et al., 1972, 1973; Gavin et al., 1974; Goldfine et al., 1973; Soll et al., 1974). In the case of antibody synthesis exposing lymphoid cells to antibodies to antigen-specific receptors themselves leads to diminished or absent responses. Previously responsive lymphoid cells fail to form immunoglobulins if they are exposed to antiimmunoglobulin sera before they interact with antigenic substances (Strayer et al., 1975; Köhler, et al., 1978; Köhler, 1975; Van Boxel et al., 1976; Huber et al., 1977; Kearney et al., 1976; Klaus et al., 1977; Raff et al., 1975; Knopf et al., 1973). Prior treatment of neoplastic cells with antibodies to certain tumor-associated antigens converts them to antibody and complement resistance (Aoki et al., 1972; Aoki and Johnson, 1972). The neoplastic cells "escape" antibody-mediated destruction as a result. The tumor-associated antigens of malignant cells from Burkitt's lymphoma patients freshly isolated from the tumor-bearing individual may be undetectable, appearing only after a period of *in vitro* growth.

Alterations in the quantities of receptors present on previously sensitive cells, following prior stimulation, have been described not only for insulin receptors but for β-adrenergic receptors (Mukherjee et al., 1975; Kebebian et al., 1975), acetylcholine receptors of mouse muscle cells (Stanley and Drachman, 1978; Heinemann et al., 1977), thyrotropin-releasing hormone receptors of rat pituitary cells (Genovesi et al., 1977), and others.

Cellular receptors for human growth hormone, like other hormone receptors, undergo down regulation (Lesniak et al., 1973). In this instance, exposing the sensitive cells to hormone concentrations as low as $10^{-10} M$ leads to a diminution in the number or receptors expressed, along with diminished sensitivity to further stimulation. It is estimated that the quantity of hormone leading to a 50% reduction in receptor number is sufficient to occupy only 10% of the receptors present, suggesting that complex cellular controls governing receptor display are operative.

The quantity of immunoglobulin-like receptors for antigen present on the cells treated with antiimmunoglobulin sera is reduced. Suppression of antibody synthesis may be specific for particular antigens if the immunoglobulin antibodies used are directed toward the antigen-combining regions of the receptor. As described by Köhler (1975), the formation of antibodies toward antigenic determinants as small as phosphorylcholine is inhibitable by treating the cells prior to antigen stimulation with antibodies for the combining regions of phosphorylcholine immunoglobulins.

Responsiveness of mouse lymphoid cells to antigenic stimulation returns after continued cellular metabolism in the absence of antibodies to immunoglobulins. This is the case if the cells are obtained from adult, but not neonatal animals. In neonates, the inhibitory effects of antiimmunoglobulins persist for indefinite periods after the antibodies are removed. Terminal differentiation of the cells, as determined by the capacity of the cells to respond to subsequent antigenic stimulation, is inhibited (Raff et al., 1975). This difference in the specialized behavior of lymphoid cells from adult and neonates following exposure to antiimmunoglobulins may be an indication that the effects of external ligand on the expression of specific receptors are an additional parameter of cellular development.

II. Antigenic Modulation as a Model System for Studying Ligand-Induced Effects on Receptor Expression

It is conceivable that cellular mechanisms leading to down regulation of hormone receptors and antigen receptors of immunoglobulin-forming cells are akin to antigenic modulation, a phenomenon described initially by E. A. Boyse, L. J. Old, and their colleagues for the thymus–leukemia (TL) system (Boyse et al., 1963, 1967; Old et al., 1968; Lamm et al., 1968). TL antigens are membrane-associated glycoproteins present on the surface membranes of thymocytes of certain mouse strains and on mouse leukemia cells. Following a period of exposure to TL antibodies, either *in vivo* or *in vitro*, cells previously sensitive are now resistant to TL antibodies and complement (C).

TL antigens disappear from the membranes of modulated cells as determined by several immunologic methods, described extensively (Old and Stockert, 1977). It is not the result of a "covering" of the antigens by noncytotoxic antibodies (it does not occur at 4°C). The fluorescence of cells stained with TL mouse antiserum and fluorescein-conjugated rabbit antimouse sera gradually diminishes, eventually disappearing, as TL antigens are gradually "stripped" from the membrane. More than 99% of the cells are involved. Further incubation of the modulated cells in medium without TL antiserum, or passage of neoplastic cells expressing TL antigens through nonimmunized genetically compatible hosts, leads to re-formation and reexpression of TL antigens by the cells (Fig. 1).

As a model to investigate the effects of specific antibodies on the expression of membrane-associated determinants, Yu and Cohen (1974a,b; Yu et al., 1975) studied formation of TL antigens of neoplastic mouse cells in the presence (and absence) of TL antibodies. It was found that antibody-induced changes in the expression of TL antigens of two lines of murine leukemia cells are an outgrowth of their metabolism. TL antigens, like other cellular constituents, are synthesized and degraded continuously; they have a metabolic half-life which is characteristic of the cells used in the investigation (vide infra). It was determined that modulation results from a more rapid rate of antigen degradation than synthesis.

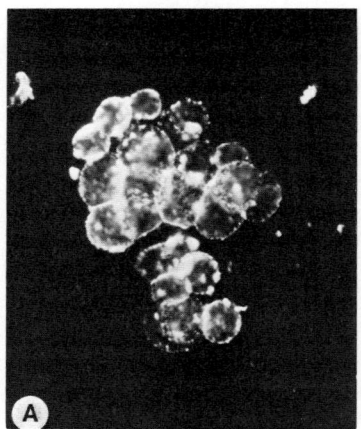
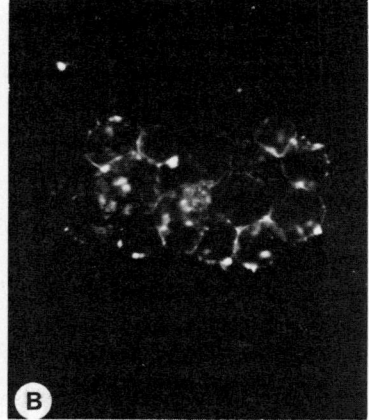

FIG. 1. Fluorescent staining of ASL-1 cells by TL 1,2,3 alloantiserum and fluorescein-conjugated rabbit anti-mouse immunoglobulins (RAM-Ig). (A) ASL-1 cells were incubated with TL 1,2,3 antiserum at 37°C for 15 minutes and then washed with phosphate-buffered saline (PBS). They were incubated a second time with fluorescein-conjugated RAM-Ig at 4°C for another 20 minutes, after which they were washed and resuspended in PBS. The stained cells were examined immediately in a microscope equipped with a UV light source. (B) ASL-1 cells were treated in the same way as those in A, except that they were preincubated at 37°C for 6 hours in medium containing TL 1,2,3 antiserum before staining.

III. The Metabolism of TL Antigens in the Presence (and Absence) of TL Antiserum

TL antigens are found on immature thymocytes of certain mouse strains (e.g., strain A) but not on others (e.g., C57BL/6). Mice phenotypically TL(−) possess genes for forming TL antigens; T cell leukemias arising in such animals are often TL(+). Their association with nonneoplastic, functionally immature mouse thymocytes classifies TL antigens as developmental rather than tumor-associated determinants. Three such determinants, associated with the same macromolecule, are identified on nonneoplastic cells (Liang and Cohen, 1976); a fourth is sometimes found with neoplastic cell types (Old and Stockert, 1977).

TL antisera may be raised in histocompatible TL(−) mice injected with TL(+) thymus or leukemia cells. As membrane-associated glycoproteins, TL and other antigens exposed to the external milieu are labeled "externally" with ^{125}I, in the presence of lactoperoxidase (Liang and Cohen, 1977a). Nonexposed cellular constituents are not labeled by this procedure. Antigens in the process of synthesis may be labeled by incubating viable cells under growth conditions in medium to which [^3H]fucose has been added (Yu and Cohen, 1974b).

Detergent-solubilized hydrophobic macromolecules retain their antigenic specificity in solutions containing nonionic detergents and may be selectively immunoprecipitated with specific antiserum (Liang and Cohen, 1977a). The metabolic turnover of TL antigens is investigated by extracting radioactively labeled cells with nonionic detergents as nonidet P-40 (NP-40) and determining the specific activity of immunoprecipitates raised at equivalence with TL antiserum. The biosynthetic rate of membrane-associated glycoproteins is determined by incubating the cells for defined periods in medium containing [^3H]fucose followed by extraction with NP-40 and recovery of the antigens by selective immunoprecipitation. The rate of antigen degradation is estimated by labeling the cells with ^{125}I followed by incubation for varying periods and recovery of remaining antigens by immunoprecipitation from NP-40 cell extracts. These techniques have been described in detail previously (Yu and Cohen, 1974a,b, Liang and Cohen, 1976).

A. Synthesis and Degradation of TL Antigens

Using these procedures, we investigated the effect of TL antiserum on the rate of synthesis and the rate of degradation of TL antigens of murine leukemia cells. We used several antiserum preparations containing antibodies that were specific for various determinants of the

TL complex. The results we obtained are summarized in Table I; they indicate the following:

1. The metabolic half-life of TL antigens of RADA-1 cells, a radiation-induced leukemia cell line of strain A mice, incubated in medium free of TL antibodies, is 16 hours. The half-life of TL antigens of RADA-1 cells incubated in medium to which TL antiserum has been added, undergoing antigenic modulation, shortens to 4 hours (Fig. 2, Insert).

The shortened half-life of TL antigens of modulating cells results from a more rapid rate of antigen disappearance from the membrane than replacement (Fig. 3). This is the case in analogous experiments involving differing antiserum concentrations with specificities for various determinants of the TL antigen complex.

The effects of specific antiserum on the rate of metabolism of TL antigens of RADA-1 cells are not unique for that cell line. Similar

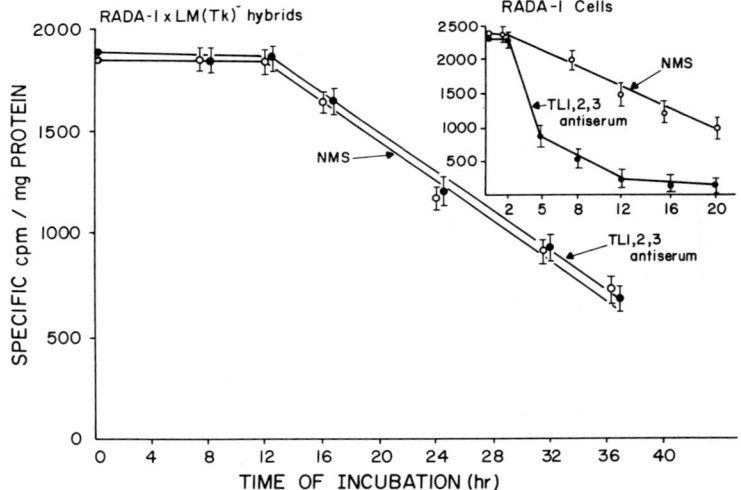

FIG. 2. Effect of TL 1,2,3 antiserum on the rate of disappearance of TL antigens of RADA-1 and RADA-1 × LM(TK)⁻ cells. RADA-1 or RADA-1 × LM(TK)⁻ cells (clone No. 11) were incubated for 5 hours in medium containing [^{3}H]fucose and then in fresh medium containing a 100-fold excess concentration of unlabeled L-fucose and TL antiserum or normal mouse serum (NMS). The incubation was continued at 37°C. At various periods, approximately 5×10^6 cells were removed from the culture, washed, and lysed with NP-40. Immunoprecipitation of TL antigens released into the supernatant was performed according to the procedures described previously. Specific activity (specific cpm/mg protein) was calculated after the specific cpm and protein concentration in the NP-40 extracts were determined. Vertical bars indicate the standard deviations of each mean value. From Liang and Cohen (1977b).

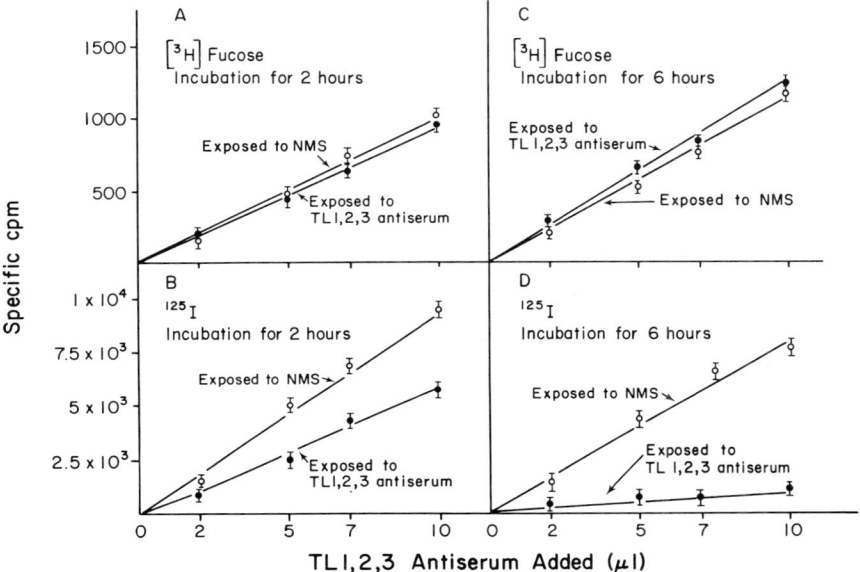

FIG. 3. Solubilization and immunoprecipitation of [³H]fucose-labeled (A and C) or ¹²⁵I-labeled (B and D) membrane antigens of ASL-1 cells preincubated at 37°C in medium containing TL 1,3 antiserum or NMS. After washing, the cells were extracted with NP-40 and TL antigens released from the cells were immunoprecipitated with various amounts of TL 1,2,3 antiserum followed with RAM-Ig. For cells labeled with ³H, [³H]fucose was added to the medium for the last 5 hours of incubation. Vertical bars indicate the experimental variation between duplicate samples. From Liang and Cohen (1977b).

effects of antiserum exposure are observed for TL antigens of ASL-1 cells, an independently arising leukemia cell line of strain A mice. The metabolic half-life of TL antigens of ASL-1 cells, 18 hours in the absence of TL antiserum and 9 hours in its presence, is distinctly different, however, than that found for TL antigens of RADA-1 cells obtained under similar conditions. In both instances, the half-life of TL antigens is shortened by antiserum exposure (Cohen and Liang, 1976).

The biosynthetic rate of TL antigens is unaffected by exposing the cells to TL antiserum; the rate of antigen synthesis is indistinguishable in cells incubated in the presence or absence of TL antiserum (Fig. 3). After formation on the membrane, TL antigen–antibody complexes are traced to endocytic vacuoles where they are degraded (Stackpole *et al.*, 1974).

2. TL antigens, like other membrane-associated determinants, are

"shed" from the cells to the surrounding medium, at least in part, in an antigenically intact form. They may be recovered by immunologic means from the medium. The rate of "shedding" of TL antigens from cells exposed to TL antiserum undergoing modulation is not detectably different than that of nonmodulating cells (Yu and Cohen, 1974a,b).

3. The half-life of $H-2^a$ antigens of both RADA-1 and ASL-1 cells is approximately 26 hours; it is distinctly different than that of TL antigens of the same cell types. $H-2^a$ antigens of these cells, unlike TL antigens, do not undergo modulation. Prior exposure of the cells to $H-2^a$ antiserum has no effect upon the cells subsequent sensitivity to $H-2^a$ antiserum and C. The metabolic half-life of $H-2^a$ antigens is unaffected by exposing the cells to $H-2^a$ antiserum (Yu and Cohen, 1974a,b; Liang and Cohen, 1976; Cohen and Liang, 1978).

4. Exposing the cells to $H-2^a$ antiserum has no effect upon the rate of turnover of TL antigens; in a similar manner, TL antibodies have no effect upon the half-lives of $H-2^a$ antigens. The metabolic half-lives and response to specific antisera of $H-2^a$ and TL antigens appear to be controlled independently of each other (Cohen and Liang, 1976).

B. TUMOR-ASSOCIATED ANTIGEN OF ASL-1 CELLS

During the course of our investigations, we detected the presence of a tumor-associated antigen of neoplastic ASL-1 cells (Yu and Cohen, 1974a,b). This determinant was detected by immunologic means using antisera raised in syngeneic strain A mice injected with mitomycin-C treated ASL-1 cells. Histocompatible TL(−) (BALB/c × C3H/He)F_1 mice injected with the same cell types form antibodies with specificity for both TL and the tumor-associated antigen. Electrophoresis of immunoprecipitates of NP-40-derived cellular extracts of ASL-1 cells or of nonneoplastic thymus cells of strain A mice distinguish the tumor antigen as an antigen separate from TL antigens (Yu and Cohen, 1974a,b).

The rate of metabolism of the tumor-associated antigen of ASL-1 cells is different than that of TL and H-2 antigens, formed by the same cells. Its half-life is 44 hours, both in the presence and absence of specific antiserum; its expression is unaffected by exposing the cells to specific antiserum. The tumor antigen of ASL-1 cells fails to undergo modulation.

Exposing ASL-1 cells to sera containing antibodies to both TL and the tumor-associated antigen leads to a progressive reduction in the expression of TL determinants with no detectable effects upon the expression of the tumor antigen. Electrophoresis of immunoprecipitates

TABLE I

METABOLIC HALF-LIFE OF SEVERAL MEMBRANE-ASSOCIATED DETERMINANTS OF PARENTAL AND HYBRID MURINE CELLS

Determinants	Cell type	Antiserum	Half-life (hours)
TL1,2,3[c]	ASL-1	(−)[a]	18
		(+)[b]	9
	ASL-1 × LM(TK)−	(−)	28–32[d]
		(+)	28–32[d]
	RADA-1	(−)	16
		(+)	4
	RADA-1 × LM(TK)−	(−)	28–32[e]
		(+)	28–32[e]
H-2[a]	ASL-1	(−)	26
		(+)	26
	ASL-1 × LM(TK)−	(−)	26[d]
		(+)	26[d]
H-2[k]	ASL-1 × LM(TK)−	(−)	26[d]
		(+)	26[d]
Tumor associated	ASL-1	(−)	44
		(+)	44
	ASL-1 × LM(TK)−	(−)	34–38[d]
		(+)	34–38[d]
	RADA-1	(−)	46
		(+)	46
	RADA-1 × LM(TK)−	(−)	34–38[e]
		(+)	34–38[e]

[a] In the absence of specific antiserum.
[b] In the presence of specific antiserum.
[c] Shared by individual molecules.
[d] Based on an analysis of 15 colonies.
[e] Based on an analysis of 11 colonies.

of NP-40 extracts of ^{125}I-labeled cells at varying periods after they are exposed to serum containing antibodies to TL and the tumor association leads to a progressive reduction in the quantities of TL antigens recovered. The tumor-associated antigen persists (Fig. 4).

That only one of two membrane-associated antigens undergoes modulation in the presence of antibodies to both is a likely indication that modulation is more than "simply" the result of an antigen–antibody reaction taking place on the surface of the cell; it is likely that complex cellular controls are involved.

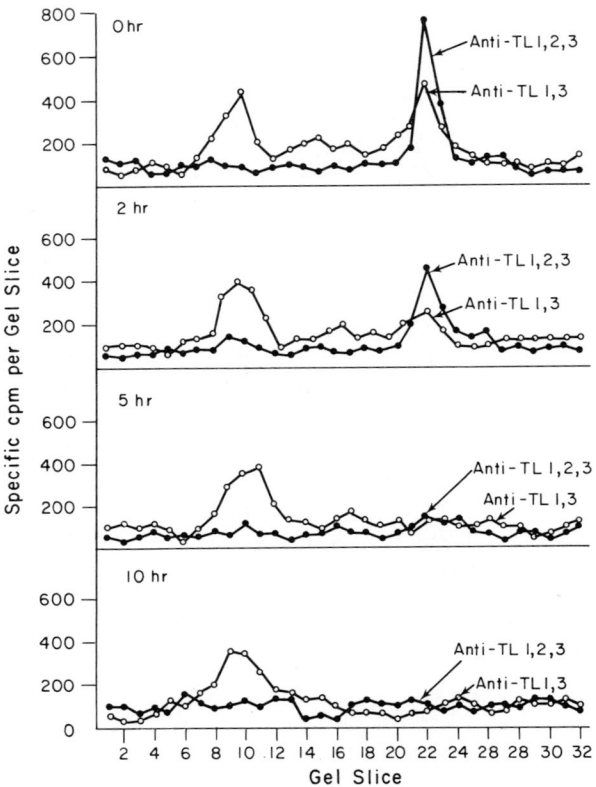

FIG. 4. SDS-polyacrylamide gel electrophoresis of immunoprecipitates, formed with TL 1,2,3 or TL 1,3 antiserum of nonidet P-40 extracts of ^{125}I-labeled RADA-1 cells undergoing modulation. TL 1,3 antiserum contains antibodies for TL antigens along with the tumor-associated antigen. From Liang and Cohen (1977b).

IV. Membrane Antigens of Somatic Hybrids of TL(+) and TL(−) Cells

The presence in differentiated cells of mechanisms involved in controlling various aspects of their specialized activities may be revealed by analyzing somatic hybrid cells. Such cells are prepared by fusing parental cells each of which is engaged in a unique differentiated function. After fusion, regulatory controls operative in one parent affect the analogous function of the other (Harris, 1970; Klebe et al., 1970; Davidson, 1974; Weiss and Chaplain, 1971; Miller et al., 1976; Croce et al., 1974; Gougne et al., 1972). Fusion, for example, of nonneoplastic cells of human origin with antibody-forming mouse myeloma cells leads to persistence of immunoglobulin synthesis (Schwaber and Cohen, 1973, 1974). Similar effects are observed in the case of membrane-associated macromolecules. Burkitt's lymphoma-derived Daudi cells, phenotypically HLA(−), form HLA antigens after they are hybridized with a second Burkitt's derived cell line, Raji. The HLA antigens expressed by the hybrid, A10, B38, and B17 are characteristic of Daudi cells which are genotypically positive for these determinants (Fellows et al., 1977). Thy 1.2 antigens are expressed after fusion of genotypically (+) but phenotypically (−) mouse strain A spleen cells with AKR mouse thymocytes. Thymocytes of AKR mice actively form Thy 1.1 determinants, immunologically distinguishable allelic forms of the *Thy-1* locus (Allison et al., 1975). In both instances, induction of repressed membrane determinants occurs.

To detect the presence of regulatory factors affecting the expression of membrane-associated antigenic determinants of murine leukemia cells, we prepared hybrids of ASL-1 cells [TL 1,2,3(+), H-2^a(+), Thy 1.2(+), and tumor antigen(+)] and LM(TK)$^-$ cells [H-2^k(+), TL(−), Thy 1.2(−), and tumor antigen(−)], and analyzed clonal isolates of hybrid cells for the expression of several membrane-associated antigens, their metabolic half-lives, and capacity to undergo antigenic modulation. ASL-1 cells do not survive *in vitro;* LM(TK)$^-$ cells, a sustained tissue culture cell line of C3H/He mouse origin, are thymidine kinase-deficient and die in medium containing hypoxanthine, aminopterin, and thymidine (HAT) (Littlefield, 1964). Hybrid cells complement the unique deficiencies of each parental source; hybrid colonies of ASL-1 and LM(TK)$^-$ are selected *in vitro* in HAT-containing medium. They possess a hybrid karyotype consisting of approximately 85 characteristic mouse chromosomes including the identification of "marker" chromosomes originating in each parental source (Liang and Cohen, 1975).

The hybrid cells form both H-2^a and H-2^k antigen complexes, characteristic of each parental source, in approximately equal proportions. Essentially 100% of the cells are killed in the presence of C and "H-2^a" antiserum (raised in H-2^k mice) or "H-2^k" antiserum (raised in H-2^a mice). Similar numbers of hybrid cells reduce by absorption H-2^a or H-2^k antisera to the same extent, indicating that the formation of H-2 antigens in these cells is codominant (Liang and Cohen, 1975).

The expression of Thy 1.2 antigens, membrane-associated determinants formed by ASL-1 but not by LM(TK)$^-$ cells, is suppressed in fused cells. Hybrid cells remain viable in the presence of Thy 1.2 (or Thy 1.1) antiserum and C; they fail to stain in immunofluorescence studies involving Thy-1 antiserum and fluorescent-conjugated rabbit antimouse sera. Under similar conditions ASL-1 cells are killed, they

TABLE II

Summary of Studies on Different Clones of Hybrid Cells

ASL-1 × LM(TK)⁻ clone number	Number chromosomes[a]	TL antigen expression	Antigenic modulation	Half-life of TL antigens incubated with TL antiserum (hours)	NMS[b]
5	86	+	−	32	32
7	85	+	−	32	32
8	88	+	−	30	30
9	85	+	−	30	30
11	86	+	−	28	28
12	84	+	−	32	32
13	83	+	−	30	30
15	82	+	−	28	28
16	88	+	−	28	28
17	83	+	−	28	28
30	85	+	−	32	30
31	85	+	−	28	30
32	85	+	−	28	28
33	84	+	−	32	32
RADA-1 × LM(TK)⁻ clone number					
1	86	+	−	32	32
2	84	+	−	28	28
3	85	+	−	30	28
4	85	+	−	28	28
5	86	+	−	28	28
8	85	+	−	30	30
10	84	+	−	32	30
11	83	+	−	32	32
13	82	+	−	30	30
14	87	+	−	28	28
15	82	+	−	32	32

[a] Based on an analysis of at least 20 metaphase spreads.
[b] Normal mouse serum.

stain positively. Similar results are reported by others (Hyman and Kelleher, 1975; Köhler *et al.*, 1977) for hybrids of Thy 1.2(+) and Thy 1.2(−) cells.

The formation of TL 1,2,3 antigens is dominant in hybrids of TL(+) and TL(−) cells. Each of 25 clonal isolates examined was susceptible to the cytolytic effects of TL antiserum and C, stained positively in immunofluorescence studies and reduced by absorption known titers of TL antiserum (Table II). TL antigens can be recovered by immunoprecipitation from ^{125}I-labeled hybrid cells extracted with NP-40. In no instance, involving continuous cultivation of ASL-1 × LM(TK)$^-$ hybrid cells for periods up to 24 months, has a clonal derivative been found failing to form TL antigenic determinants.

V. Failure of TL Antigens of Hybrid Cells to Undergo Modulation

RADA-1 × LM(TK)$^-$ hybrid cells, unlike the RADA-1 parental cells from which they are derived, fail to modulate TL antigens (Fig. 5). Hybrid cells exposed for prolonged periods to high titers of TL antiserum, conditions in excess of that required to stimulate the modulation of RADA-1 cells, remain sensitive to fresh TL antiserum and complement. After antibody exposure, the cells retain TL antigens on their surface membranes as detected by their positive staining characteristics in fluorescent antibody tests, immunoprecipitation of TL antigens from NP-40 extracts of the cells, as well as their capacity to reduce by absorption TL antisera of known titers. Sensitivity to TL antibodies and complement persists. Our attempts to stimulate modulation of the hybrid cells include exposing the cells for up to 30 hours to each of several different preparations of TL antiserum at 5-fold higher concentrations than required to modulate parental cells (parental RADA-1 cells, under similar conditions modulate in 8 hours), as well as the use of indirect methods involving the addition of rabbit antimouse immunoglobulins.

The failure of hybrid cells to undergo modulation under conditions more stringent than required to stimulate modulation of parental cells is an indication that cellular controls governing the expression of TL antigens may be distinguished from those governing adaptation to TL antiserum. It is likely that multiple cellular steps are required for modulation to take place, complex controls governing the reactions required to "strip" TL antigen–antibody conjugates from the cell surface. The precise cellular "defect" of hybrid cells, leading to failure of modulation, is unknown. Stackpole *et al.* (1978) found that the third component of complement (C3) is required for the modulation of TL antigens. Conceivably, the formation of receptors for C3, like other membrane determinants, is suppressed after fusion. Cross-linking of surface immunoglobulins of mouse lymphocytes with specific antibodies or lectins stimulates their association with cytoskeletal actin (Flanagan and Koch, 1978; Bourguignon and Singer, 1977). It is tempting to speculate that TL–actin association, required for modulation, fails to occur in hybrid cells.

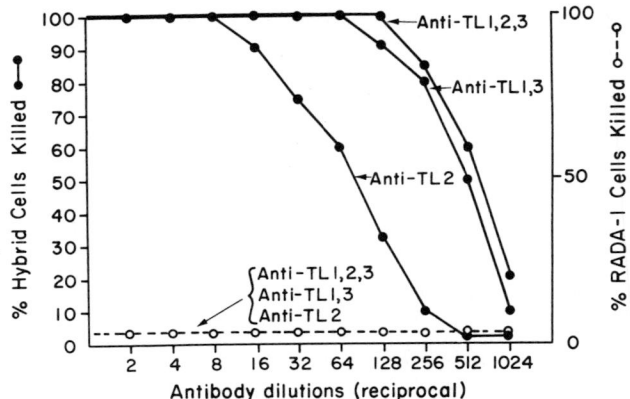

FIG. 5. ASL-1 × LM(TK)⁻ cells (clone No. 10) or ASL-1 cells (used as one parent in forming the hybrid) were incubated at 37°C for 24 hours in the presence of TL 1,2,3 antiserum (antiserum titer against A/J thymocytes, 1:512). After washing with cold minimum essential medium, the cells were tested for their susceptibility to cytotoxic effects of TL 1,3, TL 2, or TL 1,2,3 antiserum and guinea pig complement (GPC). The proportion of cells killed was determined by trypan blue dye exclusion. From Liang and Cohen (1975).

The tumor-associated antigen of ASL-1 cells is detected in each clone of hybrid cells tested. Like parental cells, it does not undergo modulation (Table II).

VI. Metabolism of Membrane Antigens of Hybrid Cells

The rate of metabolism of TL antigens of hybrid cells is different from those of parental cells. Such differences reflect the existence of control mechanisms governing, not only TL antigen expression and capacity for modulation, but its rate of metabolism as well.

The half-life of TL antigen expression of parental ASL-1 cells is 18 (± 2) hours; the half-life of TL antigens expressed by 13 clonal isolates of ASL-1 × LM(TK)⁻ hybrid cells ranges between 28 and 32 hours (Table I). The metabolic half-life of TL antigens of hybrid cells is not affected by exposing the cells to TL antisera; under similar circumstances the half-life of TL antigens of cells undergoing modulation changes significantly.

The rates of metabolism of TL antigens of parental RADA-1 cells and RADA-1 × LM(TK)⁻ hybrid cells like those of ASL-1 and ASL-1 × LM(TK)⁻ cells differ from each other as well; the metabolic half-life of TL antigens of RADA-1 cells is approximately 16 hours and that of the hybrid cells is 28–32 hours, similar to that of ASL-1 × LM(TK)⁻

hybrid cells. As described previously, the half-life of TL antigens of RADA-1 cells exposed to TL antiserum shortens from 16 to 4 hours; however, the half-life of TL antigens of the hybrid cells is unaffected by exposing the cells to TL antiserum (Fig. 2). RADA-1 × LM(TK)⁻ hybrid cells fail to undergo antigenic modulation (Cohen and Liang, 1976).

Twenty-five clones of hybrid cells were investigated in this analysis. None underwent antigenic modulation (Table II); in no instance was there an indication that TL antiserum affected either the expression or the rate of metabolic turnover of TL antigens of hybrid cells. The half-life of expression of TL antigens of each clone was essentially the same.

The total number of chromosomes of RADA-1 × LM(TK)⁻ and ASL-1 × LM(TK)⁻ hybrid cells ranged between 80 and 86; it remained essentially the same throughout the period of these experiments. Whether each chromosome of both parents was present in the hybrid cells was not determined—conceivably, certain chromosomes were lost and others duplicated, maintaining a relatively constant chromosomal number while significantly affecting important parameters of cellular physiology.

The metabolic half-life of the tumor-associated antigen of ASL-1 cells, like that of TL antigens, is different in hybrid than parental cells. The half-life of the tumor-associated antigen formed by each clone of hybrid cells is approximately 36 hours, less than the half-life of 44 hours of the analogous antigen formed by parental cells. Its metabolic half-life is unaffected by exposing either parental or hybrid cells to specific antiserum. The tumor-associated antigen of these cell types does not undergo modulation. It is of significance that the cells remain viable in the presence of tumor antiserum and active complement.

The metabolic half-lives of H-2^a antigens of ASL-1 or RADA-1 parental and ASL-1 × LM(TK)⁻ or RADA-1 × LM(TK)⁻ hybrid cells, approximately 26 hours, are equivalent. Like the tumor-associated antigen of ASL-1 cells, the half-life of H-2^a antigens is unaffected by exposing the cells to H-2^a antiserum. H-2^a antigens fail to undergo modulation (Cohen and Liang, 1978).

These results indicate the existence of complex cellular controls governing not only membrane-determinant expression but its metabolic turnover rate as well.

VII. The Quantities of TL Antigens of Parental and Hybrid Cells

Quantitative antibody absorption procedures provide a means of determining the relative amounts of specific antigenic determinants expressed by various cell types. Differences in the quantities of analogous antigens expressed between parental and hybrid cells are a likely indication of the presence of additional regulatory controls governing this aspect of cellular physiology. In this procedure, varying numbers of cells of known surface area (calculations, taken from cell diameters, are based on the assumption that the cells in suspension are spherical) are incubated with an antiserum of known titer; the number of cells required to reduce the titer as determined by subsequent analysis is determined. As usually practiced, the procedure does not indicate the absolute number of antigens expressed by the cells, nor does it indicate the presence of "buried" or "hidden" determinants which are inaccessible to the antibodies used in the analysis. To minimize antibody-induced effects upon cell membrane antigen expression, the incubation is performed at 4°C for 30 minutes. (Modulation does not occur at 4°C.) Other important aspects of membrane-determinant expression as variations in antigen density occurring during various stages of the cell cycle, or their spatial distribution on the cells' surface, are not considered by this approach.

Like antigen expression, metabolism, and modulation, the quantities of TL antigens expressed by parental and hybrid cells are quite distinct. The absorptive capacity of RADA-1 cells for TL antiserum of a known titer exceeds that of RADA-1 × LM(TK)$^-$ hybrid cells by about 2-fold (Fig. 6). Approximately 6×10^6 RADA-1 cells are required to reduce the antiserum titer by one-half; twice as many hybrid cells are

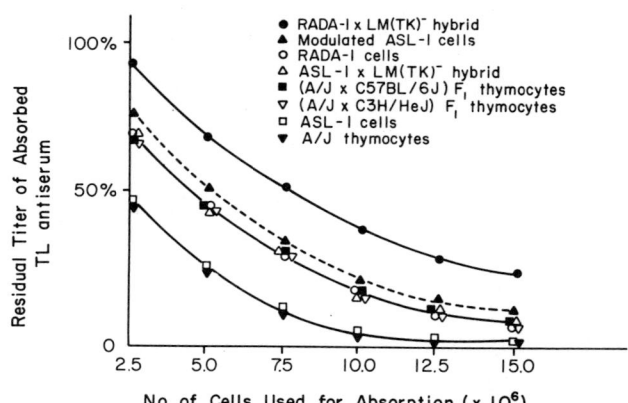

FIG. 6. The absorptive capacity of various types of TL (+) cells for TL antiserum of known titer. Appropriately diluted TL 1,2,3 antiserum was incubated at 4°C for 30 minutes with various numbers of cells. After centrifugation, the residual TL antibody titer was determined by C-mediated microcytotoxicity tests using A/J thymocytes (TL 1,2,3) as "target" cells. The percentage of cells killed was determined by trypan blue dye exclusion. From Liang and Cohen (1977a).

required to reduce the titer to the same extent. These results, however, do not take into account differences in surface area between the two cell types. Corrected for differences in size (the mean diameter of RADA-1 cells is 9.5 mμ, that of RADA-1 × LM(TK)$^-$ hybrid cells is 16.2 mμ), the relative quantity of TL antigens of parental RADA-1 cells is approximately 5-fold *greater* than that formed by RADA-1 × LM(TK)$^-$ hybrid cells.

ASL-1 cells and thymocytes of strain A mice express the highest quantities of TL antigens of any of the various cell types investigated (Fig. 6). The relative quantities of TL antigens expressed by these two cell types (corrected for size differences) is approximately 10 times greater than that of RADA-1 × LM(TK)$^-$ hybrid cells and 5 times greater than that of ASL-1 × LM(TK)$^-$ hybrid cells.

Whether the cellular controls governing expression, turnover, and quantity of TL antigen expression are coordinate has not been determined; each parameter may be controlled independently of the others. Our finding that TL antigens are formed by hybrids of TL(+) and TL(−) cells which fail to modulate may be an indication that they are not. Complementation studies in various lines of TL(+) hybrid cells would be one approach toward an understanding of these complex cellular phenomena and their possible linkage relationships.

VIII. Discussion

The data presented in this chapter reveal the existence of cellular controls governing the expression, rates of metabolism, and quantities of TL antigens formed by two lines of murine leukemia cells. TL antiserum stimulates the reversible disappearance of TL antigens from the cellular membrane, a phenomenon termed antigenic modulation which in many ways is akin to the effects of hormones and other ligands on the quantities of specific receptors formed by suitably differentiated cells.

TL antigens, like other cellular constituents, undergo a metabolic turnover, that is, they are synthesized and degraded continuously. Their half-life of expression on the membrane may be estimated by isotopic labeling procedures that are specific for previously formed "exposed" membrane proteins (^{125}I-labeled lactoperoxidase). The half-life of TL antigens of ASL-1 cells, a leukemia cell line of strain A mice, is 18 hours, changing to 9 hours in cells undergoing antigenic modulation. The half-life of analogous immunologically cross-reactive TL antigens of RADA-1 cells, and independently arising leukemia cell line of the same mouse strain, is 16 hours in the absence of TL antiserum, 4 hours in its presence.

TL antiserum stimulates a reduction in the expression of TL antigens; we considered several possible cellular mechanisms.

1. During modulation the rate of synthesis of TL antigens may diminish as the rate of antigen degradation continues, without interruption or alteration, leading to a gradual reduction in the quantities of antigens present at the membrane. This would indicate the existence of cellular control mechanisms affecting as a result of an antigen–antibody reaction occurring on the surface of the cell primary modes of protein biosynthesis. We found no evidence that the TL antiserum affected the biosynthetic rate of TL antigens. In each instance involving either ASL-1 or RADA-1 cells, the presence of TL antibodies in the cellular milieu had no detectable effect upon the rate of TL antigen formation. The rate of incorporation of [^{3}H]fucose into TL antigens of leukemia cells undergoing modulation was indistinguishable from that of cells incubated in medium without TL antibodies.

2. The association of immunoglobulins with membrane-associated antigenic determinants may trigger a nonspecific alteration in the lipoprotein structure of the cell membrane so that newly formed determinants for a time can no longer "fit" into place. Antigen degradation resulting from metabolic turnover under this scheme would continue without replacement by newly formed determinants, leading to a gradual diminution in the total quantity of antigens present at the cell surface. Alternatively, as TL antigens are glycoproteins, TL antiserum may lead to the cessation of antigen glycosylation, conceivably required for insertion of newly formed determinants into the membrane. Thy 1(−) mutants of mouse myeloma cells have been described with such a defect (Trowbridge et al., 1978). We find no evidence, however, that modulation of one determinant affects the quantity of other antigens expressed by the cells or their rates of metabolism. The expression of independently arising, antigenically distinct, nonassociated determinants appears to be governed by controls which are independent of each other (Cohen and Liang, 1976). Exposing TL(+) ASL-1 cells to antibodies to TL antigens has no effect upon the metabolism of H-2^a antigens of the same cells. Although TL and H-2 antigens share many similarities [e.g., they both possess β_2-microglobulin in their overall structure, are of similar molecular weights, and share amino acid sequence homology (Anundi et al., 1975, Davies et al., 1969)], significant undescribed differences between the two antigens may exist, affecting the capacity of newly formed TL antigens to integrate into the membranes of modulated cells.

3. Membrane-associated receptors are "shed" from the membrane to the surrounding medium (Calafat et al., 1976) in some instances, in

association with cytoskeletal actin (Koch and Smith, 1978). Modulation may occur by a shedding of receptors from the membrane to the medium, stimulated by combination with specific antiserum. Mammary tumor (Calafat et al. 1976), H-2 (Koch and Smith, 1978) and TL antigens of murine leukemia cells (Yu and Cohen, 1974b) are examples of determinants which are shed spontaneously from the cell. The "usual" rate of receptor metabolism may be unaffected by combination with antibodies; however, modulation may result from the association of antibodies with receptors that are "loosely" bound to the membrane, allowing the ligand to "pull" them off. Antibody-induced shedding has been described for the virally specified determinants of cells infected with measles virus (Rustigian, 1966) and for acetylcholine receptors of mouse muscle cells (Stanley and Drachman, 1978). We find no evidence supporting this, however. In prior experiments (Yu and Cohen, 1974b), the quantities of TL antigens recovered from the medium of antibody-exposed ASL-1 cells were indistinguishable from those of control cells. TL determinants are detected, in both instances, in the cell-free supernatant, in an antigenically intact form. It is conceivable, nevertheless, that antigenic substances pulled from the membrane by antibodies are rapidly degraded extracellularly, obscuring our ability to detect this putative mechanism of modulation. The concentration in the medium of shed TL antigens was small (Yu and Cohen, 1974b).

4. Antibodies may induce during modulation an enzyme-mediated change in the structure of membrane-associated complementary antigens. The molecule might be altered in such a way that the new form no longer combines with specific antibodies, thus becoming undetectable in immunologic assays. This scheme might involve a "seeking" out by newly formed enzyme of the various antigens involved. Conceivably, each determinant might be associated with its own enzyme, activated as a result of its interaction with antibody.

We have no evidence either to support or to exclude this possibility. Our observation that hybrid cells express TL antigens but fail to undergo modulation is an indication that cellular mechanisms involved in forming TL antigens may be distinguished from those required for conversion to antibody and complement resistance. It suggests that multiple cellular steps are involved in the modulation process, one or more of which is "defective" in hybrid cells.

We find that reduction in the quantities of TL antigens expressed by modulating leukemia cells results from a more rapid rate of antigen degradation than synthesis. This was found to be the case in repeated experiments for two lines of murine leukemia cells involving different preparations of TL antisera (Liang and Cohen, 1976; Cohen and Liang,

1976, 1978) (Figs. 2 and 3). The duration of exposure of TL antigens of cells undergoing modulation is less than one-half that of the control. It is likely that the half-life of TL antigen–antibody complexes on the cells' surface is less than that of analogous antigens which are not associated with antibodies. In the course of metabolism, newly formed surface antigens would be removed shortly after they associate with specific antibodies. It might be predicted, therefore, that TL antigens appear on the surface of modulated cells in the presence of specific antibodies. The cells' capacity to "escape" complement-mediated lysis is poorly understood; however, acquisition of resistance to TL antibodies and complement precedes the complete disappearance of TL antigens from the membrane (Liang and Cohen, 1977a). ASL-1 cells, shortly after they convert to TL antibody and complement resistance, retain greater quantities of TL antigens on their surface membranes than other lines of TL(+) cells which are sensitive (Fig. 6).

It might be inferred that modulation is a "passive" event occurring without the involvement of other aspects of cellular physiology resulting "simply" from the effects on the cells' surface of an antigen–antibody conjugate. There are several indications that this may not be the case and that more complex controls are operative. Not every membrane determinant of ASL-1 cells undergoes modulation. In spite of the presence of H-2 antigen–antibody complexes on the surface membrane and the formation of "patches" and "caps" detected by fluorescent antibody methods, leukemia cells remain sensitive to H-2 antisera (Liang and Cohen, 1976). The metabolism of H-2 antigens is unaffected by H-2 antiserum (Liang and Cohen, 1976). Antigenically distinct, nonassociated membrane glycoproteins respond to antisera containing multiple immunoglobulin specificities independently of each other. Exposure of cells to serum containing antibodies to *both* TL and the tumor-associated antigen of ASL-1 cells leads to modulation of TL antigens alone (Fig. 4). The tumor antigen persists. It is of significance that the sole presence of antibodies to tumor-associated antigens, in the presence of complement, is insufficient to lyse these neoplastic cells. Somatic hybrids of TL(+) and TL(-) cells form TL antigens but fail to undergo modulation as a further indication that complex cellular controls governing the effect of antiserum on antigen "display" are operative. Modulation may be a "luxury" cell function which as others of this category is often suppressed after fusion (Gougne et al., 1972).

IX. Similar Features of Down Regulation and Antigenic Modulation

Is the mechanism leading to reductions in the quantities of hormone receptors, an aftermath of hormone stimulation, akin to modulation of TL antigens? Several features of these two phenomena which are similar suggest that they may be related.

After their association on the cell surface, TL antigen–antibody complexes are traced intracellularly to endocytic vacuoles where they are degraded (Stackpole et al., 1974). In an analogous fashion, complexes of thyrotropin-releasing hormone and its receptor on human cells are found, after they associate, intracellularly in endocytic vacuoles where they are degraded to constituent amino acids (Hinkly and Tashyian, 1975). In the case of antigens, bivalent antibodies lead to aggregation on the cell surface before internalization. Whether complexed receptors aggregate before endocytosis is not certain. Receptors on various human cell types for low-density lipoproteins after binding to ligand are internalized and at least partially degraded (Anderson et al., 1977). In some instances, bound receptors are not degraded but are "recycled." Glucocorticoid receptors of mouse muscle cells after association with hormone behave this way (Aronow, 1978).

Acetylcholine receptors of mouse neuromuscular junctions undergo a metabolic turnover, and their half-life is shortened after they are complexed with receptor antibodies obtained from myasthenia gravis patients (Stanley and Drachman, 1978). Other ligands for acetylcholine receptors, such as bungarotoxin, do not affect the receptors turnover rate, an indication that the behavior of such receptors to different ligands differs.

In each instance, both modulation and down regulation lead to diminished quantities of specific cell surface determinants; they correlate with alterations in the "usual" cellular responsiveness to external stimuli. Genetic control manifestations of cellular development governing various parameters of membrane receptor expression appear to be critical determining factors.

X. Concluding Remarks

The quantities of TL antigens expressed by ASL-1 and RADA-1 cells, independently arising murine leukemia cell lines, diminish after they are exposed to TL antibodies; the cells ordinarily sensitive convert to TL antibody and complement resistance (antigenic modulation). The reduction in the quantities of TL antigens expressed by the cells is an outgrowth of antigen metabolism. The half-life of expression of TL antigens of ASL-1 cells changes from 18 to 9 hours in the presence of TL antibodies; for RADA-1 cells, TL antibodies leads to a shortening of the half-life of TL antigens from 16 to 4 hours. Specific antibodies do not affect the rates of metabolism of every antigen of these leukemia cells. The half-lives of H-2^a or a tumor-associated antigen of the cells (26 and 44 hours, respectively) are not affected by exposing them to specific antibodies. Antiserum with specificities for TL as well as the tumor antigen of ASL-1 cells leads to modulation of TL antigens alone; the tumor antigen persists, and its metabolic half-life is unchanged. Selectivity of modulation is taken as an indication that antibody-induced effects on membrane antigen metabolism and expression are more than simply the result of an antigen–antibody reaction occurring on the surface of the cells; more complex controls are probably involved.

In order to detect the possible presence of cellular controls governing the effects of antiserum on membrane-antigen expression, somatic hybrids of TL(+) ASL-1 cells and TL(−) LM(TK)⁻ cells, an *in vitro* maintained mouse cell line, were formed. After fusion, regulatory influences of one parent affect analogous functions of the other. Hybrid cells form H-2 antigens of both parental sources in approximately equal proportion and share a hybrid karyotype. They form TL antigens but fail to undergo antigenic modulation even under conditions more stringent than required to stimulate modulation of parental cells. The metabolic half-life of TL antigens expressed by ASL-1 × LM(TK)⁻ hybrid cells, approximately 30 hours, is distinctly different than that of parental ASL-1 cells. Like ASL-1 cells, however, the half-life of TL antigens is unaffected by antiserum exposure. The quantities of TL antigens expressed by parental cells is 5-fold greater than that of hybrid cells, another parameter of antigen expression which appears to be under cellular control. Not every antigen of ASL-1 cells is expressed in ASL-1 × LM(TK)⁻ hybrid cells. The expression of Thy 1.2 antigens, characteristic of ASL-1 cells, is suppressed after fusion. Changes from parental cells in several parameters of antigen expression in hybrid cells are taken as indications that regulatory circuits governing the expression of each several membrane determinants are controlled independently of each other.

Several possible models were considered to explain antigenic modulation. Antibody-induced shedding of membrane-bound antigens, alterations in the lipoprotein structure of the membrane resulting from antiserum exposure, and alterations in antigenic specificity induced by antiserum were rejected as possible explanations. The available evidence supports the notion that antibody-induced alterations in the quantities of membrane antigens expressed, in those instances in which they occur, are outgrowths of their metabolism. The biosynthetic rate of TL antigens is unaffected by antiserum exposure; however, the rate of degradation of TL antigens of modulating cells is faster than that of the control, leading to a progressive diminution in the quantities of antigens expressed.

REFERENCES

Allison, D. C., Meier, P., Majeune, M., and Cohen, E. P. (1975). *Cell* **6,** 521.
Anderson, R. G. W., Brown, M. S., and Goldstein, J. L. (1977). *Cell* **10,** 351.
Anundi, H., Rask, L., Ostberg, L., and Peterson, P. A. (1975). *Biochemistry* **14,** 5046.
Aoki, T., and Johnson, P. A. (1972). *J. Natl. Cancer Inst.* **49,** 183.
Aoki, T., Geering, G., Berh, E., and Old, L. J. (1972). *In* "Recent Advances in Human Tumor Virology and Immunology" (W. Nakahara, K. Nichioka, T. Hirayama, and Y. Ito, eds.), p. 425. Univ. of Tokyo Press, Tokyo.
Aronow, L., (1978). *Fed. Proc.* **37,** 162.
Bourguignon, L. Y. W., and Singer, S. J. (1977). *Proc. Natl. Acad. Sci. U.S.A.* **74,** 5031.
Boyse, E. A., Old, L. J., and Luell, S. (1963). *J. Natl. Cancer Inst.* **31,** 987.
Boyse, E. A., Stockert, E., and Old, L. J. (1967). *Proc. Natl. Acad. Sci. U.S.A.* **58,** 954.
Calafat, J., Hilgers, J., Blitterswijk, W. J., Verbeet, M., and Hageman, P. C. (1976). *J. Natl. Cancer Inst.* **56,** 1019.
Cohen, E. P., and Liang, W. (1976). *J. Supramol. Struct. (Suppl.)* **1,** 89.
Cohen, E. P., and Liang, W. (1978). *In* "Protides of the Biological Fluids" (Proceedings, XXVth Colloquium). Pergamon, New York.
Croce, C. M., Kieba, I., Koprowski, H., Malino, M., and Rothblat, G. H. (1974). *Proc. Natl. Acad. Sci. U.S.A.* **71,** 110.
Davidson, R. L. (1974). *Annu. Rev. Genet.* **8,** 195.
Davies, D. A. L., Alkins, B. J., Boyse, E. A., Old, L. J., and Stockert, E. (1969). *Immunology* **16,** 669.
Fellows, M., Kamoun, M., Wiels, J., Dausset, J., Clements, G., Zeuthen, J., and Klein, G. (1977). *Immunogenetics* **5,** 423.
Flanagan, J., and Koch, G. L. E. (1978). *Nature (London)* **273,** 278.
Gavin, J. R., Roth, J., Neville, D. M., DeMeyts, P., and Buell, D. N. (1974). *Proc. Natl. Acad. Sci. U.S.A.* **71,** 84.
Genovesi, E. V., Mark, P. A., and Wheelock, E. F. (1977). *J. Exp. Med.* **146,** 520.
Goldfine, I. D., Kahn, C. R., Neville, D. M., Jr., Roth, J., Garrison, M. M., and Bates, R. W. (1973). *Biochem. Biophys. Res. Commun.* **53,** 852.
Gougne, C., Ruiz, F., and Ephrussi, B. (1972). *Proc. Natl. Acad. Sci. U.S.A.* **69,** 330.
Harris, H. (1970). "Cell Fusion." Harvard Univ. Press, Cambridge, Massachusetts.
Heineman, S., Bevan, S., Kullberg, R., Lindstrom, J., and Rice, J. (1977). *Proc. Natl. Acad. Sci. U.S.A.* **74,** 3090.
Hinkle, P. M., and Tashyian, A. H. (1975). *Biochemistry* **14,** 3845.
Huber, B., Gershon, R. R., and Cantor, H. (1977). *J. Exp. Med.* **145,** 10.
Hyman, R., and Kelleher, R. J. (1975). *Somatic Cell Genet.* **1,** 335.
Kahn, C. R., Neville, D. M., Jr., Garden, P., Freychet, P., and Roth, J. (1972). *Biochem. Biophys. Res. Commun.* **48,** 135.
Kahn, C. R., Neville, D. M., Jr., and Roth, J. (1973). *J. Biol. Chem.* **248,** 244.
Kearney, J. F., Cooper, M. D., and Lawton, A. R. (1976). *J. Immunol.* **116,** 1664.
Kebebian, J. W., Zatz, M., Romero, J. A., and Axelrad, J. (1975). *Proc. Natl. Acad. Sci. U.S.A.* **72,** 3735.
Klaus, G. G. B., Abbas, A. K., and McElroy, P. J. (1977). *Eur. J. Immunol.* **7,** 387.
Klebe, R. J., Chen, T., and Ruddle, F. H. (1970). *Proc. Natl. Acad. Sci. U.S.A.* **66,** 1220.

Knopf, P. M., Gestree, A., and Hyman, R. (1973). *Eur. J. Immunol.* **3**, 251.
Koch, G. L. F., and Smith, M. J. (1978). *Nature (London)* **273**, 274.
Köhler, H. (1975). *Transplant. Rev.* **27**, 24.
Köhler, G., Lefkovitz, I., Elliott, B., and Coutinho, A. (1977). *Eur. J. Immunol.* **7**, 758.
Köhler, H., Richardson, B. C., and Smyk, S. (1978). *J. Immunol.* **120**, 233.
Lamm, M. G., Boyse, E. A., Old, L. J., Lisowska-Bernstein, G., and Stockert, E. (1968). *J. Immunol.* **101**, 99.
Lesniak, M. A., Roth, J., Gordon, P., and Gavin, J. R. (1973). *Nature (London), New Biol.* **241**, 20.
Liang, W., and Cohen, E. P. (1975). *Proc. Natl. Acad. Sci. U.S.A.* **72**, 1873.
Liang, W., and Cohen, E. P. (1976). *Somatic Cell Genet.* **2**, 291.
Liang, W., and Cohen, E. P. (1977a). *J. Natl. Cancer Inst.* **58**, 1079.
Liang, W., and Cohen, E. P. (1977b). *J. Natl. Cancer Inst.* **58**, 1601.
Littlefield, J. W. (1964). *Science* **145**, 709.
Miller, O. J., Miller, D., Dev, V. G., Rantravahi, R., and Croce, C. M. (1976). *Proc. Natl. Acad. Sci. U.S.A.* **73**, 4531.
Mukherjee, C., Caron, M. G., and Lefkowitz, R. I. (1975). *Proc. Natl. Acad. Sci. U.S.A.* **72**, 1945.
Old, L. J., and Stockert, E. (1977). *Annu. Rev. Genet.* **2**, 127.
Old, L. J., Stockert, E., Boyse, E. A., and Kim, J. H. (1968). *J. Exp. Med.* **127**, 523.
Raff, M. C., Owen, J. J. T., Copper, M. D., Lawton, A. R., III, Megson, M., and Gathings, W. E. (1975). *J. Exp. Med.* **142**, 1052.
Rustigian, R. (1966). *J. Bacteriol.* **92**, 1805.
Schwaber, J., and Cohen, E. P. (1973). *Nature (London)* **244**, 444.
Schwaber, J., and Cohen, E. P. (1974). *Proc. Natl. Acad. Sci. U.S.A.* **71**, 2003.
Soll, A. H., Goldfine, I. D., Roth, J., Kohn, C. R., and Neville, D. M., Jr. (1974). *J. Biol. Chem.* **245**, 41.
Stackpole, C. W., Jacobson, J. G., and Lardis, M. P. (1974). *J. Exp. Med.* **140**, 939.
Stackpole, C. W., Jacobson, J. B., and Galuska, S. (1978). *J. Immunol.* **120**, 188.
Stanley, E. F., and Drachman, D. B. (1978). *Science* **200**, 1285.
Strayer, D. S., Lee, W. M. F., Rowley, D. A., and Köhler, H. (1975). *J. Immunol.* **114**, 728.
Trowbridge, I. S., Hyman, R. I., and Mazauskas, C. (1978). *Cell* **14**, 21.
Van Boxel, J. A., Broder, S., and Waldman, T. A. (1976). *In* "Leucocyte Membrane Determinants Regulating Immune Reactivity" (V. P. Eijsvoogel, D. Roos, and W. P. Zeiklemaker, eds.), p. 297. Academic Press, New York.
Weiss, M. C., and Chaplain, M. (1971). *Proc. Natl. Acad. Sci. U.S.A.* **68**, 3026.
Yu, A. C., and Cohen, E. P. (1974a). *J. Immunol.* **112**, 1285.
Yu, A. C., and Cohen, E. P. (1974b). *J. Immunol.* **112**, 1296.
Yu, A., Liang, W., and Cohen, E. P. (1975). *J. Natl. Cancer Inst.* **55**, 229.

CHAPTER 14

TOPOGRAPHIC DISPLAY OF CELL SURFACE COMPONENTS AND THEIR ROLE IN TRANSMEMBRANE SIGNALING

Garth L. Nicolson

DEPARTMENTS OF DEVELOPMENTAL AND CELL BIOLOGY AND
PHYSIOLOGY, COLLEGE OF MEDICINE
UNIVERSITY OF CALIFORNIA
IRVINE, CALIFORNIA

I. Introduction	305
II. Plasma Membrane Organization	306
A. Membrane Proteins	306
B. Peripheral Membrane Proteins	307
C. Membrane Fluidity	308
III. Dynamics of Cell Membrane Components	310
A. Dynamics of Membrane Lipids	310
B. Cholesterol	310
C. Membrane Viscosity	311
D. Calcium Effects	311
E. Lateral Phase Separations	312
F. Lateral Mobility of Membrane Proteins and Glycoproteins	313
G. Ligand-Induced Redistribution	314
IV. Topographic Control of Cell Surface Receptors	318
A. Transmembrane Control over Membrane Organization	321
B. Cytoskeletal Transmembrane Control	321
C. Membrane-Associated Cytoskeletal Models	324
V. Transmembrane-Mediated Communication	326
A. First Messenger Communication	327
B. Ionic Communication	328
C. Second Messenger or Enzymatic Communication	329
D. Cytoskeletal-Mediated Communication	331
VI. Concluding Remarks	333
References	333

I. Introduction

Complex patterns of embryonic development in higher organisms demanded the evolution of special mechanisms to regulate cell–cell interactions, cell communication, cell positioning, cell movement, and cell proliferation. The signals which bring about and control these phenomena to an overwhelming degree operate through specific receptors at the cell surface. Each cell possesses a myriad of multiple receptor types, and the identity of each cell can be determined by its fingerprint of cell surface receptors and the dynamic display and positioning of these receptors at its surface.

The receipt of information at the cell surface occurs via interaction of soluble ligands or macromolecular structures borne by adjacent cells with specific receptors leading to direct or indirect transmembrane-mediated communication. Direct transmembrane-mediated communication is envisioned as entry of ligands such as ions and small molecules into cells. Indirect transmembrane-mediated communication is envisioned as a one- or multiple-hit process in which each receptor is triggered to transmit or transduce its information across the membrane into the cell via other molecules. Mediation of transmembrane information by an indirect route can also be envisaged as a multiple-hit receptor process that is dependent upon the topographic display or pattern and/or mobility of receptors at the cell surface. Once a specific pattern is disturbed or formed by receptor–ligand interaction, information can be mediated across the membrane.

II. Plasma Membrane Organization

General consensus has been reached over the last few years that the structures of widely divergent cellular membranes conform to a number of basic principles (for reviews, see E. D. Korn, 1969; Singer, 1971, 1974; Singer and Nicolson, 1972; Wallach, 1972; Bretscher, 1973; Nicolson, 1976; Nicolson et al., 1977a). (1) The major membrane lipids, such as the phospholipids, are arranged in a planar bilayer configuration which is predominantly in a "fluid" state under physiological conditions. (2) The lipid bilayer is not a continuous structure and is interrupted by numerous proteins which are inserted or intercalated to various degrees into the bilayer. (3) In most membranes the arrangement of lipids, glycolipids, proteins, and glycoproteins is asymmetric with respect to the distribution of specific molecules in the inner and outer halves of the bilayer. (4) The proteins and glycoproteins of cell membranes are quite heterogeneous and can be operationally divided into two main classes: integral (or intrinsic) membrane proteins and peripheral (or extrinsic) membrane proteins (Capaldi and Green, 1972; Singer and Nicolson, 1972).

A. Membrane Proteins

A great deal of information now exists which indicates that integral or intrinsic membrane proteins and glycoproteins are globular, bimodal components which interact with both the hydrophobic (hydrocarbon) inner zone of the membrane as well as the hydrophilic outer zones of the membrane. The integral class of membrane proteins is characterized by its strong interactions with membrane lipids. Integral protein–lipid interactions are driven by the favorable entropy gained through sequestering the hydrophobic portions of the integral membrane proteins away from the hydrophilic phase of the membrane as well as the surrounding aqueous environment. These interactions lead to the intercalation of the integral membrane proteins into the hydrophobic zone of the lipid bilayer which is formed by the hydrocarbon tail groups of the membrane phospholipid fatty acids. Thus, while the hydrophilic portion of the integral membrane proteins protrude from the bilayer into the external and/or internal aqueous environment, the hydrophobic portions of their structures are intercalated into the bilayer to varying degrees dependent upon their three-dimensional folding and primary sequence. There are now several examples where cell membrane integral proteins span the entire membrane and have portions of their structures protruding at both the inner and out membrane surfaces (Bretscher, 1971; Segrest et al., 1973; Morrison et al., 1974; Hunt and Brown, 1975; Gahmberg, 1977).

Integral membrane proteins and glycoproteins need not exist as simple molecular entities in the membrane. There is evidence that certain integral membrane proteins exist as oligomeric complexes (for review, see Nicolson, 1976). These complexes can be subunit in nature or heteromolecular and contain units of differing molecular types. In addition, many integral membrane proteins interact with peripheral membrane proteins at the inner (Nicolson and Painter, 1973; Elgsaeter and Branton, 1974; Ji and Nicolson, 1974; Elgsaeter et al., 1976) or outer (Hynes et al., 1976) membrane surface. These interactions (to be discussed in later sections) are important in determining the topographic displays and mobilities of cell surface receptors.

B. Peripheral Membrane Proteins

Peripheral membrane proteins are those proteins and glycoproteins which are only weakly bound to the surfaces of biological membranes (Singer and Nicolson, 1972). Peripheral membrane proteins bind to membranes via noncovalent bonds, and binding does not appear to be highly dependent upon hydrophobic interactions for stability due to the fact that peripheral components can be removed by treating membranes with high salt concentrations or divalent cation chelating agents. Thus, peripheral membrane proteins probably associate with integral membrane proteins and glycoproteins and perhaps also lipids and glycolipids by ionic interactions or by weak nonionic interactions such as van der Waals and London dispersion forces.

C. Membrane Fluidity

In its simplest form the plasma membrane can be thought of as a two-dimensional solution of a mosaic of integral membrane proteins and glycoproteins embedded in a predominantly fluid lipid bilayer (Singer and Nicolson, 1972). To this basic structure peripheral membrane proteins are loosely attached to either membrane surface. In addition, nonmembrane cytoskeletal assemblies, such as microtubules and microfilaments, are dynamically associated with the membrane at the inner surface (Fig. 1) (Nicolson, 1976; Loor, 1977; Nicolson et al., 1977a). The function of these membrane-associated cytoskeletal ele-

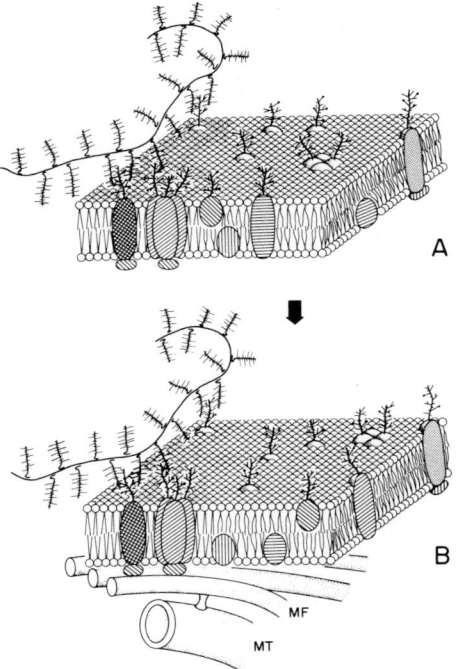

Fig. 1. Modified version of the fluid mosaic model of membrane structure showing transmembrane control over the distribution and mobility of cell surface receptors by peripheral and membrane-associated cytoskeletal components. In this hypothetical model the mobility of integral glycoprotein complexes is controlled by outer surface peripheral glycosaminoglycans and also by membrane-associated cytoskeletal elements at the inner surface. (A) Some glycoprotein complexes are shown to be sequestered into a specific lipid domain indicated by the shaded area, while others exist in a free or unaggregated state and are capable of free lateral motion. (B) Aggregation of certain glycoproteins has stimulated attachment of membrane-associated cytoskeletal elements. MF, microfilaments; MT, microtubules.

ments in controlling the topographic display and mobility of cell surface receptors will be discussed in Section IV,B.

The concept of a fluid mosaic membrane architecture has two important implications for membrane function. (1) This arrangement permits membrane components to be organized asymmetrically, thus allowing specific membrane components to be localized predominantly or exclusively in the inner or outer half of the bilayer or at the inner or outer membrane surface (Bretscher, 1973). Membrane glycoproteins and glycolipids exhibit striking asymmetry in the display of their carbohydrate sequences, and in every plasma membrane that has been isolated and investigated thus far the oligosaccharide portions of these molecules reside exclusively on the outer surface of the membrane functioning as specific cell surface receptors (Nicolson and Singer, 1974). Alternatively, certain peripheral membrane proteins, for example, the fibrous protein spectrin of the human red blood cell membrane (Marchesi *et al.*, 1969), reside exclusively at the inner surface of the membrane (Nicolson *et al.*, 1971) and are important in membrane stabilization. The striking asymmetry of the plasma membrane dictates that the components oriented to either the inner or outer surface of the membrane should not be able to rotate at appreciable rates from one side of the membrane to the other (Singer and Nicolson, 1972). The lack of appreciable transmembrane rotation in cell membranes (Bretscher, 1973; Nicolson and Singer, 1974) may be due to the tremendous thermodynamic barriers which tend to prevent hydrophilic molecules from passing through the hydrophobic matrix of the membrane (Singer and Nicolson, 1972; Singer, 1974). (2) Another important feature of the fluid mosaic model of membrane structure is that this type of structural arrangement allows rapid lateral diffusion within the plane of the membrane (for reviews, see Edidin, 1974; Singer, 1974; Cherry, 1976; McConnell, 1975b; Nicolson, 1976; Nicolson *et al.*, 1977a). This permits reversible and rapid changes in the topographic display of specific cell surface receptors. Although some cell surface components diffuse quite rapidly and apparently freely within the membrane, others move much more slowly, and their movement appears to be under restraint by a variety of mechanisms (see Section IV). The graded hierarchy of mobilities for different receptors on the surface membrane and even within specific regions on the cell surface suggests that cells contain specific controlling mechanisms that restrict the mobility of certain receptors in order to maintain specific topographic arrangements or patterns on the cell surface (Nicolson and Poste, 1976a).

Rapid and reversible changes in the topographic display of surface receptors can occur in response to stimuli from outside the cell and also from within the cell. These rapid and reversible changes in the pattern and display of receptors may serve an important function in transmembrane-mediated communication (see Section V).

III. Dynamics of Cell Membrane Components

The dynamics of membrane components are not limited to a consideration of the complex structural and topographic rearrangements exhibited by different membrane components. An equally important aspect is the synthesis, assembly, and turnover of different membrane constituents. These latter phenomena will not be considered here, and the reader is directed to a number of other reviews which deal with this subject (Poste and Nicolson, 1977).

A. Dynamics of Membrane Lipids

The forces which stabilize most artificial and biological lipid bilayer membranes are due mostly to hydrophobic effects. Biological membranes are formed predominantly from a matrix of fluid state lipids which have a viscosity similar to light machine oil (for reviews, see Edidin, 1974; Nicolson, 1976). Using nuclear magnetic resonance and electron spin resonance, the bending and flexing of fatty acid hydrocarbon chains in the lipid matrix have been estimated. There is a considerable degree of lipid acyl chain anisotropic motion in biological membranes, and this determines their bulk fluidity. Paramagnetic resonance spectral parameters indicate that fatty acid acyl chain bending increases down the methylene chains from the polar head groups (Horwitz et al., 1972; Levine et al., 1972a,b; McFarland, 1972). Interestingly, chain bending also increases from the glycerol carbons toward the polar head groups of phosphatidylcholine (Levine et al., 1972a). More recently, nuclear magnetic resonance has been used to measure the anisotropic motion of undisturbed lipid bilayers instead of bilayers doped with nitroxide spin labels. The nuclear magnetic resonance data indicate that chain flexing and bending are relatively constant down phospholipid acyl chains and that a rapid change in flexing and bending occurs only near the methylene ends of the chains (Horwitz et al., 1972; Seelig and Niederburger, 1974). At 37°C the phospholipid acyl chains undergo rapid rotation and kinking, and the kinks that form appear to fluctuate rapidly up and down the hydrocarbon chains (Träuble and Eibl, 1975).

B. Cholesterol

In most biological membranes cholesterol plays a vital role in modulating the bulk membrane fluidity. Cholesterol can interact with certain fatty acids and sphingolipids (Brockerhoff, 1974), and when this interaction occurs it serves to decrease the lateral molecular spacing and reduce the flexibility of the phospholipid acyl chains (Hubbell and McConnell, 1971; Oldfield and Chapman, 1972).

Cholesterol–phospholipid interactions appear to rigidify biological membranes at physiologic temperatures, but they fail to convert membranes to solid phase states (Oldfield and Chapman, 1971; Rottem et al., 1973; Vanderkooi et al., 1974). Cholesterol serves to create an intermediate fluid condition in either fluid or solid phases (Oldfield and Chapman, 1972); thus, most biological membranes at physiological temperatures usually show decreased phospholipid acyl chain mobility and flexing.

C. Membrane Viscosity

Are the viscosities of biological membranes determined mainly by their lipids or by their proteins or protein–lipid complexes? In the membranes that have been studied the bulk fluidity characteristics are certainly determined by lipids, because artificial lipid bilayers made from extracted membrane lipids are similar in viscosity to the membranes from which they were derived (reviews: Edidin, 1974; Nicolson et al., 1977a). Studies using electron spin resonance and fluorescent polarization have estimated biological membrane viscosity to be a few poise (Edidin, 1974; Nicolson et al., 1977a). Using techniques such as magnetic resonance spin exchange, diffusion constants varying from $D = 1 \times 10^{-8}$ to $12 \times 10^{-8} \, \text{cm}^2 \, \text{sec}^{-1}$ have been calculated depending on the type of nitroxide probe, lipid composition, divalent cation concentration, and the methods of measurement and calculation (Edidin, 1974; Nicolson et al., 1977a). These measurements indicate that the lateral diffusion rates of phospholipids are quite high; thus, a single phospholipid molecule can move a distance of a few micrometers per second when the membrane is in a fluid state (McConnell, 1975a,b).

D. Calcium Effects

Calcium ions can cause condensation of phospholipid bilayers and restrict the mobility of certain phospholipids (Schnepel et al., 1974; Butler et al., 1970; Ehrstrom et al., 1973; Sauerheber and Gordon, 1975). However, in some cases the presence of Ca^{2+} enhances the lateral mobility of membrane lipids. For example, Adams et al. (1976) introduced Ca^{2+} into red blood cell membranes during hemolysis and found that the calcium results in increased lipid mobility. In this case calcium may have induced membrane protein aggregation decreasing protein–lipid interactions (Carraway et al., 1975; Elgsaeter et al., 1976). Calcium is an important structural regulator in cell membranes and serves important functions in secretion, excitation, contraction, and membrane fusion (for reviews, see Rasmussen, 1970; Carafoli *et al.*, 1975). Calcium influx is an important transmembrane signal during activation of cells by mitogens, transmitters, and, in certain cases, hormones (Rasmussen, 1970; Carafoli *et al.*, 1975; McManus *et al.*, 1975; Berridge, 1975; Raff *et al.*, 1976). Changes in membrane Ca^{2+} binding can be modulated by a variety of extra- and intracellular molecules such as cyclic nucleotides, ATP, neurotransmitters, and certain hormones (Carafoli *et al.*, 1975; Berridge, 1975; Triggle, 1972). Since alterations in Ca^{2+} mobilization and membrane interactions occur rapidly and reversibly, this molecule is well suited to play a regulatory role in modulating membrane properties. Calcium-ion binding to membranes can modify charge effects, induce structural reorganizations, as well as cause stereospecific activation of individual membrane components.

E. Lateral Phase Separations

Phospholipids undergo endothermic phase transitions at characteristic temperatures for (reviews, see Träuble, 1972; Chapman, 1973; McConnell, 1975b; Kimelberg, 1977). The acyl chains of membrane phospholipids are in a relatively rigid, extended, and parallel packed state below their phase transition temperature; but the acyl chains become more fluid or liquid-crystalline in nature and undergo cooperative chain motion such as flexing, bending, and kinking when the temperature is at or above the phase transition temperature. Phase changes from "solid" to "fluid" state result in a reduction of bilayer thickness and an increase in the area occupied per lipid molecule (Phillips and Chapman, 1968; White, 1970), and importantly lipid molecules are capable of lateral motion in the plane of the membrane (Träuble and Eibl, 1974; McConnell, 1975b; Kimelberg, 1977).

In synthetic membranes made up of two or more types of phospholipids, lateral phase separations can occur when the phospholipids have different phase transition temperatures. Lateral phase separations require long-range lipid molecular motion in the bilayer plane, and these motions result in lipid compositional segregation. Lipid phase separations are generally not characterized by sharp temperature transitions in mixed lipid membranes. They occur over a fairly broad temperature range compared with pure lipids, and in these temperature ranges solid and fluid lipid phases coexist.

Membranes obtained from mutant lipid-requiring cells or bacteria grown on different fatty acid supplements undergo lateral phase separation similar to synthetic binary lipid mixtures (Verkleij *et al.*, 1972; Speth and Wunderlich, 1973; Kleemann and McConnell, 1974).

Freeze-fracture electron microscopy has been used to identify fluid and solid phase membrane regions in these biological systems, because the fluid membrane regions appear smooth after freeze fracture, while solid phase lipid fracture surfaces have ridges which make them appear banded. The amounts of smooth (fluid phase) and banded (solid phase) lipid regions are directly proportional to the amount of fluid versus solid phase lipid domains estimated by electron spin resonance (Grant et al., 1974). Biological membranes which have been freeze-fractured and observed by electron microscopy contain numerous intercalations or particles which are known to contain proteins and/or glycoproteins (Hong and Hubbell, 1972; Segrest et al., 1974; Vail et al., 1974). When lipid phase transitions occur in biological membranes, the intercalated particles are sequestered into the fluid phase lipids and are completely excluded from solid phase lipids (Grant and McConnell, 1974; Kleemann and McConnell, 1976).

Cholesterol can affect the formation of lipid domains. Concentrations of cholesterol up to 20 mol% do not prevent lateral lipid phase separations, but cholesterol above this level can dampen the extent of phase separation. Kleemann and McConnell (1976) have found different protein (particle) distributions in solid phase and fluid phase lipid regions in a reconstituted model membrane vesicle system containing the ATPase protein from rabbit sarcoplasmic reticulum. Below the phase transition temperature (15°C) for a 10 mol% cholesterol in dimyristroyl phosphatidylcholine mixture, two protein particle phases exist: a particle-rich phase and a particle-poor phase. When the cholesterol content was increased to 20 mol%, particles began to form stringlike linear arrays close to, but below the phase transition temperature (23°C), and these particles dispersed above the phase transition temperature. Lipid phase transitions can occur in lipid mixtures possessing up to 33 mol% cholesterol content, although these are generally quite broad (Darke et al., 1972; de Kruyff et al., 1974). In many biological membranes with broad lipid phase transition changes, local lipid domain formation leading to protein clustering, segregation, or sequestration may be important in modifying the topographic display of receptors. Also, many membrane-bound enzymes are extremely sensitive to the fluid state of the lipid immediately surrounding them (for review, see Kimelberg, 1977), and thus their activities can be controlled by changes in membrane physical states.

F. LATERAL MOBILITY OF MEMBRANE PROTEINS AND GLYCOPROTEINS

As discussed earlier, integral membrane proteins and glycoproteins are not static components of biological membranes; they undergo dynamic topographic changes with respect to other membrane components. Although the lateral mobilities of membrane proteins and glycoproteins appear to be much less than the membrane phospholipids, they are nonetheless significant. Diffusion constants for membrane proteins such as rhodopsin have been calculated to be approximately $4-5 \times 10^{-9}$ cm^2 sec^{-1}, a value at least an order of magnitude less than that found for phospholipid diffusion (Poo and Cone, 1974). Schlessinger et al. (1976) also found large differences between lipid-in-lipid and protein-in-lipid lateral diffusion, and their experiments revealed heterogeneity in the mobility of cell surface proteins and glycoproteins. At least one class of cell surface glycoproteins appeared to have its mobility regulated by the cell cytoskeletal system (see Section III,G).

Relatively low rates of lateral mobility for many cell surface glycoproteins can be attributed to the presence of peripheral and transmembrane controlling mechanisms which appear to impede movement. For example, Peters et al. (1974) found that the membrane glycoproteins of human erythrocytes have very low rates of lateral diffusion ($D = <3 \times 10^{-12}$ cm^2 sec^{-1}). Similarly, Frye and Edidin (1970) used cell surface antigen intermixing on mouse–human heterokaryons after Sendai virus-induced fusion to measure lateral mobility, and from their experiments a diffusion constant of approximately $D = 2 \times 10^{-10}$ cm^2 sec^{-1} can be calculated for the lateral diffusion of antibody–antigen complexes (Edidin and Wei, 1977). More recently, Edidin and Fambrough (1973) estimated the average rate of lateral diffusion of fluorescent Fab' monovalent antibodies bound to antigens after spot application to muscle fibers. By measuring the extent of patch spreading with time, Edidin and Fambrough were able to calculate a diffusion constant of $D = 1-2 \times 10^{-9}$ cm^2 sec^{-1} for the Fab-receptor complexes.

G. LIGAND-INDUCED REDISTRIBUTION

Multivalent ligands that bind to cell surface receptors are able to induce lateral movement and topographic rearrangements of the ligand–receptor complexes (for reviews, see Unanue and Karnovsky, 1973; Edidin, 1974; Loor, 1977; Nicolson, 1976; Nicolson et al., 1977a). Several studies have shown that the inherent topography of most, but not all, cell surface receptors is essentially random over the entire cell surface. However, there are examples of receptors which show apparent nonrandom topographic distributions. Wartiovaara et al. (1974) found that the surface peripheral glycoprotein fibronectin (LETS, CSA, SF) was distributed nonrandomly on mouse fibroblasts, and its distribution coincided closely with cellular fibrillar structures such as surface ridges and microvilli which contain submembranous microfilament assemblages. Surface immunoglobulin molecules on mouse lymphocytes also appear to exist inherently as inner connected networks rather than as random entities (Abbas et al., 1975), and the distributions of several neuronal surface antigens have been shown to be nonrandom (Rostas and Jeffrey, 1975).

Molecular segregation and regional specialization or membrane domain formation have been seen most dramatically on mammalian spermatozoa surfaces. Different species of spermatozoa have distinct discontinuous distributions of cell surface anionic sites (Bedford et al., 1972; Yanagimachi et al., 1972), which indicates that these components are not capable of free lateral mobility around the entire sperm surface. Surface lectin receptors also differ significantly in their lateral mobility at different regions of the sperm plasma membrane (Nicolson and Yanagimachi, 1974). For example, the distributions of lectin receptors on rabbit spermatozoa were found to be distinctly different between acrosomal and postacrosomal head regions (Nicolson et al., 1977b) (Fig. 2). Certain antigens on the spermatozoa are present in only specific regions of the sperm head, indicating their inability to diffuse to tail regions (Koo et al., 1973; Koehler and Perkins, 1974; Koehler, 1975a,b).

In general, the binding of multivalent ligands to cell surfaces resulting in cross-linking of adjacent ligand–receptor complexes leads to redistribution and the formation of small clusters, larger patches, and in certain situations polar redistributions to form caps. After the redistribution occurs, the patched or capped ligand–receptor complexes are often internalized by endocytosis or in some cases shed from the outer cell surface (for reviews, see Nicolson et al., 1977a; Sundqvist, 1977; de Petris, 1977). The type of ligand-induced redistribution of surface receptors, its extent, and subsequent fate are determined by the nature of the ligand and the receptor and cell type involved (Loor, 1977; Nicolson, 1976). For example, use of the same ligand, the lectin concanavalin A (Con A), to cause redistribution of Con A receptors on normal lymphocytes and untransformed fibroblasts results in cap formation on the former and patch formation on the latter. Clusters and patches, when they form on cells after ligand-induced redistribution, do not necessarily lead to caps. Phillips and Perdue (1976a,b) have found that antigens on virus-transformed chick fibroblasts can undergo ligand-induced "marginal" redistribution where the receptor–ligand complexes are swept quickly away from the cell periphery. They also found that, on certain cells, clusters which had formed coalesced into regions where endocytosis eventually occurred without cap formation (Fig. 3).

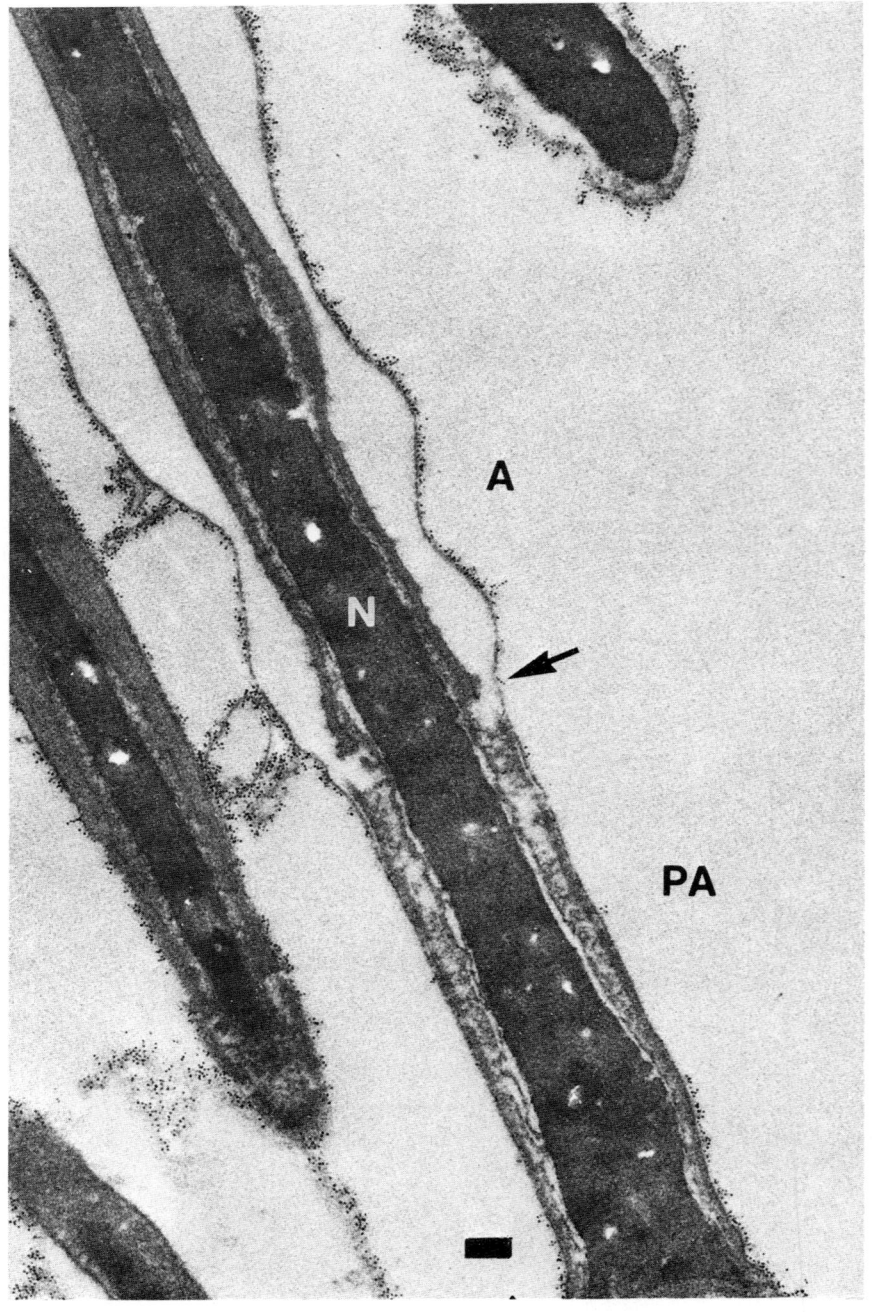

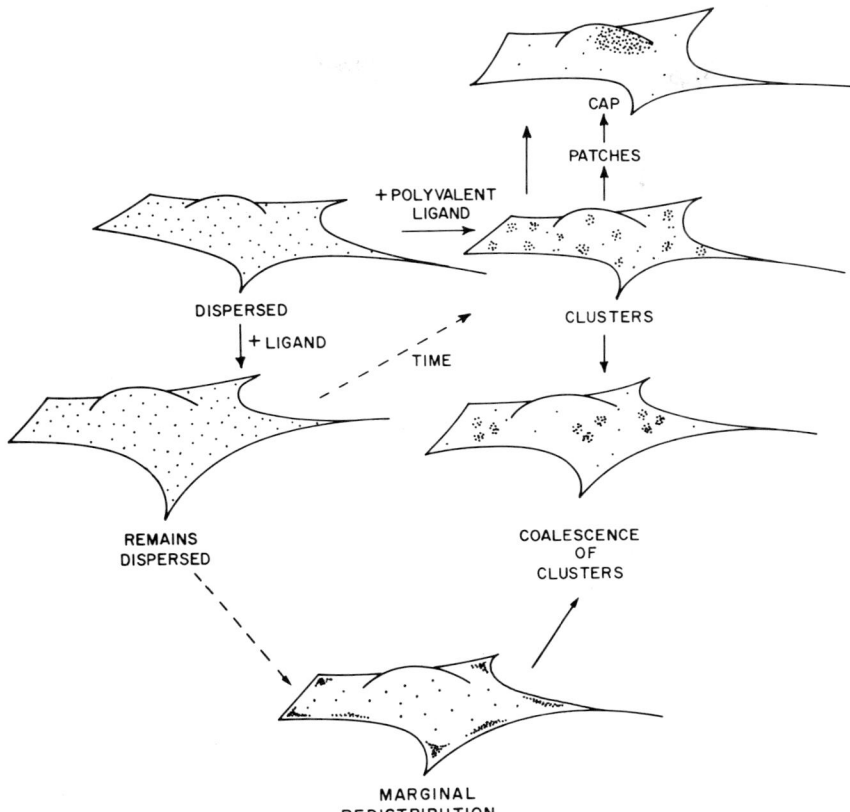

Fig. 3. Alternate pathways of ligand-induced receptor redistribution on cell surfaces. After ligand binding, initially dispersed receptors can remain dispersed or undergo ligand-induced clustering. Adapted from Nicolson and Poste (1976a).

Another important property of cell surface receptors is that most of these molecules appear to be capable of independent redistribution with the appropriate multivalent ligand. For example, Neauport-Sautes et al. (1973a,b) showed that ligand-induced redistribution of H2-D alloantigens on mouse lymphocytes was not accompanied by redistribution of H2-K alloantigens. On the other hand, Poulik et al.

Fig. 2. The distribution of lectin receptors on rabbit spermatozoa head acrosomal (A) and postacrosomal (PA) regions. Rabbit spermatozoa were labeled with ferritin-conjugated *Ricinus communis* agglutinin at 22°C. Note the absence of lateral diffusion of lectin–receptor complexes from acrosome to postacrosome area (border indicated by arrow) in sperm head. N, nucleus. Bar equals 0.2 μm.

(1973) found that HL-A histocompatibility antigens on human lymphocytes comigrated with β_2-microglobulin, indicating a possible relationship between these two components. Comigration is dependent on several criteria, the most important of which appears to be the concentration of the multivalent ligand. De Petris (1975) has shown with mouse lymphocytes that low concentrations of Con A result in redistribution of Con A receptors independent from surface immunoglobulins, but at high concentrations (>25 µg/ml Con A) comigration occurs.

IV. Topographic Control of Cell Surface Receptors

Since there is a wide body of evidence that indicates that surface receptors can have unique displays, distributions, or mobilities on the cell surface, controlling mechanisms must exist which govern cell surface dynamics and display of surface receptors. Cell surface receptors are mobile in the membrane; however, their movement can be restricted by several mechanisms (Fig. 4). Planar associations or aggregations with other molecules can serve to restrict the lateral mobilities of receptors. Sequestration of receptors into specific membrane domains or exclusion from these domains can also lead to a nonuniform display of surface receptors. Noncovalent lateral association of surface receptors into supramolecular aggregates or even paracrystalline arrays may be more common than first thought (Rash et al., 1974). For example, Friend and Fawcett (1974) have found striking regional differences in the topography of freeze-fracture intramembranous particles in spermatozoa acrosomal, postacrosomal, and tail regions. In the acrosome these components (particles) are present in a paracrystalline array suggestive of a membrane with very low capability for lateral motion. Labeling of surface receptors in this region (Nicolson and Yanagimachi, 1974) by ferritin–lectin ligands also suggests a low rate of mobility of lectin receptors in the acrosomal region.

An example of lateral association leading to aggregated membrane components and paracrystalline membrane structures is the formation of intercellular junctions between adjacent cells (for reviews, see McNutt and Weinstein, 1973; Staehelin, 1974; McNutt, 1977). The morphology and ultrastructure of various junction types will not be reviewed here; it should simply be noted that these structures share in common the property of developing from subunit components which form supramolecular aggregates in the membrane plane and also across to adjacent plasma membranes. Lateral associations between junctional subunits within the same cell membrane may be stabilized by disulfide bonds and possibly other noncovalent forces. Cell junc-

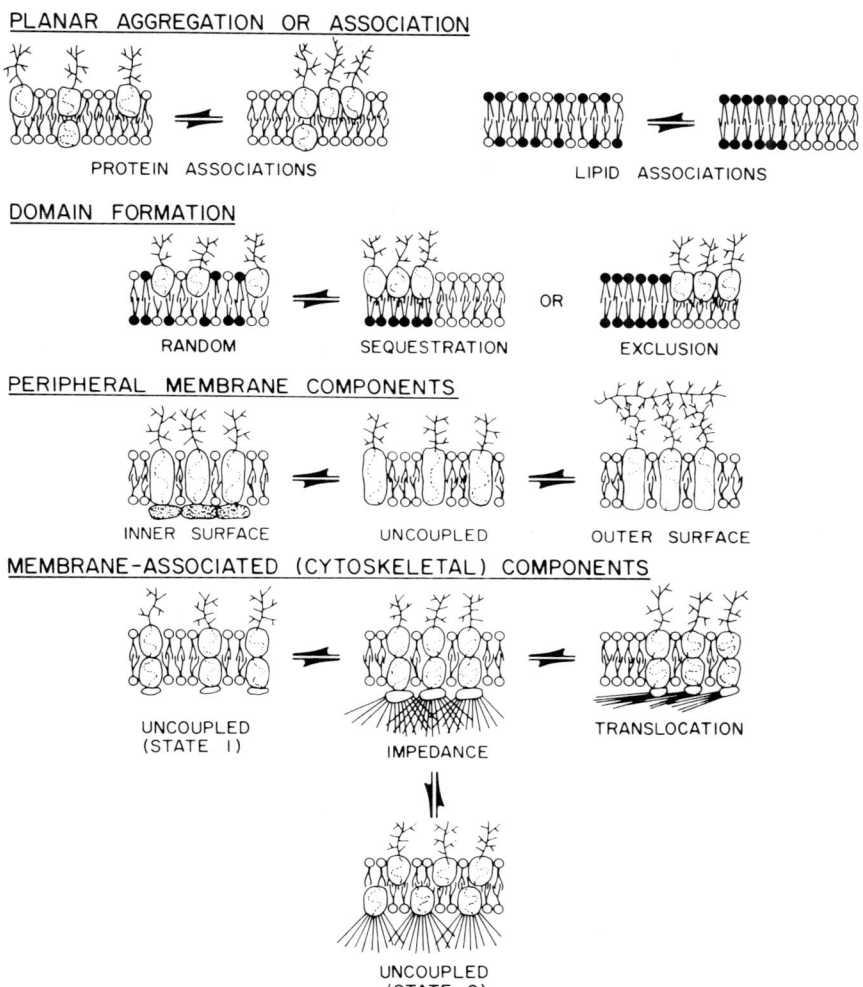

Fig. 4. Some possible restraining mechanisms on the lateral mobility of cell surface receptors. From Nicolson (1976).

tional complexes are probably capable of some restricted lateral mobility, but their size and attachments, both within the cell and between adjacent cells, restrict mobility even though they exist in a fluid lipid matrix.

The assembly of components into oligomeric complexes in nerve postsynaptic membrane endings is probably another example of lateral associations which lead to somewhat ordered arrays of surface receptors (Sandri et al., 1972). Postsynaptic components remain associated even after the removal of efferent presynaptic nerve membrane endings (Mathews et al., 1976; Kelly et al., 1976). Another class of membrane control (control by peripheral membrane components) may be involved in the stabilization of the postsynaptic complex. Indeed, the existence of a dense zone on the inner membrane surface called the postsynaptic density (observable by electron microscopy due to its heavy-metal staining characteristics; Cotman et al., 1974) may be important in stabilizing lateral associations.

There are other examples of plasma membrane lateral associations leading to the formation of tightly ordered structures, and one of the best documented is the sequestration of viral membrane components during the assembly and release of certain animal viruses. Lipid-containing animal viruses undergo their final maturation step at the cell surface where they are assembled utilizing host plasma membrane lipids by a process called "budding." An important step in this process is the clustering of virus-specified envelope proteins and glycoproteins within the plasma membrane and the exclusion of host plasma membrane components (Rott and Klenk, 1977). Complexing of virus envelope glycoproteins and proteins within the membrane involves either random insertion of virus-specified proteins into the host cell plasma membrane followed by lateral aggregation into clusters at the budding site or transport and subsequent fusion of intracellular plasma membrane precursor vesicles containing virus envelope proteins with the plasma membrane. Once the viral envelope components reach the cell surfaces, stability of their self-association is necessary to maintain the viral budding site within a fluid matrix. Of course, an additional factor may involve the association of peripheral membrane components or other membrane-associated structures at the inner surface of the prospective viral bud.

Preferential association of many plasma membrane proteins and glycoproteins may be due in part to lipid domain formation. If proteins are sequestered into fluid lipid domains, their dynamic display would be limited to these regions, and the converse would be true if they preferentially associated with solid phase lipids. The presence of different lipid phase states may be important in determining lateral associations of membrane components if, for example, membrane proteins self-associate in solid lipid phases due to preferential protein–protein interactions over protein–lipid interactions. In addition, different phase states from one lipid leaflet to the other in the lipid bilayer could modify partitioning across the bilayer and modify transmembrane interactions.

A. Transmembrane Control over Membrane Organization

Transmembrane controlling mechanisms are those in which the display, topography, and dynamics of proteins and glycoproteins situated on one surface of the cell membrane are controlled through transmembrane linkages to components on the opposite side of the membrane. In the case of cell surface receptors, transmembrane linkages to inner surface peripheral components or membrane-associated cytoskeletal elements are particularly important (Nicolson, 1976). The distinction between peripheral membrane components and membrane-associated components has been defined as the following. Peripheral membrane components are proteins, glycoproteins, glycosaminoglycans, or other components that are held to the membrane by noncovalent forces other than hydrophobic interactions within the lipid matrix, while membrane-associated components are not true membrane components in that they occur widely throughout the cell but are also attached to the inner membrane surface. Membrane-associated components are transient in nature, and their structure and organization are dependent upon cell energy systems, ion concentrations, and a variety of controlling molecules such as cyclic nucleotides. As a group of membrane-associated components, cytoskeletal elements are synonymous with microtubular, microfilament, and intermediate filament assemblages.

Transmembrane restraints are involved in controlling the display and mobility of a number of cell surface receptors. One of the first documented was the human erythrocyte membrane where the lateral mobilities of the major glycoproteins are under restraint and are unable to undergo free lateral diffusion (Loor *et al.*, 1972; Peters *et al.*, 1974). In this case an inner surface peripheral protein network involving the fibrous protein, spectrin, and at least one other component—erythrocyte membrane actin—serve to restrict the mobility of the membrane glycoproteins which are known to be transmembrane (Bretscher, 1971; Morrison *et al.*, 1974). These two peripheral components are capable of forming associations in solution (Tilney and Detmers, 1975), and they probably form a dense network on the inner erythrocyte membrane surface where they restrict the lateral mobility of the transmembrane glycoproteins (Nicolson and Painter, 1973; Ji and Nicolson, 1974; Elgsaeter and Branton, 1974; Elgsaeter *et al.*, 1976).

B. Cytoskeletal Transmembrane Control

Transmembrane control over the display, topography, and dynamics of cell surface receptors by membrane-associated cytoskeletal assemblages such as microtubules and microfilaments may be one of the most common mechanisms of cell surface receptor control. Microtubules have been found often in close association with the plasma membrane (Weber et al., 1974; Brinkley et al., 1975), and they are thought to perform "anchoring" or "restricting" roles in maintaining cell surface receptor topography (Yahara and Edelman, 1972, 1973; Edelman, 1974, 1976; Edelman et al., 1973; Berlin et al., 1974). Cytoplasmic microtubules are capable of rapid, reversible assembly and disassembly, and their state of polymerization is controlled by a number of factors such as temperature, Ca^{2+} concentration, pH, and cyclic nucleotides (Kirschner and Williams, 1974).

Another important class of membrane-associated cytoskeletal elements, the actin-containing microfilaments, are also often associated with plasma membrane inner surfaces. In addition, microfilaments occur in a variety of organizational states such as "bundles" that penetrate deep into the cell from the plasma membrane inner surface or "lattice" or "network" filament structures which are present in the cortical regions of the cell extending up to the plasma membrane (Goldman, 1972, 1975; Weber et al., 1974; Lazarides, 1975). Microfilament assemblages may be joined to the plasma membrane in regions containing α-actinin (Lazarides and Burridge, 1975; Tilney and Detmers, 1975). Microfilaments are thought to perform important muscle-like contractile functions in cells such as cytoplasmic streaming, cell shape maintenance, and locomotion, as well as movements or impedence of movements of plasma membrane components (Poste et al., 1975b; Goldman et al., 1976; Nicolson, 1976; Nicolson et al., 1977a).

There are several examples where cell surface receptor topography, display, and dynamics appear to be under regulatory control by the membrane-associated cytoskeletal components (for reviews, see Unanue and Karnovsky, 1973; Edelman, 1976; Loor, 1977; Nicolson, 1976; Nicolson et al., 1977a). The most documented example of cytoskeletal transmembrane control is in the immunoglobulin (Ig)-bearing B lymphocyte. The attachment of the multivalent ligand, anti-Ig, to B lymphocytes causes Ig surface receptors to undergo ligand-induced redistribution to form clusters, patches, and eventually caps (Taylor et al., 1971; Yahara and Edelman, 1972; Loor et al., 1972). This process will not be discussed in detail here; instead, the reader is referred to recent reviews by Unanue and Schreiner (1977) and de Petris (1977). The role of the microfilament system in cap formation is thought to be active (contractile), and these elements may provide the stress necessary to translocate ligand–receptor clusters into patches and caps. On the other hand, Edelman (1974, 1976) has proposed that microtubules play an important role in anchoring or immobilizing cell surface receptors, but during the capping process microtubules may function as skeletal components allowing directional translocation of the patches in a cap at one end of the cell (Nicolson and Poste, 1976a; de Petris, 1977).

There are a variety of drug and environmental conditions which prevent lymphocyte cap formation (Nicolson, 1976; de Petris, 1977). Cap formation requires cellular energy (Taylor *et al.*, 1971; Loor *et al.*, 1972; Yahara and Edelman, 1972; Unanue *et al.*, 1973). Drugs that disrupt microfilament organization, such as the cytochalasins, prevent cap formation to various degrees (Yahara and Edelman, 1972; Unanue *et al.*, 1973; de Petris, 1974, 1975; Poste *et al.*, 1975a,b), while drugs that cause microtubule depolymerization have little effect (Unanue *et al.*, 1973; Poste *et al.*, 1975a) or may actually enhance cap formation (Yahara and Edelman, 1972; de Petris, 1975; Poste *et al.*, 1975b). Combinations of these drugs, cytochalasin B plus colchicine or vinblastine, when administered to lymphocytes (de Petris, 1974, 1975; Poste *et al.*, 1975a,b) or polymorphonuclear leukocytes (Ryan *et al.*, 1974a), completely or almost completely block cap formation. These data suggest that both microfilaments and microtubules are involved in the capping process; the microfilament system is thought to provide active contractile movement necessary to translocate the ligand–receptor clusters and patches, while the microtubules probably provide the skeletal structure necessary for the contractile apparatus as well as proper orientation to ensure a coalescence of the receptor–ligand aggregates at one end of the cell.

Tertiary amine local anesthetics have been used to obtain additional insight into the role of membrane-associated cytoskeletal assemblages in regulating the dynamics of surface receptors. At low anesthetic concentrations where membrane fluidity is unaffected (Papahadjopoulos *et al.*, 1975), tertiary amine local anesthetics produce a variety of cellular effects such as inhibition of cell spreading (Rabinovitch and de Stefano, 1974), cell movement (Gail and Boone, 1972), adhesion (O'Brien, 1962; Rabinovitch and de Stefano, 1973), and fusion (Poste and Reeve, 1972). Local anesthetics also modify the dynamics of ligand-induced receptor redistribution (Poste *et al.*, 1975a,b; Nicolson and Poste, 1976b). Local anesthetics enhance lectin-induced agglutination of untransformed fibroblasts (Poste *et al.*, 1975a,b; Nicolson and Poste, 1976b) and reverse the inhibition of lectin-induced agglutination of transformed fibroblastic cells by colchicine and vinblastine (Poste *et al.*, 1975b). Lymphocyte capping is dramatically prevented by local anesthetics (Ryan *et al.*, 1974b; Poste *et al.*, 1975b), and local anesthetics induce the breakdown of preformed caps on lymphocytes (Poste et al., 1975b). The effects of local anesthetics can be duplicated by treatment of cells with colchicine and vinblastine together with cytochalasin B. The fact that colchicine or cytochalasin B alone will not suffice to duplicate the effects of local anesthetics suggests that microtubules and microfilaments are both involved in the maintenance of surface receptor distribution and dynamics (for reviews, see Nicolson, 1976; Edelman, 1976; de Petris, 1977; Nicolson et al., 1977a; Loor, 1977). In this scheme the topography of membrane receptors at any time reflects interplay between the microfilaments and microtubule system, and this may also be indirectly controlled by the intermediate filament system (Albertini and Anderson, 1977).

C. Membrane-Associated Cytoskeletal Models

In Fig. 5 the basic structure of the plasma membrane is envisioned as an elaboration of the fluid mosaic membrane model (Singer and Nicolson, 1972) with added peripheral and membrane-associated components. On the extracellular side of the membrane, glycosaminoglycans or mucopolysaccharides are shown to be associated with integral membrane glycoproteins, although certain parts of their structures may anchor hydrophobically in the membrane lipid matrix (Kraemer, 1975). Many of the integral membrane proteins or glycoproteins or their oligomeric complexes have been presented as asymmetric transmembrane oriented. Peripheral membrane components have been placed at the inner plasma membrane surface, and some of these have been shown to interact with membrane-associated cytoskeletal elements.

Parallel arrays of microfilaments are shown extending from the tip of a microvillus where they are thought to be attached by linkage molecules (Mooseker and Tilney, 1975) deep within the cell cytoplasm, and they are also found running parallel to the plasma membrane (Fig. 5). Adjacent microfilament arrays may be attached by myosin molecules similar to the attachment of microfilaments to the terminal web of intestinal endothelial cells by thick filaments (Mooseker and Tilney, 1975). In addition, a variety of bridgelike molecules have been illustrated as being present in the cytoskeletal system, evidence provided by electron microscopy (Mooseker and Tilney, 1975; Hepler and Palevitz, 1974).

In order to satisfy the proposal that the membrane-associated microfilaments and microtubules play coordinating role in the regulation and the movement and distribution of cell surface receptors (Poste et

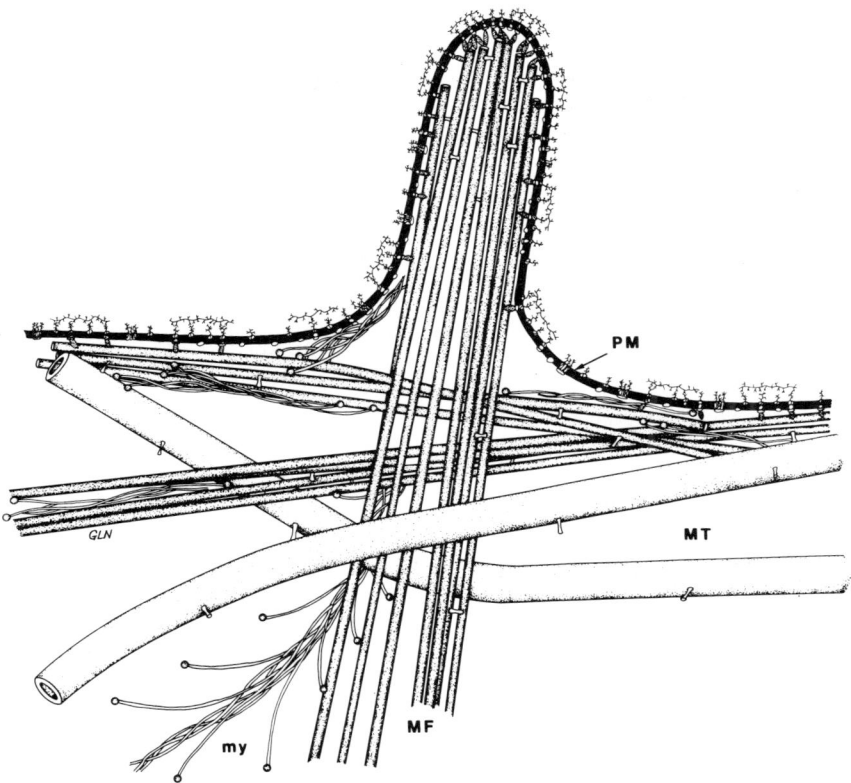

FIG. 5. Hypothetical interactions between membrane-associated microtubule (MT) and microfilament (MF) systems involved in transmembrane control over cell surface receptor mobility and distribution. This model envisages opposite but a coordinated role for the microfilaments (contractile) and microtubules (skeletal and directional) and suggests that they are linked to one another or to the same plasma membrane (PM) inner surface components. This linkage may occur through myosin molecules or through cross-bridging molecules such as α-actinin. In addition, peripheral membrane components linked at the inner or outer plasma membrane surface may extend this control over specific membrane domains. From Nicolson et al. (1977a).

al., 1975b; Nicolson, 1976; Nicolson and Poste, 1976a), microfilaments and microtubules must be connected to one another or to similar inner surface plasma membrane components, and in this scheme both possibilities are shown (Fig. 5). The transmembrane proteins and glycoproteins which are linked to the cytoskeletal assemblages are thought not to be large enough macromolecular complexes to be visualized by freeze-fracture electron microscopy (Karnovsky et al., 1972; Matter and Bonnet, 1974; Pinto da Silva et al., 1975). Here they are thought to be smaller transmembrane structures or structures linked together at the interface between halves to the lipid bilayer (Nicolson, 1976), so that these transmembrane linkage components would not be visualized by freeze-fracture techniques. An important concept of this structure advanced by Edelman (1974, 1976) is the idea that the cytoskeletal system exerts modulatory control over the cell surface by existing in "free" or "membrane-attached" equilibrium states, so that at any one time certain classes of transmembrane-linked surface receptors are under cytoplasmic control. In fact, only a few transmembrane-linked components could control the distribution and mobility of a large number of cell surface receptors through peripheral membrane interactions as suggested by Hynes (1974). In Fig. 5 this is illustrated by the associations of peripheral glycoproteins and glycosaminoglycans with a variety of surface components. In addition, it should be mentioned that not all cell surface components are probably coupled to one of the transmembrane systems (in Fig. 5) or the restraint mechanisms discussed previously (Fig. 4).

V. Transmembrane-Mediated Communication

The selective cellular response to the diverse array of extracellular signals in the form of ions, hormones, neurotransmitters, antigens, and other recognition structures is dependent upon a variety of transmembrane signaling systems. The receipt of the proper signal by the proper cell usually results in a highly specific stimulus culminating in a selective cellular response. The specificity and diversity in the signal, the receptor component, and the transmembrane transduction of the signal, and finally initiation of an intracellular event forms the basic concept of transmembrane communication (Greaves, 1976). One of the most important properties of this system is that configurational complementarity between the signal and the receptor is an important requirement. Second, the interaction of the signal with the receptor can result in a graded response determined by the number of signals and by the number of receptors receiving them as well as by the transmembrane communication system used to transmit the information and the type of response that the cell will undergo. This, and the fact that each cell receives a variety of diverse signals simultaneously, allows for each cell type to respond in a different manner to the information it receives. In this section some of the more obvious mechanisms for transmembrane communication will be briefly discussed as well as a few highly speculative models.

A. First Messenger Communication

The simplest communication system could be envisioned as one that allows direct entry of the signal into the cells. This can obviously occur via a variety of mechanisms such as simple diffusion, facilitated diffusion, active transport, and endocytosis. The first possible mode of entry, simple diffusion, where the signal simply partitions between the extra- and intracellular phases, will not be discussed further. Although in certain instances this form of communication may be important, it does not allow the fine selectivity inherent in communication systems which utilize specific receptors. Other mechanisms of transport of small or large molecules require some type of stereospecific recognition at the cell surface. This can eventually lead to entry of the molecule into the cell where the molecule itself or an altered form of the molecule becomes the important messenger. An interesting variation on this scheme is shown in Fig. 6, where the entry of the message into the cell is dependent upon its dynamic interaction with cell surface receptors. An example of this form of communication is the binding and

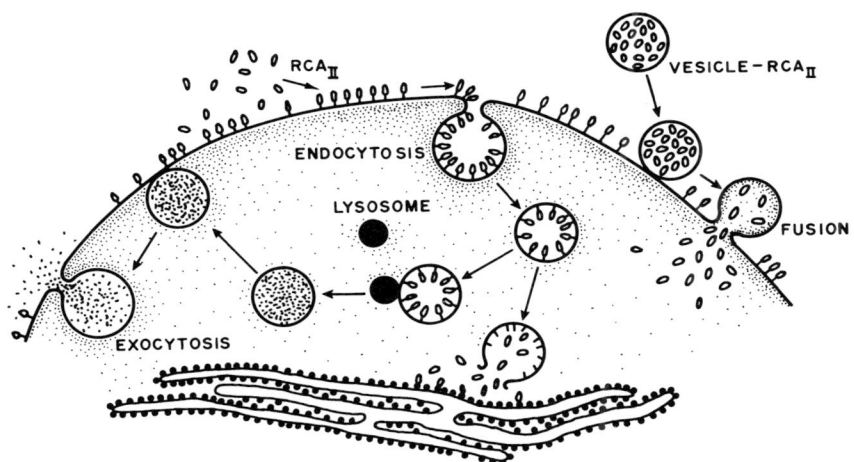

FIG. 6. Possible routes of entry of toxins into mammalian cells. Normal route of entry is (a) binding to the cell surface, (b) ligand-induced clustering, (c) endocytosis of the ligand–receptor complexes, (d) rupture of some of the endocytotic vesicles. Other endocytotic vesicles containing ligand–receptor complexes may fuse with lysosomes and subsequently excrete degraded toxin molecules. Another possible entry mechanism utilizes lipid vesicle-encapsulated toxin molecules. This latter route bypasses the normal entry mechanism by fusion of the vesicles with the plasma membrane releasing the encapsulated toxin directly into the cell cytoplasm. RCA_{II} = *Ricinus communis II* agglutinin. From Nicolson and Poste (1978).

cellular entry of certain plant toxins. The protein synthesis inhibiting toxin from *Ricinus communis* must bind and enter the cell to exert its toxic effect, and several studies have shown that this toxin binds to cell surface glycoproteins and stimulates an endocytotic event. Once the toxin molecules have been endocytosed, they escape some of the endocytotic vesicles and exert their toxic effect directly on the protein synthesis machinery in the cytoplasm (Refsnes *et al.*, 1974; Nicolson *et al.*, 1975, 1978; Nicolson and Poste, 1978).

Another possible mechanism for cell-to-cell communication at long range is the release of small vesicles containing the messenger molecule. If these vesicles bear the correct complementary informational molecules, then they could undergo specific binding to the target cell via its surface receptors. Once this specific binding has taken place, nonspecific fusion between the membrane of the vesicle and the plasma membrane of the cell is much more likely resulting in the entry of the messenger molecule into the target cell (Fig. 6). Mechanisms similar to this have been proposed for embryonic induction (Slavkin *et al.*, 1972; Saxén *et al.*, 1976).

B. Ionic Communication

There are a variety of ways in which ions and small molecules can enter cells to trigger specific events. Some acceptable and more speculative models appear in Fig. 7. For simplicity all of these models are based on pore entry mechanisms for ion flux. In Fig. 7A a ligand binds to the exterior surface at a pore receptor complex. This event activates the gating mechanism and allows free entry of the ion into the cell. In this case the ion becomes the messenger. An alternative process has as an intermediate step the assembly of the pore macromolecular complex (Fig. 7B). Binding of the ligand to one or more subunits of the pore complex results in a configurational change which allows assembly and finally activation of the pore resulting in ion entry into the cell. Ion and small molecule communication between cells can occur through the formation of intercellular junctions (Fig. 7C). In this example a gap junction is shown in its preassembled and assembled state, the latter allowing cell-to-cell passage of small molecules and ions. The last two mechanisms are more speculative and involve membrane phase changes. In Fig. 7D and E an annulus of solid phase lipid around the assembled pore (Fig. 7D) similar to the Ca^{2+} transport protein of sarcoplasmic reticulum (Bennett *et al.*, 1978) or around the disassembled subunits of the pore (Fig. 7E) prevents this system from allowing ion or small molecule communication across the membrane. After a lipid phase transition which could be caused by

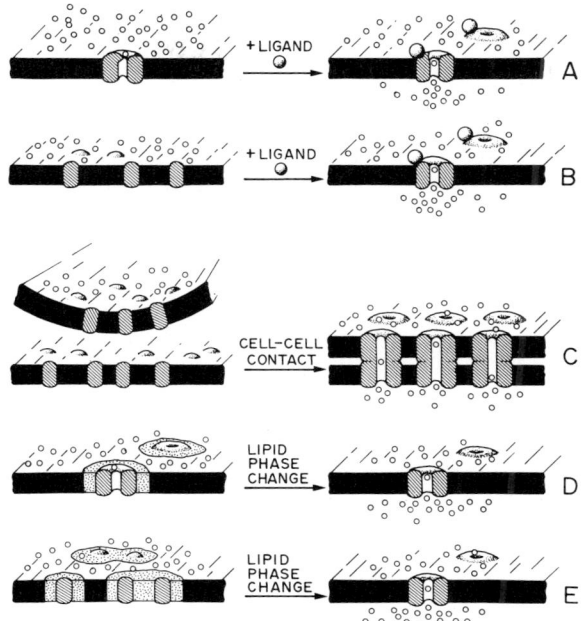

FIG. 7. Hypothetical models on control of ion communication and entry into cells. See text for explanation.

a change in membrane physical state due to temperature modifications or biosynthetic or enzymatic compositional alterations, the annulus of solid phase lipids is converted to the fluid state resulting in activation of the complex and ion or small molecule passage. The existence of plasma membrane enzymes and transport systems requiring fluid state lipids for activity has been well documented in a variety of biological systems (for review, see Kimelberg, 1977). It should be noted that these latter models are particularly speculative, since it remains to be proven that lipid phase changes are involved in transmembrane communication controlling systems.

C. Second Messenger or Enzymatic Communication

One of the classical systems for transmembrane communication is the second messenger model (see discussions by Rodbell, 1972; Goldberg *et al.*, 1974; Resch, 1976; Rasmussen, 1977). In its most generalized scheme (Fig. 8A) the second messenger model theorizes that the binding of an external ligand (first message) to a cell surface receptor results in a conformational transduction of this information across

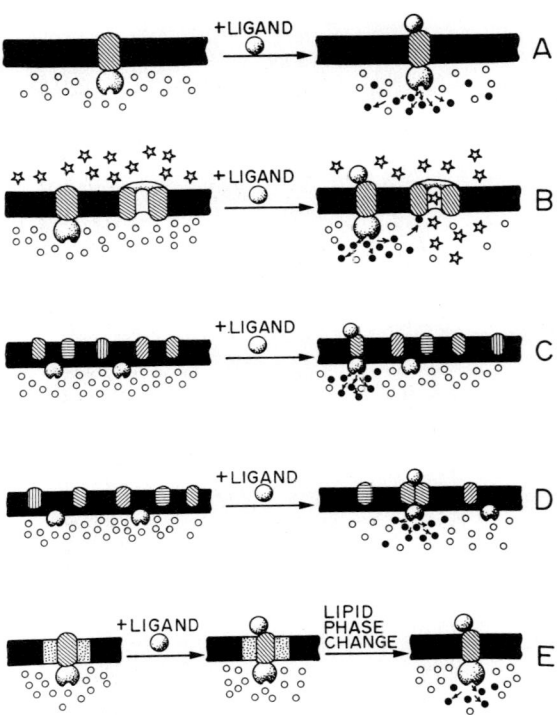

FIG. 8. Some hypothetical models for second messenger or enzymatic transmembrane communication. See text for explanation.

the membrane to the inner membrane surface where an enzyme is activated. The enzyme product becomes the second messenger. The main advantage of this system over direct entry of the first messenger is that amplification of the signal can occur. Thus, for every external message received by a cell, a large number of second messengers can be produced in the cytoplasm. Variations on this scheme can lead to a third messenger hypothesis whereby the second messenger activates a number of cellular compounds or enzymes resulting in rapid biosynthesis and membrane replacement or enzymatic modification of existing plasma membrane components. In Fig. 8B the second messenger is shown enzymatically modifying a transmembrane ion channel component allowing the third messenger, a specific ion, to enter the cell. The entering ion then becomes the effector.

Not every receptor may have a specific enzyme such as an adenylcyclase located immediately across the membrane at all times. Cuatrecasas (1974) has proposed that hormone–receptor complexes might diffuse laterally in the membrane until proper contact is made with an adenylcyclase molecule at the inner membrane surface (Fig. 8C). This mechanism was introduced, in part, to explain the time lage between receipt of the first message and measurement of a specific event occurring inside the cell. In addition, it has obvious economic advantages for the cell, since only one or a few second messenger enzymes are necessary to interact with a large number of cell surface receptors. A variation of this model (Fig. 8D) requires that the receptors undergo ligand-induced redistribution to form complexes which are necessary in order to activate the enzyme at the inner membrane surface. Finally, another alternative would be similar to the lipid phase change models presented in Fig. 7. In this case (Fig. 8E) binding of the ligand would initiate a change in lipid phase state or composition immediately surrounding the receptor. As the annulus of lipid around the receptor is changed to fluid state lipids, the ligand–receptor complex is stimulated to activate an enzyme at the inner membrane surface.

In the foregoing models for ligand–receptor–enzyme activation resulting in a second messenger, it has been assumed that the inactive enzyme(s) are inherently present in the membrane or bound to the inner membrane surface as a peripheral membrane protein. This is not a necessary condition for any of the models in Fig. 8. Soluble cytoplasmic enzymes could easily substitute for membrane enzymes, if an intermediate step in the activation process is the binding of the soluble enzyme to an activated ligand–receptor complex.

D. Cytoskeletal-Mediated Communication

Control over cellular cytoskeletal systems is thought to be important in transmembrane communication (Edelman, 1976) and cellular differentiation (Wessells et al., 1971). There are several possible mechanisms for cytoskeletal-mediated transmembrane communication (Fig. 9). The binding of an external ligand to the cell surface could result in a transmembrane signal which releases membrane-associated cytoskeletal elements (Fig. 9A). In an alternative process the ligand could be bound to a transmembrane ion–pore complex which, when activated by the binding of an external ligand, allows ion penetration into the cell. In this case the change in ion concentration near the inner surface of the membrane triggers release of the membrane-associated cytoskeletal components (Fig. 9B). In an interesting experiment Edidin and Wei (1977) found that redistribution of antigen–antibody complexes on Sendai virus-fused heterokaryons is dependent upon membrane potential and can be modified by non-

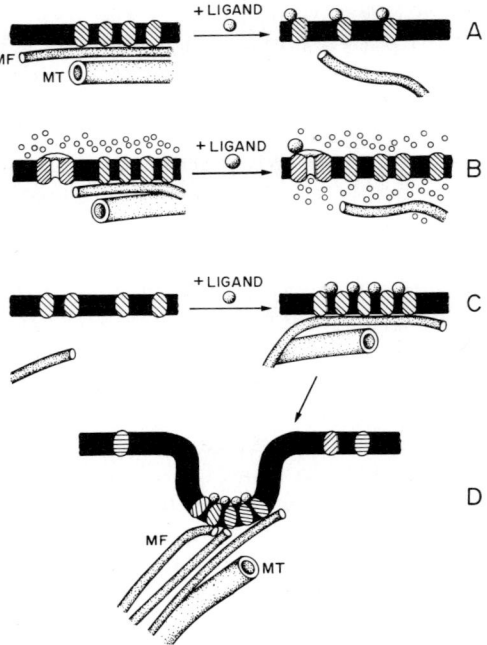

Fig. 9. Some hypothetical models for transmembrane control involving cytoskeletal systems such as microfilaments (MF) and microtubules (MT). See text for explanation.

physiological concentrations of potassium ions or drugs which affect the plasma membrane Na^+–K^+-ATPase. The uncoupling and resultant increase in diffusional freedom of antigen–antibody complexes on cells with high cytoplasmic K^+ ion concentrations could be due to such a release mechanism.

A common event in many types of cells after binding of an external ligand is ligand-induced redistribution to form clusters and patches. When this aggregation is transmembrane communicated into the cell, polymerization or attachment of cytoskeletal components could occur (Fig. 9C). Contraction of the microfilament system along oriented microtubular pathways could also result in selective endocytosis of a specific membrane region (Fig. 9D). Once membrane invagination has occurred, this region of the membrane can pinch off and reseal leaving the ligand–receptor complexes inside the endocytotic vesicle.

After endocytosis has occurred, the endocytotic vesicles appear to be transported to specific cell regions for storage, breakdown, or fusion with lysozomal particles or Golgi compartments (Gonatus et al., 1977).

Modulation of cell surface receptors may follow one of these routes. Indeed, the results of Steinman et al. (1976) suggest that the entire plasma membrane is internalized every few hours and implies that the cell surface is regenerated at approximately the same rate, presumably because of transmembrane feedback mechanisms. It has been hypothesized that the degradation of endocytotic vesicles is linked to membrane recycling, because rates of endocytosis are much greater than the half-lives of plasma membrane proteins, supporting the concept that endocytosis need not necessarily result in endocytotic membrane degradation (Cook, 1977).

VI. Concluding Remarks

The cell surface is no longer considered simply to be an interfacial boundary for the cell cytoplasm. Its importance in selecting for each cell a felicitous set of external signals for transmembrane communication cannot be disregarded. Indeed, without transmembrane signals in proper discrimination, sequences, or amplification, organisms would fail to undergo precise cellular programs of growth, division, differentiation, movement, communication, sensation, and ultimately death. Since there are innumerable signals and communications that must be selectively received by the appropriate cells, it is fitting that there are probably a multitide of different ways in which cells can transmit signals from their exterior to their interior environments. Unfortunately, our ignorance concerning these transmembrane communication systems has limited to a certain degree molecular explanations for the cellular events above. This has not, however, dampened our enthusiasm for speculation as is evident in the final sections of this brief review.

REFERENCES

Abbas, A. K., Ault, K. A., Karnovsky, M. J., and Unanue, E. R. (1975). *J. Immunol.* **114**, 1197–1204.
Adams, D., Markes, M. E., Leivo, W. J., and Carraway, K. L. (1976). *Biochim. Biophys. Acta* **426**, 38–45.
Albertini, D. F., and Anderson, E. (1977). *J. Cell Biol.* **73**, 111–127.
Ali, I. U., Mautner, V., Lanza, R., and Hynes, R. O. (1977). *Cell* **11**, 115–126.
Bedford, J. M., Cooper, G. W., and Calvin, H. I. (1972). *In* "The Genetics of the Spermatozoa" (R. A. Beatty and S. Gluecksohn-Waelsch, eds.), pp. 69–89. Bogtrykeriet Forum, Copenhagen.
Bennett, J. P., McGill, K. A., and Warren, G. B. (1978). *Nature (London)* **274**, 823–825.
Berlin, R. D., Oliver, J. M., Ukena, T. E., and Yin, H. H. (1974). *Nature (London)* **247**, 45–46.
Berridge, M. J. (1975). *In* "Advances in Cyclic Nucleotide Research" (P. Greengard and G. A. Robison, eds.), Vol. 6, pp. 1–98. Raven, New York.

Bretscher, M. S. (1971). *J. Mol. Biol.* **59,** 351–357.
Bretscher, M. S. (1973). *Science* **181,** 622–629.
Brinkley, B. R., Fuller, G. M., and Highfield, D. P. (1975). *Proc. Natl. Acad. Sci. U.S.A.* **72,** 4981–4985.
Brockerhoff, H. (1974). *Lipids* **9,** 645–650.
Butler, K. W., Dugas, H., Smith, I. C. P., and Schneider, H. (1970). *Biochem. Biophys. Res. Commun.* **40,** 770–776.
Capaldi, R. A., and Green, D. E. (1972). *FEBS Lett.* **25,** 205–209.
Carafoli, E., Clementi, F., Drabikowski, W., and Margreth, A., eds. (1975). "Calcium Transport in Contraction and Secretion." North-Holland Publ., Amsterdam.
Carraway, K. L., Triplett, R. B., and Anderson, D. R. (1975). *Biochim. Biophys. Acta* **379,** 571–581.
Chapman, D. (1973). *In* "Biological Membranes" (D. Chapman and D. F. H. Wallach, eds.), Vol. 1, pp. 91–144. Academic Press, New York.
Cherry, R. J. (1976). *In* "Biological Membranes" (D. Chapman and D. F. H. Wallach, eds.), Vol. 3, pp. 47–102. Academic Press, New York.
Cook, G. M. W. (1977). *In* "The Synthesis and Turnover of Cell Surface Components" (G. Poste and G. L. Nicolson, eds.), Vol. 4 of "Cell Surface Reviews," pp. 85–163. North-Holland Publ., Amsterdam.
Cotman, C. W., Banker, G., Churchill, L., and Taylor, D. (1974). *J. Cell Biol.* **63,** 441–455.
Cuatrecasas, P. (1974). *Annu. Rev. Biochem.* **43,** 169–214.
Darke, A., Finer, E. G., Flook, A. G., and Phillips, M. C. (1972). *J. Mol. Biol.* **63,** 265–279.
De Kruyff, B., Van Dijck, P. W. M., Demel, R. A., Schuijff, A., Brants, G., and Van Deenen, L. L. M. (1974). *Biochim. Biophys. Acta* **356,** 1–7.
de Petris, S. (1974) *Nature (London)* **250,** 54–56.
de Petris, S. (1975) *J. Cell Biol.* **65,** 123–146.
de Petris, S. (1977) *In* "Dynamic Aspects of Cell Surface Organization" (G. Poste and G. L. Nicolson, eds.), Vol. 3 of "Cell Surface Reviews," pp. 643–728. North-Holland Pub., Amsterdam.
Edelman, G. M. (1974). *In* "Control of Proliferation in Animal Cells" (B. Clarkson and R. Baserga, eds.), Vol. 1 of "Cold Spring Harbor Conferences on Cell Proliferation," pp. 357–377. Cold Spring Harbor Laboratory, New York.
Edelman, G. M. (1976). *Science* **192,** 218–226.
Edelman, G. M., Yahara, I., and Wang, J. L. (1973). *Proc. Natl. Acad. Sci. U.S.A.* **70,** 1442–1446.
Edidin, M. (1974). *Annu. Rev. Biophys. Bioeng.* **3,** 179–201.
Edidin, M., and Fambrough, D. (1973). *J. Cell Biol.* **57,** 27–37.
Edidin, M., and Wei, T. (1977). *J. Cell Biol.* **75,** 483–489.
Ehrström, M., Eriksson, L. E. G., Israelachvili, J., and Ehrenberg, A. (1973). *Biochem. Biophys. Res. Commun.* **55,** 396–402.
Elgsaeter, A., and Branton, D. (1974). *J. Cell Biol.* **63,** 1018–1036.
Elgsaeter, A., Shotton, D. M., and Branton, D. (1976). *Biochim. Biophys. Acta* **426,** 101–122.
Friend, D. W., and Fawcett, D. W. (1974). *J. Cell Biol.* **63,** 641–652.
Frye, L. D., and Edidin, M. (1970). *J. Cell Sci.* **7,** 319–333.
Gahmberg, C. G. (1977). *In* "Dynamic Aspects of Cell Surface Organization" (G. Poste and F. L. Nicolson, eds.), Vol. 3 of "Cell Surface Reviews," pp. 371–471. North-Holland Publ., Amsterdam.
Gail, M. H., and Boone, C. W. (1972). *Exp. Cell Res.* **63,** 252–255.

Goldberg, N. D., Haddox, M. K., Dunham, E., Lopez, C., and Hadden, J. W. (1974). In "Control of Proliferation in Animal Cells" (B. Clarkson and R. Baserga, eds.), Vol. 1 of "Cold Spring Harbor Conferences on Cell Proliferation," pp. 609–625. Cold Spring Harbor Laboratory, New York.
Goldman, R. D. (1972). *J. Cell Biol.* **52**, 246–254.
Goldman, R. D. (1975). *J. Histochem. Cytochem.* **23**, 529–542.
Goldman, R., Pollard, T., and Rosenbaum, J., eds. (1976). "Cell Motility," Vol. 3 of "Cold Spring Harbor Conferences on Cell Proliferation," Cold Spring Harbor Laboratory, New York.
Gonatus, N. K., Kim, S. U., Stieber, A., and Avrameas, S. (1977) *J. Cell Biol.* **73**, 1–13.
Grant, C. W. M., and McConnell, H. M. (1974) *Proc. Natl. Acad. Sci. U.S.A.* **71**, 4653–4657.
Grant, C. W. M., Wu, S. H.-W., and McConnell, H. M. (1974). *Biochim. Biophys. Acta* **363**, 151–158.
Greaves, M. F. (1976). *In* "Receptors and Recognition" (P. Cuatrecasas and M. F. Greaves, eds.), Vol. 1, Series A, pp. 1–32. Chapman & Hall, London.
Hepler, P. K., and Palevitz, B. A. (1974). *Annu. Rev. Plant Physiol.* **25**, 309–362.
Hong, K., and Hubbell, W. L. (1972). *Proc. Natl. Acad. Sci. U.S.A.* **69**, 2617–2621.
Horwitz, A. F., Horsley, W. J., and Klein, M. P. (1972). *Proc. Natl. Acad. Sci. U.S.A.* **69**, 590–593.
Hubbell, W. L., and McConnell, H. M. (1971). *J. Am. Chem. Soc.* **93**, 314–326.
Hunt, R. C., and Brown, J. C. (1975). *J. Mol. Biol.* **97**, 413–422.
Hynes, R. O. (1974). *Cell* **1**, 147–156.
Hynes, R. O., Destree, A. T., and Mautner, V. (1976). *In* "Membranes and Neoplasia" (V. T. Marchesi, ed.), pp. 189–202. Liss, New York.
Ji, T. H., and Nicolson, G. L. (1974) *Proc. Natl. Acad. Sci. U.S.A.* **71**, 2212–2216.
Karnovsky, M. J., Unanue, E. R., and Leventhal, M. (1972). *J. Exp. Med.* **136**, 907–917.
Kelly, P., Cotman, C. W., Gentry, C., and Nicolson, G. L. (1976). *J. Cell Biol.* **71**, 487–496.
Kimelberg, H. K. (1977). *In* "Dynamic Aspects of Cell Surface Organization" (G. Poste and G. L. Nicolson, eds.), Vol. 3 of "Cell Surface Reviews," pp. 205–293. North-Holland Publ., Amsterdam.
Kirschner, M. W., and Williams, R. C. (1974). *J. Supramol. Struct.* **2**, 412–428.
Kleemann, W., and McConnell, H. J. (1974). *Biochim. Biophys. Acta* **345**, 220–230.
Kleemann, W., and McConnell, H. M. (1976). *Biochim. Biophys. Acta* **219**, 206–222.
Koehler, J. K. (1975a). *J. Cell Biol.* **67**, 647–659.
Koehler, J. K. (1975b). *In* "The Biology of the Male Gamete" (J. G. Duckett and P. A. Races, eds.), pp. 337–342. Academic Press, London.
Koehler, J. K., and Perkins, W. D. (1974). *J. Cell Biol.* **60**, 789–795.
Koo, G. C., Stackpole, C. W., Boyse, E. A., Hämmerling, U., and Lardis, M. P. (1973). *Proc. Natl. Acad. Sci. U.S.A.* **70**, 1502–1505.
Korn, E. D. (1969). *Annu. Rev. Biochem.* **38**, 263–288.
Kraemer, P. M. (1975). *In* "Mammalian Cells: Probes and Problems" (C. R. Richmand, D. F. Petersen, P. F. Mullaney, and E. C. Anderson, eds.), pp. 242–245. USERDA Technician Information Center, Oak Ridge, Tennessee.
Lazarides, E. (1975). *J. Histochem. Cytochem.* **23**, 507–528.
Lazarides, E., and Burridge, K. (1975). *Cell* **6**, 289–298.
Levine, Y. K., Birdsall, N. J. M., Lee, A. G., and Metcalfe, J. G. (1972a). *Biochemistry* **11**, 1416–1421.
Levine, Y. K., Partington, P., Roberts, G. C. K., Birdsall, N. J. M., Lee, A. G., and Metcalfe, J. C. (1972b). *FEBS Lett.* **23**, 203–207.

Loor, F. (1977). *Prog. Allergy* **23**, 1–153.
Loor, F., Forni, L., and Pernis, B. (1972). *Eus. J. Immunol.* **2**, 203–212.
Marchesi, V. T., Steers, E., Jr., Tillack, T. W., and Marchesi, S. L. (1969). *In* "The Red Cell Membrane, Structure and Function" (G. A. Jamieson and T. J. Greenwalt, eds.), pp. 117–130. Lippincott, Philadelphia, Pennsylvania.
Mathews, R. A., Johnson, T. C., and Hudson, J. E. (1976). *Biochem. J.* **154**, 57–64.
Matter, A., and Bonnet, C. (1974). *Eur. J. Immunol.* **4**, 704–707.
McConnell, H. M. (1975a). *In* "Functional Linkage in Biomolecular systems" (F. O. Schmitt, D. M. Schneider, and D. M. Crothers, eds.), pp. 123–131. Reven, New York.
McConnell, H. M. (1975b). *In* "Cellular Membranes and Tumor Cell Behavior" (E. F. Walborg, ed.), pp. 61–80. Williams & Wilkins, Baltimore, Maryland.
McFarland, B. G. (1972). *Chem. Phys. Lipids* **8**, 303–313.
McManus, J. P., Whitfield, J. F., Boynton, A. L., and Rixon, R. H. (1975). *In* "Advances in Cyclic Nucleotide Research" (G. I. Drummond, P. Greengard, and G. A. Robison, eds.), Vol. 5, pp. 719–734. Raven, New York.
McNutt, N. S. (1977). *In* "Dynamic Aspects of Cell Surface Organization" (G. Poste and G. L. Nicolson, eds.), Vol. 3 of "Cell Surface Reviews," pp. 75–126. North-Holland Publ., Amsterdam.
McNutt, N. S., and Weinstein, R. S. (1973). *Prog. Biophys. Mol. Biol.* **26**, 45–101.
Mooseker, M. S., and Tilney, L. G. (1975). *J. Cell Biol.* **67**, 725–743.
Morrison, M., Mueller, T. J., and Huber, C. T. (1974). *J. Biol. Chem.* **249**, 2658–2660.
Neauport-Sautes, C., Lilly, F., Silvestre, E., and Kourilsky, F. M. (1973a). *J. Exp. Med.* **137**, 511–526.
Neauport-Sautes, C., Silvestre, D., and Lilly, F. (1973b). *Transplant. Proc.*, **5**, 443–446.
Nicolson, G. L. (1976). *Biochim. Biophys. Acta* **457**, 57–108.
Nicolson, G. L., and Painter, R. G. (1973). *J. Cell Biol.* **59**, 395–406.
Nicolson, G. L., and Poste, G. (1976a). *New Engl. J. Med.* (Pt. I) **295**, 197–203; (Pt. II) **295**, 253–258.
Nicolson, G. L., and Poste, G. (1976b). *J. Supramol. Struct.* **5**, 65–72.
Nicolson, G. L., and Poste, G. (1978). *J. Supramol. Struct.* **8**, 235–245.
Nicolson, G. L., and Singer, S. J. (1974). *J. Cell Biol.* **60**, 236–248.
Nicolson, G. L., and Yanagimachi, R. (1974). *Science* **184**, 1294–1296.
Nicolson, G. L., Marchesi, V. T., and Singer, S. J. (1971). *J. Cell Biol.* **51**, 265–272.
Nicolson, G. L., Lacorbiere, M., and Hunter, T. R. (1975). *Cancer Res.* **35**, 144–155.
Nicolson, G. L., Poste, G., and Ji, T. H. (1977a). *In* "Dynamic Aspects of Cell Surface Organization" (G. Poste and G. L. Nicolson, eds.), Vol. 3 of "Cell Surface Reviews," pp. 1–73. North-Holland Publ., Amsterdam.
Nicolson, G. L., Usui, N., Yanagimachi, R., Yanagimachi, H., and Smith, J. R. (1977b). *J. Cell Biol.* **74**, 950–962.
Nicolson, G. L., Smith, J. R., and Hyman, R. H. (1978). *J. Cell Biol.* **78**, 565–576.
O'Brien, J. R. (1962). *J. Clin. Pathol.* **15**, 446–455.
Oldfield, E., and Chapman, D. (1971). *Biochem. Biophys. Res. Commun.* **43**, 610–616.
Oldfield, E., and Chapman, D. (1972). *FEBS Lett.* **23**, 285–297.
Papahadjopoulos, D., Jacobson, K., Poste, G., and Sheperd, G. (1975). *Biochim. Biophys. Acta* **394**, 504–519.
Peters, R., Peters, J., Tews, K. H., and Bähr, W. (1974). *Biochim. Biophys. Acta* **367**, 282–294.
Phillips, E. R., and Perdue, J. F. (1976a). *J. Cell Sci.* **20**, 459–477.
Phillips, E. R., and Perdue, J. F. (1976b). *J. Supramol. Struct.* **4**, 27–44.

Phillips, M. C., and Chapman, D. (1968). *Biochim. Biophys. Acta* **163**, 301–313.
Pinto da Silva, P., Martinez-Palomo, A., and Gonzalez-Robles, A. (1975). *J. Cell Biol.* **64**, 538–550.
Poo, M., and Cone, R. A. (1974). *Nature (London)* **247**, 438–440.
Poste, G., and Nicolson, G. L., eds. (1977). "The Synthesis, Assembly and Turnover of Cell Surface Components," Vol. 4 of "Cell Surface Reviews." North-Holland Publ., Amsterdam.
Poste, G., and Reeve, P. (1972). *Exp. Cell Res.* **72**, 556–560.
Poste, G., Papahadjopoulos, D., Jacobson, K., and Vail, W. J. (1975a). *Biochim. Biophys. Acta* **394**, 520–539.
Poste, G., Papahadjopoulos, D., and Nicolson, G. L. (1975b). *Proc. Natl. Acad. Sci. U.S.A.* **72**, 4430–4434.
Poulik, M. D., Bernoco, M., Bernoco, D., and Ceppellini, R. (1973). *Science* **182**, 1352–1355.
Rabinovitch, M., and De Stefano, M. J. (1973). *J. Cell Biol.* **59**, 165–176.
Rabinovitch, M., and De Stefano, M. J. (1974). *Exp. Cell Res.* **88**, 153–162.
Raff, M. C., Freedman, M., and Gomperts, B. (1976). *In* "Membrane Receptors of Lymphocytes" (M. Seligman, J. L. Preud'homme, and F. M. Kourilsky, eds.), pp. 393–398. North-Holland Publ., Amsterdam.
Rash, J. E., Staehelin, L. A., and Ellisman, M. H. (1974). *Exp. Cell Res.* **86**, 187–190.
Rasmussen, H. (1970). *Science* **170**, 404–412.
Rasmussen, H. (1977). *In* "Cell and Tissue Interactions" (J. W. Lash and M. M. Burger, eds.), pp. 243–268. Raven, New York.
Refsnes, K., Olsnes, S., and Pihl, A. (1974). *J. Biol. Chem.* **249**, 3557–3562.
Resch, K. (1976). *In* "Receptors and Recognition" (P. Cuatrecasas and M. F. Greaves, eds.), Vol. 1, Series A, pp. 61–117. Chapman & Hall, London.
Rodbell, M. (1972). *In* "Current Topics in Biochemistry" (C. B. Anthenson, R. F. Goldberger, and A. N. Schecter, eds.), pp. 187–218. Academic Press, New York.
Rostas, J. A. P., and Jeffrey, P. L. (1975). *Neurosci. Lett.* **1**, 47–53.
Rott, R., and Klenk, H-D. (1977). *In* "Virus Infection and the Cell Surface" (G. Poste and G. L. Nicolson, eds.), Vol. 2 of "Cell Surface Reviews," pp. 47–82. North-Holland Publ., Amsterdam.
Rottem, S. J., Yashouv, J., Ne'eman, Z., and Razin, S. (1973). *Biochim. Biophys. Acta* **323**, 496–508.
Ryan, G. B., Borysenko, J. S., and Karnovsky, M. J. (1974a). *J. Cell Biol.* **62**, 351–365.
Ryan, G. B., Unanue, E. R., and Karnovsky, M. J. (1974b). *Nature (London)* **250**, 56–57.
Sandri, C., Akert, K., Livingston, R. E., and Moor, H. (1972). *Brain Res.* **41**, 1–16.
Sauerheber, R. D., and Gordon, L. M. (1975). *Proc. Soc. Exp. Biol. Med.* **150**, 28–31.
Saxén, L., Karkinin-Jääskeläinen, M., Lehtonen, E., Nordling, S., and Wartiovaara, J. (1976). *In* "The Cell Surface in Animal Embryogenesis and Development" (G. Poste and G. L. Nicolson, eds.), Vol. 1 of "Cell Surface Reviews," pp. 331–407. North-Holland Publ., Amsterdam.
Schlessinger, J., Koppel, D. E., Axelrod, D., Jacobson, K., Webb, W. W., and Elson, E. L. (1976). *Proc. Natl. Acad. Sci. U.S.A.* **73**, 2409–2413.
Schnepel, G. H., Hegner, D., and Schummer, U. (1974). *Biochim. Biophys. Acta* **367**, 67–74.
Seelig, J., and Niederberger, W. (1974). *Biochemistry* **13**, 1585–1588.
Segrest, J. P., Kahne, I., Jackson, R. L., and Marchesi, V. T. (1973). *Arch. Biochem. Biophys.* **155**, 167–183.

Segrest, J. P., Gulik-Krzywicki, T., and Sardet, C. (1974). *Proc. Natl. Acad. Sci. U.S.A.* **71,** 3294–3298.
Singer, S. J. (1971). *In* "Structure and Function of Biological Membranes" (L. I. Rothfield, ed.), pp. 145–222. Academic Press, New York.
Singer, S. J. (1974). *Annu. Rev. Biochem.* **43,** 805–833.
Singer, S. J., and Nicolson, G. L. (1972). *Science* **175,** 720–731.
Slavkin, H. C., Croissant, R., and Bavetta, L. A. (1972). *Mech. Age Dev.* **1,** 139–169.
Speth, V., and Wunderlich, F. (1973). *Biochim. Biophys. Acta* **291,** 621–628.
Staehelin, L. A. (1974). *Int. Rev. Cytol.* **39,** 191–283.
Steinman, R. M., Brodie, S. E., and Cohn, Z. A. (1976). *J. Cell Biol.* **68,** 665–687.
Sundqvist, K. G. (1977). *In* "Dynamic Aspects of Cell Surface Organization" (G. Poste and G. L. Nicolson, eds.), Vol. 3 of "Cell Surface Reviews," pp. 551–600. North-Holland Publ., Amsterdam.
Taylor, R. B., Duffus, W. P. H., Raff, M. C., and de Petris, S. (1971). *Nature (London) New Biol.* **233,** 225–229.
Tilney, L. G., and Detmers, P. (1975). *J. Cell Biol.* **66,** 508–520.
Träuble, H. (1972). *Biomembranes* **3,** 197–227.
Träuble, H., and Eibl, H. (1975). *In* "Functional Linkage in Biomolecular Systems" (F. O. Schmitt, D. M. Schneider, and D. M. Crothers, eds.), pp. 59–90. Raven, New York.
Triggle, D. J. (1972). *In* "Progress in Surface and Membrane Science" (J. F. Danielli, M. D. Rosenberg, and D. A. Cadenhead, eds.), Vol. 5, pp. 267–331. Academic Press, New York.
Unanue, E. R., and Karnovsky, M. J. (1973). *Transplant Rev.* **14,** 184–210.
Unanue, E. R., and Schreiner, G. F. (1977). *In* "Dynamic Aspects of Cell Surface Organization" (G. Poste and G. L. Nicolson, eds.), Vol. 3 of "Cell Surface Reviews," pp. 619–641. North-Holland Publ., Amsterdam.
Unanue, E. R., Karnovsky, M. J., and Engers, H. D. (1973). *J. Exp. Med.* **137,** 675–689.
Vail, W. J., Papahadjopoulos, D., and Moscarello, M. A. (1974). *Biochim. Biophys. Acta* **345,** 463–467.
Vanderkooi, J., Fischkoff, S., Chance, B., and Cooper, R. A. (1974). *Biochemistry* **13,** 1589–1595.
Verkleij, A. J., Ververgaert, P. H. J., Van Deenen, L. L. M., and Elbers, P. F. (1972). *Biochim. Biophys. Acta* **288,** 326–332.
Wallach, D. F. H. (1972). "The Plasma Membrane: Dynamic Perspectives, Genetics and Pathology." Springer-Verlag, Berlin and New York.
Wartiovaara, J., Linder, E., Ruoslahti, E., and Vaheri, A. (1974). *J. Exp. Med.* **140,** 1522–1533.
Weber, K., Lazarides, E., Goldman, R. D., Vogel, A., and Pollack, R. (1974). *Cold Spring Harbor Symp. Quant. Biol.* **39,** 363–369.
Wessells, N. K., Spooner, B. S., Ash, J. F., Bradly, M. O., Luduena, M. A., Taylor, E. L., Wrenn, J. T., and Yamada, K. M. (1971). *Science* **171,** 135–143.
White, S. H. (1970). *Biophys. J.* **10,** 1127–1148.
Yahara, I., and Edelman, G. M. (1972). *Proc. Natl. Acad. Sci. U.S.A.* **69,** 608–612.
Yahara, I., and Edelman, G. M. (1973). *Exp. Cell Res.* **81,** 143–155.
Yanagimachi, R., Noda, Y. D., Fujimoto, M., and Nicolson, G. L. (1972). *Am. J. Anat.* **135,** 497–520.

SUBJECT INDEX

A

Acrosome reaction, 32
Adhesion, intercellular
  neural tissue, 226, 253
  sea urchin cells, 199–201
Affinity chromatography, 267
Agglutination assays, 205
γ-Aminobutyric acid, 250
Aminopterin, 225
Anesthetics, effects on membranes, 323
Antigenic modulation, 283
  failure of in hybrid cells, 293
  possible cellular mechanism, 298–300
  similarities with down regulation, 300
Antigens, cell surface
  A_L in human erythrocytes, 96
  antibody-induced redistribution, 190
  coded by human chromosome 7, 149
  corpus callosum, 265
  egg, 142
  F9
    on early embryos, 147–149
    on embryonal carcinoma cells, 125
    immunochemistry of, 128
    relationship to T/t complex, 127, 174
    on sperm, 4
  Forssman, 159
  H-2
    on early mouse embryos, 147, 169
    on sperm, 4, 170
  H-3, 171
  H-6, 171
  H-Y
    on preimplantation embryos, 171
    on sperm, 4
  Ia
    on mouse embryos, 147
    on sperm, 4, 170
  N18 (neuroblastoma) surface polypeptides, 218
  NS (nervous system), 223, 259–277
    developmental expression, 268–273
    on preimplantation embryos, 145
    on sperm, 4
    subcellular distribution in nervous tissue, 273
  newt embryonic germ layers, 54
  PCC4
    on blastocysts, 135
    on sperm, 4
  Ran-1, 239
  redistribution of, 315
  shedding of, 288
  stage specific embryonic antigen-1, 153–158
  SV40 related, 149–150
  testicular teratoma, 55
  Thy-1, 239, 245–251
  thymus leukemia, 283
Antigens, quantitation of
  on neuroblastoma cells, 217
  on sperm cell surface, 22
Antisera
  effects on
    embryonic development, 161, 175, 190, 229
    mouse embryogenesis, 132
    sea urchin gastrulation, 210
  reactivity with preimplantation embryos, 188–189
Antisera to
  bindin, 33
    cross reactivity with different species, 40
    immunochemical characterization, 34
    immunoperoxidase localization, 35–40
  cell adhesion material, 221
  cerebellar cells, 145
  Drosophila chromosome proteins
    5.2/21, 78
    band 2, 79
    NHC protein fraction, 78
    p, 80
  embryonal carcinoma (F9) cells, 173

erythrocytes, human, 93
fibronectin, 187
galactocerebroside, 243
glial fibrillary acidic protein, 241
glycophorin, 104
human cells, 92
human chorionic gonadotropin, 182
hybrid cells, mouse–human, 149
interspecies neural membrane antigens, 221
mouse blastocyst, 142, 176
mouse egg, Pronase-treated, 175–176
mouse embryos, 142–160
mouse L cells, 185
mouse placenta, 177–182
mouse spermatogenic cells, 10
  adult seminiferous epithelium, 10
  adult pachytene spermatocytes, 10
  B spermatogonia, 17–24
  round spermatids, 10
  sperm from vas deferens, 10
N_1 (teratoma) cell line, 175
neuroblastoma cells, 216, 221
retina cells, 276
RNA polymerase, 84
Schwann cells, 238
sea urchin gastrula membranes, 204
slime mold, 209
teratoma 402AX cells, 172
zona pellucida, 144
Ascidians
  early development, 50–52
  sperm–egg interaction, 63
ASL-1 cell line, 287
Astrocytes, 241, 249

B

Bindin, 32, 62
  distribution of sea urchin sperm, 36
  membrane fusion, at site of, 38
Blood–testis barrier, 10
Bromodeoxyuridine, preparation of auxotrophic mutants, 91

C

Calcium ion
  in cell membranes, 311
  in morula compaction, 133

Capping, of surface antigens, 190
  inhibition of, 323
  role of microfilaments, 322
Cerebellar cells, 248
Cholesterol, 310, 313
Chorion, 63
Complementation analysis, 102–104
Chromatin, 76
Chromosomes
  human-7, cell surface antigen, 149
  human-11, 93
  polytene
    immunofluorescent staining of, 72–76
    preparation of, 72
Cytochalasin B
  effect on capping, 323
  inhibition of morula compaction, 132
Cytosine arabinoside, 253
Cytotoxicity, complement-mediated, on sperm, 12

D

Deoxyribonuclease I, 82
2′-Deoxy-5-fluorouridine (FUdR), 248
Dibutyryl cyclic AMP, 253
  effect on Chinese hamster brain cell dendritic extension, 112
Down regulation, 282
Drosophila, polytene chromosomes, 71

E

Endocytic vacuoles, 301
Endocytosis, 332

F

Fab fragments
  anti-F9 sera
    effect on embryonal carcinoma cells, 133
    effect on mouse embryogenesis, 132, 191
  antisera to Arbacia eggs, 190
  antisera to sea urchin gastrula membranes, 210
Fc receptors, on embryonic cells, 160
Fertilization, 59
Fertilizin, 60
Fibronectin

antisera to, 187
on astrocytes, 242
development of mouse embryo, 56
on neural tissue cells, 265
Fluorescence-activated cell sorting, 253, 267
Forssman antigen, 159
Fusion of plasma membranes, sperm and egg, 38

G

G_{M1}, 271
Galactocerebroside, 243, 265
Ganglia
mouse superior cervical, 218
rat dorsal root, 239
Gastrulation, in sea urchins, 199–203
Germ layers, embryonic
antisera to sea urchin tissues, 208
Glial fibrillary acidic protein, 241
Glycoproteins, cell surface, 314
Granule cells, 249
G_{M1} distribution on, 271

H

Human chorionic gonadotropin
antisera to, 182
distribution on mouse blastocyst, 182–183
Hybrid cells
human–Chinese hamster ovary, preparation of, 90
TL(+) and TL (−) tumor cells, 290

I

IgA, antisera to sperm, 175
IgG, quantitative binding assays, 22, 217
IgM
antisera to stage specific embryonic antigen-1, 155
antiserum to human chromosome 7-cell surface antigen, 149
binding to Fc receptors on embryonic cells, 160
Immunofluorescence
general comments, 205
staining techniques
polytene chromosomes, 75

Immunoperoxidase
A_L antigens, 106–109
sea urchin fertilization, 33
Immunosurgery, 266
Intercellular communication, 326–333
Intercellular junctions, 318, 328
desmosomes, 3
gap junctions, 2
septate junctions, 2
tight junctions, in morula to blastocyst transformation, 168

L

Lactate dehydrogenase
A, 99
X, 4
Lectins
concanavalin A, 57, 64
on embryonal carcinoma cells, 123
on preimplantation embryos, 124
Dolichos biflorum, binding to muscle cells, 52
fucus, 47, 64
on embryonal carcinoma cells, 123
on preimplantation embryos, 124
inhibition of cytotoxic anti-A_L sera, 105
peanut agglutinin
on embryonal carcinoma cells, 123
on preimplantation embryos, 124
Ricinus communis on rabbit sperm, 316
wheat germ agglutinin
on embryonal carcinoma cells, 123
on preimplantation embryos, 124
LETS protein, see Fibronectin
Luteinizing hormone, 185

M

Membrane fluidity, 308–310
Membrane lipids, 310
Membrane proteins
intrinsic, 306
peripheral, 307
Membrane recycling, 333
Microfilaments, 322
β_2-Microglobulin, on preimplantation embryos, 171
Microtubules, 322
Monoclonal antibody
to embryonal carcinoma cells, 130

to Forsmann antigen, 159
to stage specific embryonic antigen-1, 153–158
Monovalent antibody, see Fab fragments
Mouse development
　advantages in the study of, 139–141
　preimplantation development, 167–168
Muscle
　cells, and interaction with nerves, 227
　lectins, 52

N

Nerve growth factor, effects on PC12 cells in culture, 220
Neural cells
　developmental expression of NS antigens, 268–273
　subcellular distribution of NS antigens, 273–276
Neural tumors, 261
Neuraminidase, to unmask trophectoderm antigens, 170
Neuroblastoma
　antisera to
　　clone N18, 216
　　human cells, 221
　differentiation in vitro, 224
　mouse C1300 derivatives, 216
Nuclear magnetic resonance, 311

O

Oligodendroglia, 243, 265

P

Patching, of surface antigens, 190
Phospholipids, 312
Placenta, antisera to, 177–182
Plasma membranes, isolation from germ cells, 24
Polarity, of the egg, 46–49
Postsynaptic membrane endings, 319
Preimplantation embryos, lectin binding, 124
Proteolysis, of sperm membrane molecules, 9

R

RADA-1 cell line, 286

S

Schwann cells, antisera to, 238
Sea urchins, 30–44, 199–214
　egg–sperm interaction, 32
　gastrulation, 199
　intercellular adhesion, 201
Segregation of cell lines, 49–54
Seminiferous cells, isolation of, 6–9
Sertoli cells, 2
Spermatogenesis, 1–6
　intercellular junctions, 2–3
　spermatocytogenesis, 3
　spermiogenesis, 4
Spermatozoa
　antigenic similarities with nervous system, 270
　cell surface antigens, 4, 316
　LDH-X isozyme, 4
　plasma membrane
　　fluidity of, 5, 15
　　partitioning of membrane particles, 6, 17
　sea urchin, 32
SV40 related surface antigens, 149–150

T

Teratocarcinoma, embryonal, 118
　glycopeptides, 118–122
　　isolation, 119
　　tissue distribution, 120–121
　lectin receptors, 123–125
　monoclonal antibodies to, 130
　PCC3 cell line, 123
　surface antigens, 125–127
Teratoma
　ovarian, 118
　testicular, 55, 118
Tetanus toxin, 241, 249
Thymus leukemia system, 283
　synthesis and turnover of antigens, 285–288
Transmembrane signaling, 305–333
T/t complex, relationship to F9 antigen, 127

V

Viral budding, 320
Vitelline coat, 67

CONTENTS OF PREVIOUS VOLUMES

Volume 1

Remarks
 Joshua Lederberg
On "Masked" Forms of Messenger RNA in Early Embryogenesis and in Other Differentiating Systems
 A. S. Spirin
The Transcription of Genetic Information in the Spiralian Embryo
 J. R. Collier
Some Genetic and Biochemical Aspects of the Regulatory Program for Slime Mold Development
 Maurice Sussman
The Molecular Basis of Differentiation in Early Development of Amphibian Embryos
 H. Tiedemann
The Culture of Free Plant Cells and Its Significance for Embryology and Morphogenesis
 F. C. Steward, Ann E. Kent, and Marion O. Mapes
Genetic and Variegation Mosaics in the Eye of *Drosophila*
 Hans Joachim Becker
Biochemical Control of Erythroid Cell Development
 Eugene Goldwasser
Development of Mammalian Erythroid Cells
 Paul A. Marks and John S. Kovach

Genetic Aspects of Skin and Limb Development
 P. F. Goetinck
Author Index—Subject Index

Volume 2

The Control of Protein Synthesis in Embryonic Development and Differentiation
 Paul R. Gross
The Genes for Ribosomal RNA and Their Transaction during Amphibian Development
 Donald D. Brown
Ribosome and Enzyme Changes during Maturation and Germination of Castor Bean Seed
 Erasmo Marrè
Contact and Short-Range Interaction Affecting Growth of Animal Cells in Culture
 Michael Stoker
An Analysis of the Mechanism of Neoplastic Cell Transformation by Polyoma Virus, Hydrocarbons, and X-Irradiation
 Leo Sachs
Differentiation of Connective Tissues
 Frank K. Thorp and Albert Dorfman
The IgA Antibody System
 Mary Ann South, Max D. Cooper, Richard Hong, and Robert A. Good
Teratocarcinoma: Model for a Developmental Concept of Cancer
 G. Barry Pierce

Cellular and Subcellular Events in Wolffian Lens Regeneration
  Tuneo Yamada
Author Index—Subject Index

Volume 3

Synthesis of Macromolecules and Morphogenesis in *Acetabularia*
  J. Brachet
Biochemical Studies of Male Gametogenesis in Liliaceous Plants
  Herbert Stern and Yasuo Hotta
Specific Interactions between Tissues during Organogenesis
  Etienne Wolff
Low-Resistance Junctions between Cells in Embryos and Tissue Culture
  Edwin J. Furshpan and David D. Potter
Computer Analysis of Cellular Interactions
  F. Heinmets
Cell Aggregation and Differentiation in *Dictyostelium*
  Günther Gerisch
Hormone-Dependent Differentiation of Mammary Gland *in Vitro*
  Roger W. Turkington
Author Index—Subject Index

Volume 4

Genetics and Genesis
  Clifford Grobstein
The Outgrowing Bacterial Endospore
  Alex Keynan
Cellular Aspects of Muscle Differentiation *in Vitro*
  David Yaffe
Macromolecular Biosynthesis in Animal Cells Infected with Cytolytic Viruses
  Bernard Roizman and Patricia G. Spear
The Role of Thyroid and Growth Hormones in Neurogenesis
  Max Hamburgh
Interrelationships of Nuclear and Cytoplasmic Estrogen Receptors
  Jack Gorski, G. Shyamala, and D. Toft

The Biological Significance of Turnover of the Surface Membrane of Animal Cells
  Leonard Warren
Author Index—Subject Index

Volume 5

Developmental Biology and Genetics: A Plea for Cooperation
  Alberto Monroy
Regulatory Processes in the Maturation and Early Cleavage of Amphibian Eggs
  L. D. Smith and R. E. Ecker
On the Long-Term Control of Nuclear Activity during Cell Differentiation
  J. B. Gurdon and H. R. Woodland
The Integrity of the Reproductive Cell Line in the Amphibia
  Antonie W. Blackler
Regulation of Pollen Tube Growth
  Hansferdinand Linskens and Marianne Kroh
Problems of Differentiation in the Vertebrate Lens
  Ruth M. Clayton
Reconstruction of Muscle Development as a Sequence of Macro-Molecular Synthesis
  Heinz Herrmann, Stuart M. Heywood, and Ann C. Marchok
The Synthesis and Assembly of Myofibrils in Embryogenic Muscle
  Donald A. Fischman
The T-Locus of the Mouse: Implications for Mechanisms of Development
  Salome Gluecksohn-Waelsch and Robert P. Erickson
DNA Masking in Mammalian Chromatin: A Molecular Mechanism for Determination of Cell Type
  J. Paul
Author Index—Subject Index

Volume 6

The Induction and Early Events of Germination in the Zoospore of *Blastocladiella emersonii*
  Louis C. Truesdell and Edward C. Cantino

Steps of Realization of Genetic Information in Early Development
  A. A. Neyjakh
Protein Synthesis during Amphibian Metamorphosis
  J. R. Tata
Hormonal Control of a Secretory Tissue
  H. Yomo and J. E. Varner
Gene Regulation Networks: A Theory for Their Global Structure and Behaviors
  Stuart Kauffman
Positional Information and Pattern Formation
  Lewis Wolpert
Author Index—Subject Index

Volume 7

The Structure of Transcriptional Units in Eukaryotic Cells
  G. P. Georgiev
Regulation of Sporulation in Yeast
  James E. Haber and Harlyn O. Halvorson
Sporulation of Bacilli, a Model of Cellular Differentiation
  Ernst Freese
The Cocoonase Zymogen Cells of Silk Moths: A Model of Terminal Cell Differentiation for Specific Protein Synthesis
  Fotis C. Kafatos
Cell Coupling in Developing Systems: The Heart-Cell Paradigm
  Robert L. DeHaan and Howard G. Sachs
The Cell Cycle, Cell Lineages, and Cell Differentiation
  H. Holtzer, H. Weintraub, R. Mayne, and B. Mochan
Studies on the Development of Immunity: The Response to Sheep Red Blood Cells
  Robert Auerbach
Author Index—Subject Index

Volume 8: Gene Activity and Communication in Differentiating Cell Populations

Reception of Immunogenic Signals by Lymphocytes
  Michael Feldman and Amiela Globerson
Determination and Pattern Formation in the Imaginal Discs of *Drosophila*
  Peter J. Bryant
Studies on the Control of Differentiation of Murine Virus-Induced Erythroleukemic Cells
  Charlotte Friend, Harvey D. Preisler, and William Scher
Concepts and Mechanisms of Cartilage Differentiation
  Daniel Levitt and Albert Dorfman
Cell Determination and Biochemical Differentiation of the Early Mammalian Embryo
  M. C. Herbert and C. F. Graham
Differential Gene Activity in the Pre- and Postimplantation Mammalian Embryo
  Robert B. Church and Gilbert A. Schultz
Neuronal Specificity Revisited
  R. K. Hunt and Marcus Jacobson
Subject Index

Volume 9: Experimental Systems for Analysis of Multicellular Organization

Histones, Chromatin Structure, and Control of Cell Division
  E. M. Bradbury
Control of Gene Expression during the Terminal Differentiation of Erythroid Cells
  A. Fantoni, M. Lunadei, and E. Ullu
Changing Populations of Reiterated DNA Transcripts during Early Echinoderm Development
  H. R. Whiteley and A. H. Whiteley
Regulation of Messenger RNA Translation during Insect Development
  Joseph Ilan and Judith Ilan
Chemical and Structural Changes within Chick Erythrocyte Nuclei Introduced into Mammalian Cells by Cell Fusion
  R. Appels and Nils R. Ringertz
Drosophila Antigens: Their Spatial and Temporal Distribution, Their Function and Control
  David B. Roberts
Subject Index

Volume 10: Experimental Systems for Analysis of Multicellular Organization

Experiments with Junctions of the Adhaerens Type
  Jane Overton
The Extracellular Matrix: A Dynamic Component of the Developing Embryo
  Francis J. Manasek
The Role of the Golgi Complex during Spermiogenesis
  Baccio Baccetti
Phenomena of Cellular Recognition in Sponges
  G. Van de Vyver
Freshwater Sponges as a Material for the Study of Cell Differentiation
  R. Rasmont
Differentiation of the Golgi Apparatus in the Genetic Control of Development
  W. G. Whaley, Marianne Dauwalder, and T. P. Leffingwell
Subject Index

Volume 11: Pattern Development

Interactions of Lectins with Embryonic Cell Surfaces
  Steven B. Oppenheimer
Biological Features and Physical Concepts of Pattern Formation Exemplified by Hydra
  Alfred Gierer
Molecular Aspects of Myogenesis
  John Paul Merlie, Margaret E. Buckingham, and Robert G. Whalen
Origin and Establishment of Embryonic Polar Axes in Amphibian Development
  P. D. Nieuwkoop
An Old Engine: The Gray Crescent of Amphibian Eggs
  J. Brachet
Control of Plant Cell Enlargement by Hydrogen Ions
  David L. Rayle and Robert Cleland
Subject Index

Volume 12: Fertilization

Patterns in Metazoan Fertilization
  C. R. Austin
Biochemistry of Male Germ Cell Differentiation in Mammals: RNA Synthesis in Meiotic and Postmeiotic Cells
  V. Monesi, R. Geremia, A. D'Agostino, and C. Boitani
Cell Communication, Cell Union, and Initiation of Meiosis in Ciliate Conjugation
  Akio Miyake
Sperm–Egg Association in Mammals
  R. Yanagimachi
Sperm and Egg Receptors Involved in Fertilization
  Charles B. Metz
Transformations of Sperm Nuclei upon Insemination
  Frank J. Longo and Mel Kunkle
Mechanisms of Activation of Sperm and Egg during Fertilization of Sea Urchin Gametes
  David Epel
Subject Index

DATE DUE